COMPUTER
CIRCUIT ANALYSIS

COMPUTER CIRCUIT ANALYSIS

Theory and Applications

FRANK A. ILARDI
Technical Career Institutes

PRENTICE-HALL, INC., *Englewood Cliffs, New Jersey*

Library of Congress Cataloging in Publication Data

ILARDI, FRANK A date
    Computer circuit analysis.

    1. Electronic digital computers—Circuits. 2. Logic circuits. 3. Pulse circuits. I. Title.
TK7888.4.I4   621.3819′58′35   75-26529
ISBN 0-13-165357-1

To my sons FRANK B., MICHAEL, and STEVEN

© 1976 by Prentice-Hall, Inc.
Englewood Cliffs, New Jersey

All rights reserved. No part of this book may be
reproduced in any form or by any means
without permission in writing from the publisher.

10 9 8 7 6 5 4 3 2 1

Printed in the United States of America

PRENTICE-HALL INTERNATIONAL, INC., *London*
PRENTICE-HALL OF AUSTRALIA, PTY. LTD., *Sydney*
PRENTICE-HALL OF CANADA, LTD., *Toronto*
PRENTICE-HALL OF INDIA PRIVATE LIMITED, *New Delhi*
PRENTICE-HALL OF JAPAN, INC., *Tokyo*
PRENTICE-HALL OF SOUTHEAST ASIA (PTE.) LTD., *Singapore*

D
621.3819′5835
ILA

CONTENTS

PREFACE   ix

1 NETWORK ANALYSIS   1

    1-1 Thévenin's Theorem, *1*
    1-2 Norton's Theorem, *7*
    1-3 Thévenin to Norton and Norton to Thévenin Conversion, *10*
    1-4 Millman's Theorem, *12*
    1-5 Transient Response of RC Circuits, *16*
    1-6 RC Circuits with Step Function Voltage Inputs, *24*
    1-7 RC Attenuators, *29*

2 SEMICONDUCTOR SWITCHING DEVICES   33

    2-1 The Semiconductor Diode, *33*
    2-2 Signal Characteristics of Diodes, *39*
    2-3 The Transistor Amplifier, *40*
    2-4 The Transistor Switch, *48*
    2-5 Transient Response of the Transistor Switch, *54*
    2-6 Overdrive, *56*
    2-7 Nonsaturating Techniques, *62*
    2-8 Reducing Delay Time t_d, *68*

2-9   Field Effect Transistors (FETs), *68*
2-10 The FET as a Switch, *76*
2-11 Summary, *77*

3   COMPUTER MATH AND LOGIC   83

3-1   Number Systems, *84*
3-2   Conversion Techniques, *86*
3-3   Binary Arithmetic, *93*
3-4   Complements, *98*
3-5   Decimal Codes, *102*
3-6   Boolean Algebra, *106*
3-7   Basic Logical Design, *116*
3-8   *NAND*, *NOR*, *INHIBIT*, and *IMPLICATION* Gates, *121*
3-9   Summary, *124*

4   DIODE LOGIC   128

4-1   Static Analysis of Diode *AND* Gates, *128*
4-2   A Practical Diode *AND* Gate, *131*
4-3   Transient Analysis of *AND* Gates, *136*
4-4   Static Analysis of Diode *OR* Gates, *139*
4-5   A Practical Diode *OR* Gate, *142*
4-6   Transient Analysis of *OR* Gates, *145*
4-7   Static Analysis of Cascaded Diode Gates, *148*
4-8   Level Shifting, *149*
4-9   Transient Analysis of Cascaded Diode Gates, *150*
      Appendix 4-A: AND Gate Rise Time, *153*
      Appendix 4-B: OR Gate Fall Time, *154*

5   THE INVERTER (*NOT* Gate)   159

5-1   The Basic Inverter, *159*
5-2   Steady-state Analysis of the Basic Inverter, *161*
5-3   A Saturation Mode Inverter with Collector Clamping, *165*
5-4   The Load (FAN-IN, FAN-OUT, and the UNIT LOAD Concept), *167*
5-5   Dc Noise Margins, *167*

6 RESISTOR-TRANSISTOR LOGIC RTL AND RESISTOR-CAPACITOR TRANSISTOR LOGIC RCTL   170

6-1 RTL *NOR* Gates, *171*
6-2 Analysis of the Multitransistor RTL *NOR* Gate, *173*
6-3 Resistor-Capacitator-Transistor Logic RCTL, *177*
6-4 Summary, *177*

7 DIODE-TRANSISTOR LOGIC DTL EMITTER-COUPLED LOGIC ECL WIRED LOGIC   180

7-1 The Modern Integrated Circuit Version of Diode-Transistor Logic, *180*
7-2 Emitter-Coupled Logic ECL, *187*
7-3 Wired Logic, *195*
7-4 Summary, *200*

8 TRANSISTOR-TRANSISTOR LOGIC TTL OR T²L WITH OUTPUT CIRCUIT VARIATIONS   204

8-1 Construction of the Multi-emitter Transistor, *204*
8-2 Steady-state Analysis of a TTL *NAND* Gate, *207*
8-3 Transient Response of the TTL *NAND* Gate, *222*
8-4 The TTL *NOR* Gate, *227*
8-5 The TTL Inverter, *229*
8-6 The TTL *AND* Gate, *230*
8-7 The TTL *OR* Gate, *233*
8-8 The *AND-OR-INVERT* Circuit, *233*
8-9 Output Circuit Variations, *236*
8-10 Summary, *252*

9 ASTABLE AND MONOSTABLE MULTIVIBRATORS   256

9-1 The Collector-base Coupled Astable Multivibrator, *257*
9-2 Circuit Analysis, *261*
9-3 A Modified Collector-coupled Astable MV, *268*
9-4 Synchronization of the Collector-coupled Astable MV and Frequency Division, *270*
9-5 The *pnp* Collector-coupled Astable MV, *271*
9-6 The Collector-base Coupled Monostable Multivibrator, *273*
9-7 An Application of the Monostable MV, *284*
9-8 Summary, *286*

10 THE BASIC BISTABLE AND MODERN LOGIC FLIP-FLOP — 288

10-1 The Basic Flip-flop, *288*
10-2 The Basic Logic Flip-flop, *293*
10-3 Types of Logic Flip-flops, *294*
10-4 Practical Logic Flip-flop Circuits, *297*
10-5 The Timing Problem, *301*
10-6 A D-type Flip-flop, *312*
10-7 The Relationships Between the Various Flip-flops, *314*

11 SPECIAL-PURPOSE CIRCUITS — 320

11-1 Internal Gates, *320*
11-2 A Schottky-clamped TTL Gate, *324*
11-3 Simple Lamp Drivers, *325*
11-4 Relay Drivers, *327*
11-5 A Transmission Line Driver and Terminator, *328*
11-6 A Crystal-controlled Clock Pulse Generator, *328*
11-7 The Schmitt Trigger, *330*
11-8 Analysis of the Schmitt Trigger, *333*
11-9 Fast Analysis of the Schmitt Trigger, *341*
11-10 FET Logic Circuits, *342*
11-11 CMOS Logic, *346*
11-12 Three-state Logic TSL, *350*
11-13 Interface Elements, *352*
11-14 Summary, *354*

12 PACKAGING AND APPLICATIONS OF DIGITAL CIRCUITS — 356

12-1 TTL Packages, *356*
12-2 A Monostable Package, *372*
12-3 Summary, *374*
12-4 Complex Arrays, *374*

INDEX — 401

PREFACE

This book was written after extensive discussions with representatives of the electronic industry. These people, who do the hiring for their firms, said they look for technical school graduates with good knowledge of electronics fundamentals. They feel that they can then teach them their particular system. It was noted that when these representatives interviewed prospective graduates, they tested them on their understanding of circuit theory similar to that included in this text.

This book evolved from classroom lecture notes used in courses in pulse and digital circuits at Technical Career Institutes (formerly RCA Institutes). These courses were highly successful in training students for employment with nearly every company in the electronics industry.

The first three chapters are included in this book because in some schools much of this material is not taught until the courses in pulse and digital circuits are given. To learn these subjects, a thorough understanding of network theory, semiconductor switching devices, and computer math and logic is essential. These chapters can be used by students studying the above-mentioned topics for the first time or as an excellent review by those who have had courses covering this material.

The various circuits used in computers are discussed in Chapters 4 through 11. Chapter 12 is included to answer questions that have so often been asked by students of the author: How are these circuits used? How are they packaged? What determines the number of circuits that can be included in an integrated-circuit package?

The material presented in this book should also be of great value to anyone already working in the electronics field. Because of the information covered in the first three chapters, a good background in electronics fundamentals is the only prerequisite to the use of this book. The most modern pulse and digital circuits are discussed, both a qualitative and a quantitative analysis is presented.

The author wishes to thank Sprague Electric Company, Fairchild Semiconductor, Signetics Corp., and Texas Instruments, Inc. for their cooperation. The information provided by these companies allowed the most up-to-date material to be included in this text.

Special thanks is given to Mr. William Brecher (Instructor, Computer Department, Technical Career Institutes) from whom the author learned a great deal about computers, and to Jacqueline Ilardi who typed the manuscript and helped with other details.

FRANK A. ILARDI

1

NETWORK ANALYSIS

To analyze pulse and digital circuits, it is often necessary to replace the circuit with a network of components equivalent to the actual circuit. The equivalent circuit is obtained by applying one or more network theorems. It is then analyzed by using either standard techniques or other network theorems. In this chapter only Thévenin's, Norton's, and Millman's theorems, which are used extensively throughout the text, are covered. RC time constant theory is also discussed because an understanding of these principles is essential in pulse and digital circuit analysis.

1-1 Thévenin's Theorem

Thévenin's theorem states that any two-terminal *linear* network, no matter how complex, may be replaced by a single voltage source in series with a single impedance. The value of this *Thévenin equivalent voltage* V_{TH} is the same as the voltage that appears across the load terminals if the load is replaced by an open circuit. The value of the *Thévenin impedance* Z_{TH} is the same as the impedance seen by the load when all sources are replaced by their internal impedance.

This theorem is extremely useful when a complex circuit with a changing (variable) load is analyzed. For example, every time that load resistance R_L changes in Fig. 1-1(a) all of the voltage drops and currents change. If the load voltage V_{R_L} must be determined for many different values of R_L, the entire circuit must be analyzed each time. In Fig. 1-1(b) that part of the circuit

2 / *Network Analysis*

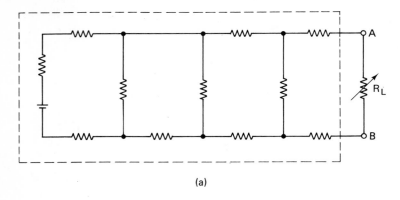

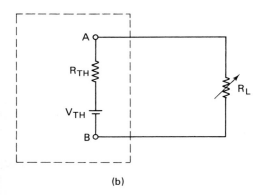

Fig. 1-1 (a) A complex series parallel circuit with a varying load. (b) The unchanging part of (a) replaced by its Thévenin equivalent.

which does *not* change is replaced by its Thévenin equivalent circuit. Now each time that R_L changes the voltage across R_L is easily calculated by

$$V_{R_L} = V_{TH}\left(\frac{R_L}{R_L + R_{TH}}\right) \tag{1-1}$$

Thévenin's theorem is also very useful when the output voltage of a circuit containing several voltage or current sources must be determined. In Fig. 1-2 a circuit containing two voltage sources is replaced by its Thévenin equivalent to produce a simple series circuit. The values of V_{TH} and R_{TH} in Fig. 1-2 are calculated by using the rules stated in the theorem, as follows:

Determine V_{TH} by removing R_L and calculating the open-circuit voltage from A to B ($V_{AB_{oc}}$). Figure 1-3(a) shows the circuit that must be analyzed to find V_{TH}.

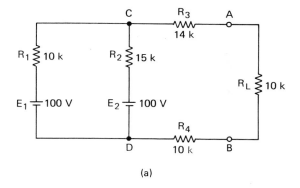

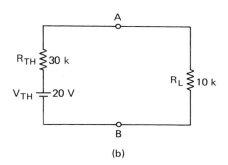

Fig. 1-2 (a) A circuit containing more than one source simplified in (b) by changing everything to the left of A and B to a Thévenin equivalent.

If $R_{AB} = \infty$,

$$I_{AB} = 0$$

and

$$V_{R_3} = V_{R_4} = 0$$

If $V_{R_3} = 0$,

$$V_A = V_C$$

and if $V_{R_4} = 0$,

$$V_B = V_D$$

Hence

$$V_{AB} = V_{CD}$$

and

$$V_{CD} = V_{R_2} + E_2 = V_{R_1} + E_1$$

4 / Network Analysis

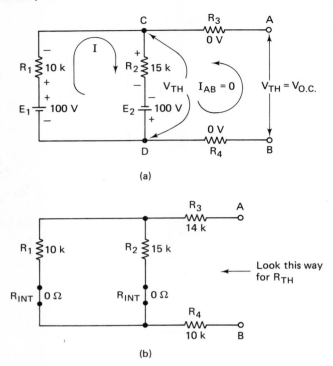

Fig. 1-3 (a) The circuit that must be analyzed to find V_{TH} in Fig. 1-2. (b) The circuit that must be analyzed to find R_{TH} in Fig. 1-2.

Therefore, to find V_{TH}, solve for either V_{R_1} or V_{R_2} and add it to the appropriate voltage source. With $R_{AB} = \infty$, the only current I flows through the series circuit shown in Fig. 1-3(a). Since E_1 and E_2 are series aiding, the total voltage in this loop is

$$E_T = E_1 + E_2$$
$$= 100 + 100$$
$$E_T = 200 \text{ V}$$

and

$$V_{R_1} = E_T \frac{R_1}{R_1 + R_2}$$
$$= 200 \frac{10}{25}$$
$$V_{R_1} = 80 \text{ V}$$

Therefore,
$$V_{CD} = V_{R_1} + E_1$$
$$= (-80) + (100)$$
$$V_{CD} = 20 \text{ V}$$

and
$$V_{TH} = 20 \text{ V}$$

The polarities for V_{R_1} and E_1 are those seen at the point C side of R_1 and E_1 when finding the voltage at C with respect to D.

Now solve for R_{TH} by replacing both sources with their internal resistance and by calculating the resistance seen by R_L. Fig. 1-3(b) shows the circuit that must be analyzed to find R_{TH}. Unless otherwise stated, assume that voltage sources have zero internal resistance and that current sources have infinite internal resistance. Then

$$R_{TH} = R_3 + R_4 + \frac{R_1 R_2}{R_1 + R_2}$$
$$= 14 + 10 + \frac{(10)(15)}{25}$$
$$R_{TH} = 30 \text{ k}\Omega$$

An alternate method of calculating V_{TH} is now shown.

$$V_{R_2} = E_T \frac{R_2}{R_1 + R_2}$$
$$= 200 \frac{15}{25}$$
$$V_{R_2} = 120 \text{ V}$$

Hence,
$$V_{CD} = V_{R_2} + E_2$$
$$= (120) + (-100)$$
$$V_{CD} = 20 \text{ V}$$

In either case
$$V_{TH} = 20 \text{ V}$$

The Thévenin equivalent of more complex networks is found section by section. The circuit is broken (opened) so that a Thévenin equivalent can be determined for a circuit no more complex than the one illustrated in Fig.

1-3(a). Then the remaining circuit is replaced one section at a time. After each section is replaced, a new partial Thévenin equivalent is calculated until finally the entire circuit has been included. This is illustrated in Ex. 1-1 in which the Thévenin equivalent of Fig. 1-4(a) is calculated.

Example 1-1: Find the Thévenin equivalent of the circuit shown in Fig. 1-4(a).

Solution: If the circuit is broken at the points marked X, the circuit arrangement to the left of these points is exactly the same as Fig. 1-3(a), and since the same values are used,

$$V_{TH_{xx}} = 20 \text{ V}$$

and

$$R_{TH_{xx}} = 30 \text{ k}\Omega$$

This results in Fig. 1-4(b). By removing R_L and solving the remaining circuit [Fig. 1-4(c)] for V_{TH} and R_{TH} the overall Thévenin equivalent is found.

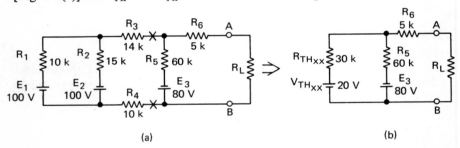

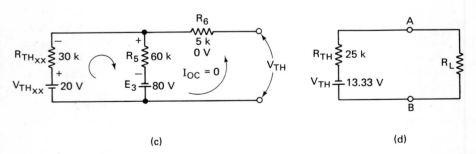

Fig. 1-4 (a) A more complex network containing three sources. (b) Fig. (a) partially simplified by finding the Thévenin equivalent of everything to the left of the points marked X. (c) Fig. (b) showing the only current with R_L removed. (d) The overall Thévenin equivalent circuit seen by R_L.

Solving for V_{TH}, it is determined that

$$V_{TH} = V_{AB_{oc}} = V_{R_5} + E_3$$

where

$$V_{R_5} = (V_{TH_{XX}} + E_3)\frac{R_5}{R_5 + R_{TH_{XX}}}$$

$$= 100\frac{60}{90}$$

$$V_{R_5} = 66.67 \text{ V}$$

Therefore,

$$V_{TH} = (+66.67) + (-80)$$

$$V_{TH} = -13.33 \text{ V}$$

Solving for R_{TH}, it is determined that

$$R_{TH} = R_6 + \frac{R_5 R_{TH_{XX}}}{R_5 + R_{TH_{XX}}}$$

$$= 5 + \frac{(60)(30)}{90}$$

$$R_{TH} = 25 \text{ k}\Omega$$

Because of the complexity of some circuits the Thévenin equivalent cannot always be found directly. In such cases other theorems are used to first simplify the circuit; then V_{TH} and R_{TH} are calculated.

1-2 Norton's Theorem

Norton's theorem states that any two-terminal linear network, no matter how complex, can be replaced by a single current source in parallel with a single impedance. The value of this *Norton equivalent current* I_N is the same as the current that flows between the load terminals if the load is replaced by a short circuit. The value of the *Norton impedance* Z_N is the same as the impedance seen by the load when all sources are replaced by their internal impedance. Note that Z_N *is the same as* Z_{TH}.

Norton's theorem is very useful when the load impedance $Z_L \ll Z_N$. For example, in Fig. 1-5, R_L varies between 10 and 100 Ω. When $R_L = 10$ Ω

$$I_L = I_N \frac{R_N}{R_L + R_N}$$

$$= 0.5 \text{ mA} \frac{10,000 \text{ }\Omega}{10,010 \text{ }\Omega}$$

$$I_L \approx 0.5 \text{ mA}$$

8 / Network Analysis

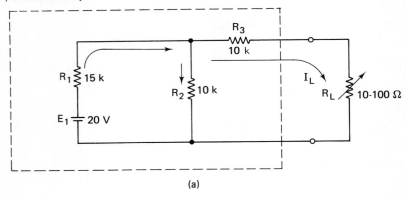

(a)

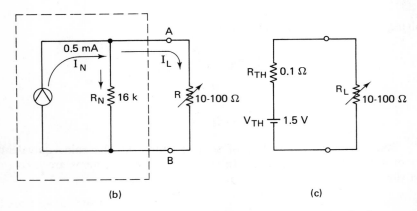

(b)   (c)

Fig. 1-5 (a) A two-terminal linear network with a varying load. (b) Its Norton equivalent circuit. (c) The same load connected to a Thévenin circuit with $R_{TH} \ll R_L$.

When

$$R_L = 100 \, \Omega$$

$$I_L = 0.5 \text{ mA} \frac{10{,}000 \, \Omega}{10{,}100 \, \Omega}$$

$$I_L \approx 0.5 \text{ mA}$$

Therefore, whenever $R_L \ll R_N$, $I_L \approx I_N$.

Thévenin's theorem is more useful when R_L is much greater than R_N or R_{TH}. This condition exists in Fig. 1-5(c). Now, regardless of the value of R_L, $V_{R_L} \approx V_{TH}$.

The values of I_N and R_N in Fig. 1-5(b) are calculated by using the rules stated in the theorem.

First, determine I_N by replacing R_L with a short circuit and calculating

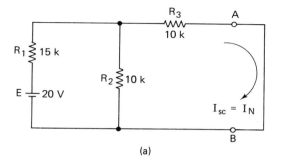

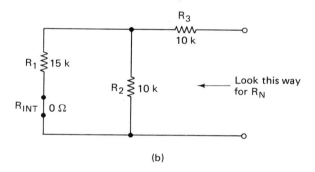

Fig. 1-6 (a) The circuit that must be analyzed to find I_N in Fig. 1-5. (b) The circuit that must be analyzed to find R_{TH} in Fig. 1-5.

the short-circuit current from A to B ($I_{AB_{sc}}$). Figure 1-6(a) shows the circuit that must be analyzed in order to find I_N.

$$I_N = I_{AB_{sc}} = I_{R_3}$$

If $R_{AB} = 0$,

$$R_T = R_1 + \frac{R_2 R_3}{R_2 + R_3}$$
$$= 15\text{ k} + 5\text{ k}$$
$$R_T = 20\text{ k}$$

If the source E_T sees 20 k,

$$I_T = \frac{E_T}{R_T}$$
$$= \frac{20}{20\text{ k}}$$
$$I_T = 1\text{ mA}$$

Using the current division ratio, it can be seen that

$$I_N = I_{R_3} = \frac{R_2}{R_2 + R_3}(I_T).$$

$$= \frac{10k}{20k}(1 \text{ mA})$$

$$I_N = 0.5 \text{ mA}$$

The arrow in the Norton current generator points in the direction of conventional current. Current must leave the generator so that it flows through R_L in the same direction as in the original circuit. As shown in Fig. 1-5(b), the arrow must point upward.

Now solve for R_N by replacing the source with its internal resistance and calculating the resistance seen by R_L. Figure 1-6(b) shows the circuit that must be analyzed to find R_N. Assuming that the voltage source has zero internal resistance, it can be seen that

$$R_N = R_3 + \frac{R_1 R_2}{R_1 + R_2}$$

$$= 10k + 6k$$

$$R_N = 16k$$

1-3 Thévenin to Norton and Norton to Thévenin Conversion

It is often advantageous to change a current source equivalent to a voltage source equivalent or vice versa. As seen by the following examples, the conversions are simple. First, the Thévenin circuit of Fig. 1-7(a) is converted to a Norton or current source equivalent.

Shorting the load terminals

$$I_N = I_{AB_{sc}} = \frac{V_{TH}}{R_{TH}} \tag{1-2a}$$

$$= \frac{20 \text{ V}}{10 \text{ k}}$$

$$I_N = 2 \text{ mA}$$

By replacing the V_{TH} source with $R_{INT} = 0$, it is seen that

$$R_N = R_{TH} = 10 \text{ k} \tag{1-2b}$$

The Norton circuit of Fig. 1-8(a) is now converted to a Thévenin or voltage-

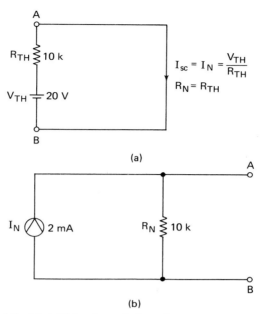

Fig. 1-7 (a) A Thévenin equivalent circuit. (b) Its Norton equivalent circuit.

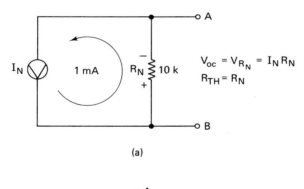

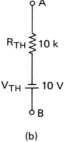

Fig. 1-8 (a) A Norton equivalent circuit. (b) Its Thévenin equivalent circuit.

source equivalent. Open-circuiting the load terminals gives

$$V_{AB_{oc}} = V_{TH} = I_N R_N$$
$$= (1 \text{ mA})(10 \text{ k})$$
$$V_{TH} = 10 \text{ V}$$

As indicated on R_N, V_{TH} is negative on top. Finally, replacing the I_N source with $R_{INT} = \infty$,

$$R_{TH} = R_N = 10 \text{ k}$$

1-4 Millman's Theorem

Millman's theorem is a very useful tool in pulse and digital circuit analysis. The theorem applies to circuits such as Fig. 1-9 in which there are only parallel branches with no resistors in the series lines. It does not lend itself to a circuit such as Fig. 1-4(a) because of R_3, R_4, and R_6.

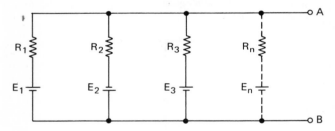

Fig. 1-9 A circuit that may be analyzed with Millman's theorem.

Referring to Fig. 1-9, we see that the theorem states that

$$V_{AB} = \frac{\dfrac{E_1}{R_1} + \dfrac{E_2}{R_2} + \dfrac{E_3}{R_3} \cdots + \dfrac{E_n}{R_n}}{\dfrac{1}{R_1} + \dfrac{1}{R_2} + \dfrac{1}{R_3} \cdots + \dfrac{1}{R_n}} \qquad (1\text{-}3)$$

where the voltages E_1, E_2, E_3, ..., E_n carry polarity signs.

If the source is positive on the A side, the voltage is positive; if the source is negative on the A side, the voltage is negative.

Derivation of Millman's Theorem. The Millman equation may be developed by applying Norton's theorem to Fig. 1-9. Figure 1-10(a) shows the current paths when the output terminals are short-circuited. The Norton current I_N is the sum of the short-circuit currents. Note that the negative

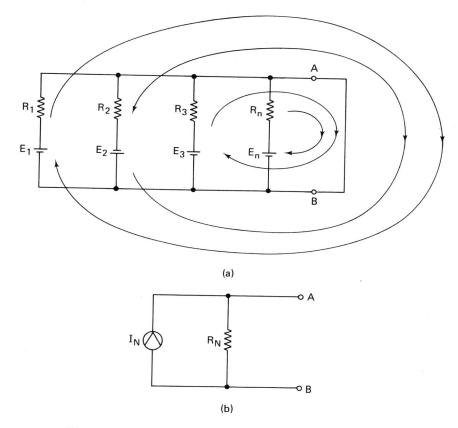

Fig. 1-10 (a) The current paths in the circuit of Fig. 1-9 with the output terminals short-circuited. (b) Its Norton equivalent circuit.

voltage source E_2 causes a current in opposition to the others. This is accounted for when the sign of E_2 is carried in the Millman equation. The short-circuit current is

$$I_N = \frac{E_1}{R_1} + \frac{E_2}{R_2} + \frac{E_3}{R_3} \cdots + \frac{E_n}{R_n}$$

and

$$R_N = \frac{1}{\frac{1}{R_1} + \frac{1}{R_2} + \frac{1}{R_3} \cdots + \frac{1}{R_n}}$$

Referring to Fig. 1-10(b), it is seen that the voltage across the A and B terminals equals the voltage developed by I_N flowing through R_N.

14 / Network Analysis

$$V_{AB} = I_N R_N$$

$$= \left(\frac{E_1}{R_1} + \frac{E_2}{R_2} + \frac{E_3}{R_3} \cdots + \frac{E_n}{R_n}\right) \times \left(\frac{1}{\frac{1}{R_1} + \frac{1}{R_2} + \frac{1}{R_3} \cdots + \frac{1}{R_n}}\right)$$

$$V_{AB} = \frac{\frac{E_1}{R_1} + \frac{E_2}{R_2} + \frac{E_3}{R_3} \cdots + \frac{E_n}{R_n}}{\frac{1}{R_1} + \frac{1}{R_2} + \frac{1}{R_3} \cdots + \frac{1}{R_n}}$$

An application of Millman's theorem is given in Ex. 1-2.

Example 1-2: Find the output voltage of the circuit in Fig. 1-11.

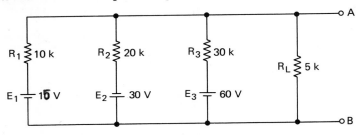

Fig. 1-11 The circuit for the Millman's theorem example.

Solution: Using Millman's theorem, it is determined that

$$V_{AB} = \frac{\frac{E_1}{R_1} + \frac{E_2}{R_2} + \frac{E_3}{R_3} + \frac{E_L}{R_L}}{\frac{1}{R_1} + \frac{1}{R_2} + \frac{1}{R_3} + \frac{1}{R_L}}$$

$$= \frac{\frac{15}{10\,k} + \frac{-30}{20\,k} + \frac{60}{30\,k} + \frac{0}{5\,k}}{\frac{1}{10\,k} + \frac{1}{20\,k} + \frac{1}{30\,k} + \frac{1}{5\,k}}$$

The common denominators, which are always equal, are in this case 60 k. Hence,

$$V_{AB} = \frac{\frac{90 - 90 + 120 + 0}{60\,k}}{\frac{6 + 3 + 2 + 12}{60\,k}}$$

$$= \frac{120}{23}$$

$$V_{AB} = 5.22 \text{ V}$$

Finding the Thévenin Voltage by Using Millman's Theorem. The Thévenin voltage of a circuit that contains only parallel branches may be found in one step by applying Millman's theorem. For example, if the Thévenin equivalent circuit of Fig. 1-11 is desired, open-circuit R_L and find $V_{TH} = V_{AB_{oc}}$ by Millman's theorem.

$$V_{TH} = \frac{\frac{E_1}{R_1} + \frac{E_2}{R_2} + \frac{E_3}{R_3}}{\frac{1}{R_1} + \frac{1}{R_2} + \frac{1}{R_3}} \quad (1\text{-}3)$$

$$= \frac{\frac{15}{10\text{ k}} + \frac{-30}{20\text{ k}} + \frac{60}{30\text{ k}}}{\frac{1}{10\text{ k}} + \frac{1}{20\text{ k}} + \frac{1}{30\text{ k}}}$$

The common denominators are 60 k. Hence,

$$V_{TH} = \frac{\frac{90 - 90 + 120}{60\text{ k}}}{\frac{6 + 3 + 2}{60\text{ k}}}$$

$$= \frac{120}{11}$$

$$V_{TH} = 10.9 \text{ V}$$

R_{TH} is found in the usual way.

Special Equation for V_{TH} of a Two-branch Circuit. When the Thévenin equivalent of more complex circuits [such as Fig. 1-4(a)] is desired, V_{TH} is calculated in steps as explained in Ex. 1-1. The circuit is always broken so that the Thévenin voltage of each section is calculated by finding the open-circuit voltage across a simple two-branch circuit. Millman's theorem is now used to develop a special equation that gives both the magnitude and polarity of V_{TH} for a two-branch circuit.

Referring to Fig. 1-12, it can be seen that

$$V_{TH} = \frac{\frac{E_1}{R_1} + \frac{E_2}{R_2}}{\frac{1}{R_1} + \frac{1}{R_2}}$$

The common denominators are $R_1 R_2$

$$= \frac{\frac{E_1 R_2 + E_2 R_1}{R_1 R_2}}{\frac{R_2 + R_1}{R_1 R_2}}$$

16 / Network Analysis

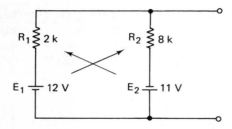

Fig. 1-12 Circuit for the development of the special Thévenin voltage Eq. (1-4).

The common denominators cancel leaving

$$V_{TH} = \frac{E_1 R_2 + E_2 R_1}{R_1 + R_2} \tag{1-4}$$

This relation is easily remembered as the sum of the cross products, as indicated in Fig. 1-12, divided by the sum of the two resistors. The voltages may be positive, negative, or zero.

$$V_{TH} = \frac{\text{sum of the cross products}}{\text{sum of the resistors}}$$

Example 1-3: Calculate the Thévenin equivalent voltage of Fig. 1-12 by using Eq. (1-4).
Solution:

$$V_{TH} = \frac{(12)(8) + (-11)(2)}{2 + 8}$$

$$= \frac{96 - 22}{10}$$

$$V_{TH} = 7.4 \text{ V}$$

It is obvious that Eq. (1-4) provides a simple and direct method of calculating both the magnitude and polarity of V_{TH}. It is used throughout this text to solve two-branch circuits.

1-5 Transient Response of RC Circuits

Resistor-capacitor circuits have many applications in pulse and digital circuits. For example, they are often used to change the shape of pulses, to filter out undesired frequencies, or to introduce desired time delays. However, they also often cause undesired time delays that limit the speed of the computer. A thorough understanding of RC circuit theory is necessary before pulse and digital circuits can be analyzed.

RC Time Constants. A capacitor is defined as a component that opposes a change in voltage; thus, at the instant the switch in Fig. 1-13 is

Transient Response of RC Circuits / 17

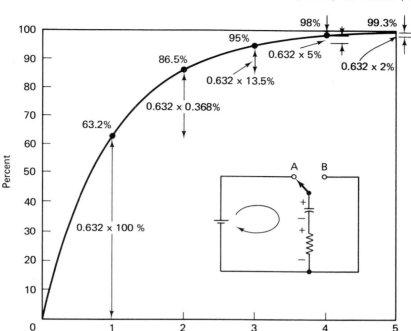

Fig. 1-13 Curve showing charging action of a capacitor in an RC circuit.

placed in position A, v_C remains 0 V and $v_R = E - v_C = E$. Then during each time interval equal to R times C, v_C increases by 63.2 percent of the difference between the applied voltage and the voltage already on C. As shown in the graph of Fig. 1-13, v_C starts at 0 V and reaches 63.2 percent of E after 1 RC interval. It then increases by 63.2 percent of the remaining 36.8 percent in the second RC interval. It continues in this manner and, in theory, v_C never fully reaches E. However, for all practical purposes, a capacitor is considered fully charged after 5 RC periods. Because v_C increases by the same 63.2 percent of the remaining voltage during each RC interval, this time is called the *time constant*.

A *universal time constant chart* is given in Fig. 1-14. It shows v_R decreasing as v_C increases. When C is fully charged, $v_R = 0$. If the switch is then thrown into position B, the capacitor starts to discharge. Since the capacitor is now the source, it applies the full voltage E to the resistor. Note that the polarity of v_R is now opposite of that caused by the charging current. As v_C gets smaller, v_R decreases. When C is fully discharged, $v_R = 0$. The same curve represents the resistor voltage during charge or discharge time because v_R decreases in either case.

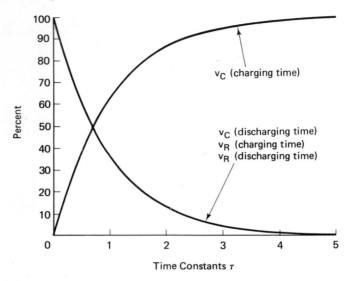

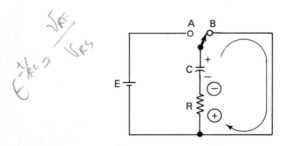

Fig. 1-14 The universal time constant chart.

This curve is a plot of $\epsilon^{-t/RC}$ where

t = the time the capacitor has to charge or discharge;
RC = the time constant of the charging or discharging circuit;
t/RC = the number of time constants the capacitor has to charge or discharge;
ϵ = 2.718;
$\epsilon^{-t/RC}$ = the ratio of v_{RF}, the voltage across the charging or discharging resistance at the *finish* (F) of t, to v_{RS}, the voltage across the same resistance at the *start* (S) of t.

The voltage across each component and the current can be calculated by using Ohm's law, Kirchhoff's voltage law, and the following equation:

$$v_{RF} = v_{RS}\epsilon^{-t/RC} \qquad (1\text{-}5)$$

This equation may be changed to permit the use of common (base 10) logarithms.

$$v_{RF} = v_{RS}\epsilon^{-t/RC}$$

$$= v_{RS} \frac{1}{\epsilon^{t/RC}}$$

$$= \frac{v_{RS}}{\log^{-1}(t/RC)\log\epsilon}$$

$$= \frac{v_{RS}}{\log^{-1}(0.4343t/RC)}$$

But multiplying by 0.4343 is the same as dividing by 2.3. Therefore,

$$v_{RF} = \frac{v_{RS}}{\log^{-1}(t/2.3RC)} \tag{1-6}$$

Very often the voltages v_{RS} and v_{RF} are known, but R, C, or the time t is not known. Solving Eq. (1-6) for each results in

$$t = 2.3RC \log(v_{RS}/v_{RF}) \tag{1-7}$$

$$R = \frac{t}{2.3C \log(v_{RS}/v_{RF})} \tag{1-8}$$

and

$$C = \frac{t}{2.3R \log(v_{RS}/v_{RF})} \tag{1-9}$$

Applications of these equations are given in Exs. 1-4 through 1-8 below.

Example 1-4: The voltage on the capacitor in Fig. 1-15(a) is zero when the switch is placed into position A. The capacitor then starts to charge toward 100 V. Determine v_C, v_R, and i after 2.5 μs.
Solution: It takes time for a capacitor to charge (or discharge) so, at the instant the switch is thrown, v_C remains zero.
    Hence,

$$v_{RS} = E - v_{CS}$$
$$= 100 - 0$$
$$v_{RS} = 100 \text{ V}$$

The condition of the circuit at this instant is shown in Fig. 1-15(b).
    After 2.5 μs, v_C has increased, which causes i and v_R to decrease.

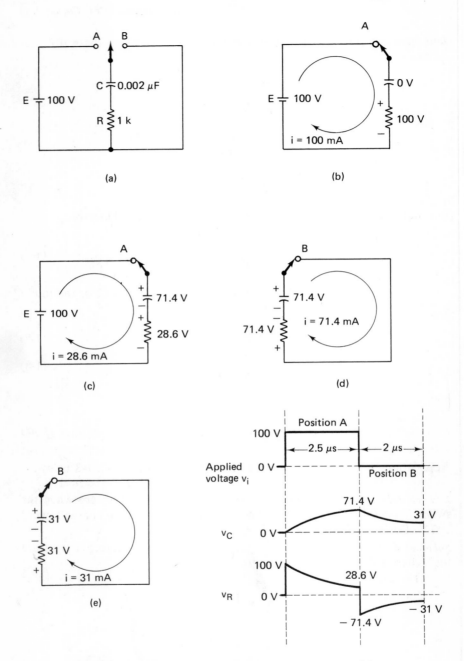

Fig. 1-15 The circuits and wave forms for Exs. 1-4 and 5.

By applying Eq. (1-6), it is determined that

$$v_{RF} = \frac{v_{RS}}{\log^{-1}(t/2.3RC)}$$

$$= \frac{100}{\log^{-1}(2.5 \times 10^{-6}/2.3 \times 10^3 \times 2 \times 10^{-9})}$$

$$= \frac{100}{\log^{-1} 0.543} = \frac{100}{3.49}$$

$$v_{RF} = 28.6 \text{ V}$$

The current is

$$i_F = \frac{v_{RF}}{R} = \frac{28.6}{1k}$$

$$= \frac{28.6}{1k}$$

$$i_F = 28.6 \text{ mA}$$

and

$$v_{CF} = E - v_{RF}$$
$$= 100 - 28.6$$
$$v_{CF} = 71.4 \text{ V}$$

The circuit with these values is shown in Fig. 1-15(c).

Example 1-5: If the switch in Fig. 1-15(a) is now thrown into position B, the capacitor discharges. Calculate the value of v_C, v_R, and i that will be present after 2 μs.

Solution: It takes time for the capacitor to discharge so, at the instant the switch is placed in position B, v_C remains 71.4 V. Since the applied voltage is zero in position B,

$$v_{RS} = E - v_{CS}$$
$$= 0 - 71.4$$
$$v_{RS} = -71.4 \text{ V}$$

and

$$i_S = \frac{v_{RS}}{R} = \frac{-71.4}{1k}$$

$$i_S = -71.4 \text{ mA}$$

The negative sign for i indicates that the discharge current flows in the opposite direction and causes v_R to reverse polarity. Figure 1-15(d) shows the circuit at this time. After 2 μs

$$v_{RF} = \frac{-71.4}{\log^{-1}(2 \times 10^{-6}/2.3 \times 10^3 \times 2 \times 10^{-9})} \tag{1-6}$$

$$= \frac{-71.4}{\log^{-1} 0.4343} = \frac{-71.4}{2.3}$$

$$v_{RF} = -31 \text{ V}$$

$$i_F = \frac{v_{RF}}{R}$$

$$= \frac{-31}{1 \text{ k}}$$

$$i_F = 31 \text{ mA}$$

and

$$v_{CF} = E - v_{RF}$$
$$= 0 - (-31)$$
$$v_{CF} = 31 \text{ V}$$

This is to be expected because the 31 V on R is being supplied by the capacitor. Figure 1-15(e) shows the circuit at this time.

Example 1-6: If the switch is now thrown back to position A, the capacitor starts to charge from 31 V toward 100 V. Calculate the time that it takes for C to charge to 70 V.

Solution: As shown in Fig. 1-16(a), at the instant the switch is thrown back to position A, v_C remains 31 V. Then

$$v_{RS} = E - v_{CS}$$
$$= 100 - 31$$
$$v_{RS} = 69 \text{ V}$$

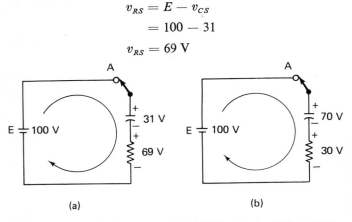

(a) (b)

Fig. 1-16 The circuits for Ex. 1-6. (a) The circuit of Fig. 1-15(a) at the instant the switch is thrown back to position A. (b) After C charges to 70 V.

and

$$i_s = \frac{v_{RS}}{R}$$
$$= \frac{69}{1k}$$
$$i_s = 69 \text{ mA}$$

After some time t, C has charged to 70 V and

$$v_{RF} = E - v_{CF}$$
$$= 100 - 70$$
$$v_{RF} = 30 \text{ V}$$

The circuit with this condition is shown in Fig. 1-16(b). The time it takes for v_C to go from 31 V to 70 V is determined by solving Eq. (1-7).

$$t = 2.3RC \log (v_{RS}/v_{RF}) \qquad (1\text{-}7)$$
$$= 2.3 \times 10^3 \times 2 \times 10^{-9} \log (69/30)$$
$$= 4.6 \times 10^{-6} \log 2.3$$
$$= 4.6 \times 10^{-6} \times 0.362$$
$$t = 1.665 \text{ } \mu\text{s}$$

Example 1-7: Referring to Fig. 1-17(a), calculate the value of R that will permit C to charge from 0 V to 50 V while the switch is in position A for 1 μs.
Solution:

$$v_{RS} = E - v_{CS}$$
$$= 100 - 0$$
$$v_{RS} = 100 \text{ V}$$
$$v_{RF} = E - v_{CF}$$
$$= 100 - 50$$
$$v_{RF} = 50 \text{ V}$$

$$R = \frac{t}{2.3C \log (v_{RS}/v_{RF})} \qquad (1\text{-}8)$$
$$= \frac{1 \times 10^{-6}}{2.3 \times 2 \times 10^{-9} \log (100/50)}$$
$$= \frac{1 \times 10^{-6}}{4.6 \times 10^{-9} \times 0.301}$$
$$R = 722 \text{ } \Omega$$

24 / Network Analysis

Example 1-8: If the capacitor in Fig. 1-17(b) has 20 V at the instant the switch is thrown into position A, what value of C will charge to 80 V in 1.7 μs?

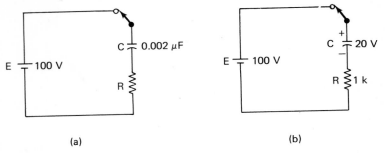

(a) (b)

Fig. 1-17 (a) The circuit for Ex. 1-7. (b) The circuit for Ex. 1-8.

Solution:

$$v_{RS} = E - v_{CS}$$
$$= 100 - 20$$
$$v_{RS} = 80 \text{ V}$$
$$v_{RF} = E - v_{CF}$$
$$= 100 - 80 \text{ V}$$
$$v_{RF} = 20 \text{ V}$$

$$C = \frac{t}{2.3R \log (v_{RS}/v_{RF})} \tag{1-9}$$

$$= \frac{1.7 \times 10^{-6}}{2.3 \times 10^3 \log (80/20)}$$

$$= \frac{1.7 \times 10^{-6}}{2.3 \times 10^3 \times 0.602}$$

$$C = 1230 \text{ pF}$$

1-6 RC Circuits with Step Function Voltage Inputs

The waveshaping and filtering action of an RC circuit is illustrated in Fig. 1-18. In this diagram the response of an RC circuit is shown for square wave input voltages of 5 kHz and 5 MHz. The circuit has a time constant τ equal to

$$RC = (10^3)(10^{-8})$$
$$= 10^{-5}$$
$$RC = 10 \ \mu\text{s}$$

An analysis of the circuit's response to each input signal follows.

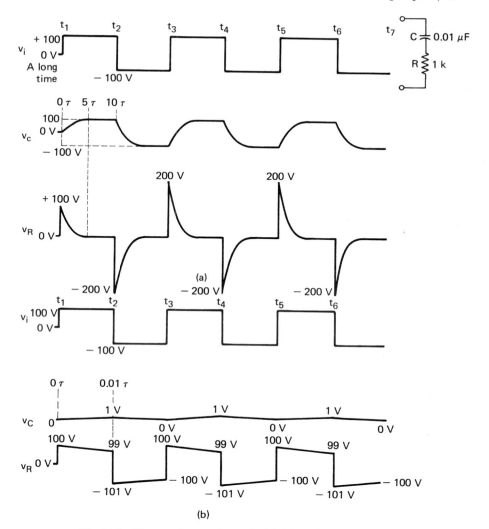

Fig. 1-18 The waveforms of v_R and with (a) a 5-kHz square-wave signal input and (b) a 5-MHz input. Note: The time just before the input voltage changes is called $-t_n$ and the time just after the input voltage changes is called t_n+.

Response to the 5-kHz Input [Fig. 1-18(a)]. The time for one cycle of the 5-kHz signal is

$$\frac{1}{F} = \frac{1}{5 \times 10^3}$$
$$= 2 \times 10^{-4}$$
$$\frac{1}{F} = 200 \ \mu s$$

During this time v_i is 100 V for one-half of the cycle and is -100 V for the other half; thus, each voltage is applied for 100 μs or 10 τ.

At $-t_1$ (Just Before the Input Rises to $+100$ V at t_1). If it is assumed that v_i has been 0 V for a long time (at least 5 τ), the capacitor has completely discharged and all voltages are zero.

At $t_1 +$ (Just After the Input Rises to $+100$ V at t_1). If it is assumed that v_i rises instantaneously, v_C remains zero and

$$v_R = v_i - v_C$$
$$= 100 - 0$$
$$v_R = 100 \text{ V}$$

Hence, the instantaneous 100-V rise in v_i appears entirely across the resistor.

At $-t_2$ (Just Before v_i Drops to -100 V at t_2). The input has been 100 V for 100 μs or 10 τ. After 50 μs or 5 τ the capacitor is fully charged to v_i, which reduces v_R to zero. These voltages remain the same through $-t_2$.

At $t_2 +$ (Just After v_i Drops to -100 V at t_2). Again, if it is assumed that the change in v_i is instantaneous, v_C remains unchanged. Hence, at t_2+, $v_C = 100$ V and

$$v_R = v_i - v_C$$
$$= (-100) - (100)$$
$$v_R = -200 \text{ V}$$

The instantaneous drop in v_i (from $+100$ V to -100 V) is also entirely across the resistor. All instantaneous changes in applied voltage are always felt on the resistance because it takes time for C to charge or discharge.

At $-t_3$ (Just Before v_i Rises to $+100$ V at t_3). The input voltage has been -100 V for 100 μs or 10 τ. The capacitor discharges the 100 V already on it and charges to the new v_i in 5 τ or 50 μs. Then $v_C = -100$ V and

$$v_R = v_i - v_C$$
$$= (-100) - (-100)$$
$$v_R = 0 \text{ V}$$

These voltages remain the same through $-t_3$.

At $t_3 +$ (Just After v_i Rises to 100 V at t_3). Again, the change in v_i (from -100 V to $+100$ V) is instantaneous, so v_C remains the same, and the 200-V rise appears across R_1 at t_3+, $v_C = -100$ V and

$$v_R = v_i - v_C$$
$$= (100) - (-100)$$
$$v_R = 200 \text{ V}$$

RC Circuits with Step Function Voltage Inputs / 27

At $-t_4$ (Just Before v_i Drops to -100 V at t_4). The capacitor has discharged the -100 V and charged to the new v_i in 50 μs or 5 τ. At this time $v_C = 100$ V and

$$v_R = v_i - v_C$$
$$= (-100) - (-100)$$
$$v_R = 0 \text{ V}$$

These voltages remain the same through $-t_4$.

This action continues, with v_C aiding v_i when the capacitor discharges and opposing v_i when it charges. When the capacitor begins discharging, v_R is either $+200$ V or -200 V. When it is fully charged to the new applied voltage, $v_R = 0$.

As shown in Fig. 1-18(a), the voltage across R in a *short time-constant** RC circuit consists of very narrow positive and negative voltage spikes. These short duration pulses are often used to trigger another circuit into (or out of) conduction. When the output voltage is taken across the resistor of a short time-constant RC circuit, it is called a *differentiator* circuit.

Response to the 5-MHz Input [Fig. 1-18(b)]. The time for one cycle of the 5-MHz signal is

$$\frac{1}{F} = \frac{1}{5 \times 10^6}$$
$$= 2 \times 10^{-7}$$
$$\frac{1}{F} = 0.2 \text{ μs}$$

During this time v_i is 100 V for 0.1 μs, or one-hundredth of a time constant, and -100 V for an equal time. In 0.01 τ the change in v_C is only approximately 1 percent of the difference between the applied voltage and the voltage already on the capacitor. Hence, the same circuit now has a *long time constant* when compared to the time for each alternation of the 5-MHz signal.

At $-t_1$ (Just Before v_i Rises to $+100$ V at t_1). If it is assumed that v_i has been 0 V for at least 5 τ, the capacitor has completely discharged and all voltages are zero.

At $t_1 +$ (Just After v_i Rises to $+100$ V at t_1). If it is assumed that v_i rises instantaneously, v_C remains zero and

$$v_R = v_i - v_C$$
$$= 100 - 0$$
$$v_R = 100 \text{ V}$$

---
*In general, a circuit is said to have a *short* time constant if its time constant is equal to or less than one-tenth of the time a voltage is applied, a *long* time constant if it is ten times the period a voltage is applied, and a *medium* time constant if it lies between these limits.

At $-t_2$ (Just Before v_i Drops to -100 V at t_2). The input has been 100 V for only 0.1 μs or 0.01 τ, during which C charges to approximately 0.01 v_i. Therefore, at $-t_2$, $v_C \approx 1$ V and

$$v_R = v_i - v_C$$
$$\approx 100 - 1$$
$$v_R \approx 99 \text{ V}$$

At t_2+ (Just After v_i Drops to -100 V at t_2). Again, if it is assumed that the change in v_i is instantaneous, v_C remains the same. Hence, at $t_2 + v_C \approx 1$ V and

$$v_R = v_i - v_C$$
$$= (-100) - (1)$$
$$v_R \approx 101 \text{ V}$$

Note that the 200-V drop in v_i appears across R.

At $-t_3$ (Just Before v_i Rises to $+100$ V at t_3). The input has been -100 V for only 0.1 μs or 0.01 τ. The capacitor tries to discharge the $+1$ V already on it and charge to the -100 V applied, but in 0.01 τ it only discharges by about 1 percent. Therefore, the change in v_C is approximately $0.01[(1) - (-100)] \approx 1$ V. Hence, at $-t_3$, v_C drops from 1 V to 0 V and $v_R = v_i = -100$ V.

At t_3+ (Just After v_i Rises 100 V at t_3). The change in v_i is instantaneous, so v_C remains zero and the 200-V rise appears across R. Therefore, at t_3+, $v_C \approx 0$ V and

$$v_R = v_i - v_C$$
$$\approx 100 - 0$$
$$v_R \approx 100 \text{ V}$$

The circuit conditions are now the same as those found at t_1+. Therefore, the calculations for the succeeding times will be the same as those made for the times $-t_2$ through t_3+. As shown in Fig. 1-18(b), the long time constant causes v_C to remain near zero at all times and makes v_R approximately the same as v_i. When the output voltage is taken across the capacitor of a long time-constant RC circuit, it is called an *integrator* circuit.

If the 5-kHz and 5-MHz signals are applied simultaneously and if the output is taken across the capacitor, the circuit is a *low-pass filter*. v_C is nearly the same as the 5-kHz input, but almost none of the 5-MHz signal appears across the capacitor.

1-7 RC Attenuators

Another application of RC circuits is in *compensated attenuators*. An example of a compensated attenuator is the low-capacitance probe (LCP) used with cathode-ray oscilloscopes (CROs). Under certain conditions, which will be explained later, distortion, even with the presence of shunt capacitance, is eliminated.

The circuit of Fig. 1-19(a) shows an LCP connected to the vertical input of a CRO. With perfect compensation, the waveform at the output of the probe v_o is the same as the waveform at its input v_i. This is shown in Fig. 1-19(b). Since at t_0+ the resistors are effectively open circuits compared to the capacitors, only a capacitive voltage divider exists. At t_0+

$$v_o = v_i \frac{C_1}{C_1 + C_i} \tag{1-10}$$

Since the time from t_0+ to $-t_1$ equals ∞, the capacitors are fully charged and are therefore open circuits across the resistors. Hence, at $-t_1$ only a resistive voltage divider exists and

$$v_o = v_i \frac{R_i}{R_1 + R_i} \tag{1-11}$$

In order for v_o at $-t_1$ to equal v_o at t_0+, Eq. (1-11) must equal Eq. (1-10). Hence, for perfect compensation

$$v_i \frac{C_1}{C_1 + C_i} = v_i \frac{R_i}{R_1 + R_i}$$

Dividing both sides by v_i and then cross multiplying yields

$$R_1 C_1 + R_i C_1 = R_i C_1 + R_i C_i$$

And by subtracting $R_i C_1$ from each side, it is found that for perfect compensation

$$R_i C_i = R_1 C_1$$

That is, the time constants must be equal.

The waveform in Fig. 1-19(c) is the result of overcompensation. The output at t_0+ is greater than v_o at $-t_1$. Although this is not desired in the LCP, it is frequently introduced to speed up the turn-on and turn-off times of transistor switches. Transient response of the transistor switch is explained in Sec. 2-5. *Overdrive* and the *speed-up capacitor* are explained in Sec. 2-6. Referring to Fig. 1-19(c), it is seen that for v_o at t_0+ to be greater than v_o at $-t_1$, Eq. (1-10) must be greater than Eq. (1-11). Hence, for overcompensation

$$v_i \frac{C_1}{C_1 + C_i} > v_i \frac{R_i}{R_1 + R_i}$$

30 / *Network Analysis*

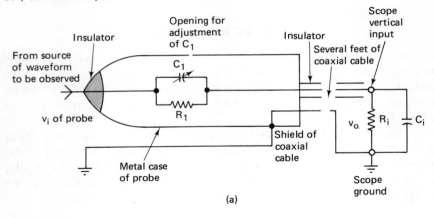

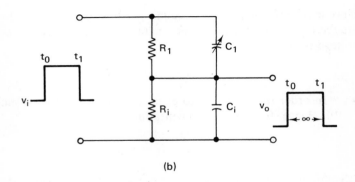

Fig. 1-19 (a) A compensated attenuator (low-capacitance probe) connected to the vertical input of an oscilloscope. (b) The input and output waveforms for perfect compensation, $R_1C_1 = R_iC_i$. (c) The output waveform with overcompensation, $R_1C_1 > R_iC_i$. (d) The output waveform with undercompensation, $R_1C_1 < R_iC_i$.

from which it is found that the overcompensated output occurs when $R_1C_1 > R_iC_i$.

The undercompensated output, shown in Fig. 1-19(d), is due to Eq. (1-11) being greater than Eq. (1-10). Hence, the undercompensated output occurs when $R_1C_1 < R_iC_i$.

Problems

1-1 Find the Thévenin equivalent of the circuit shown in Fig. 1-20(a), (b), and (c).

1-2 Find the Norton equivalent of each of the circuits shown in Fig. 1-20.

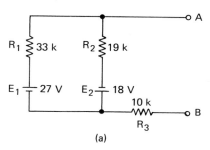

(a)

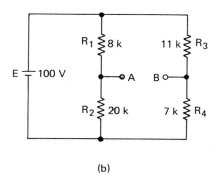

(b)

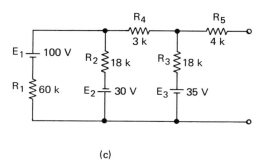

(c)

Fig. 1-20

1-3 The circuit in Fig. 1-21 is the equivalent of a diode *AND* gate with all inputs *A*, *B*, and *C* high. Calculate the output voltage by using Millman's theorem.

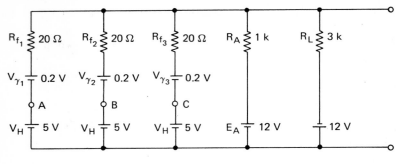

Fig. 1-21

1-4 Find the output voltage of the circuit shown in Fig. 1-22 by using Millman's theorem. Then find the current through each resistor.

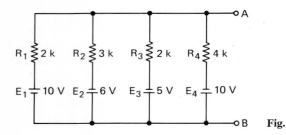

Fig. 1-22

1-5 Using Fig. 1-15(a) with $R = 10$ k, $C = 500$ pF, and $E = 100$ V:
  (a) Find v_C 8.5 µs after the switch is placed in position *A*. Assume that $v_C = 0$ V before the switch is thrown.
  (b) Assume that *C* has charged to 45 V while the switch is in position *A*. How long will it take to discharge to 10 V after the switch is thrown into position *B*?

1-6 A rectangular wave that swings between $+25$ V and -75 V at a frequency of 100 kHz with a mark-to-space ratio M/S $= 3$ is applied to the RC circuit of Fig. 1-18. Draw the waveforms of the applied voltage v_i and draw the resistor and capacitor voltages v_R and v_C for three cycles ($-t_1$ to $-t_7$) in a ladder diagram to scale. Indicate all voltages that exist just before and just after the changes in v_i.

1-7 The probe of Fig. 1-19(a) is designed to attenuate the input v_i by a factor of 10. If $R_i = 1$ M and $C_i = 50$ pF, what values of R_1 and C_1 must be used in the compensated attenuator?

2

SEMICONDUCTOR SWITCHING DEVICES

Digital computers use semiconductor diodes and transistors as high-speed switches. They are switched between their two extreme regions of operation by the voltages used to represent the two logic levels in the machine. In one extreme the semiconductor device is in its low conduction or cutoff state (open) and in the other it conducts heavily (closed). These and the amplifying region are explained in detail in this chapter. The limitations on switching speed and methods of improving the transient response are also discussed.

2-1 The Semiconductor Diode

The junction diode consists of a single *pn* junction. Its volt-ampere characteristic and schematic symbol are shown in Fig. 2-1. The arrowhead indicates the *p* section (anode) and the bar indicates the *n* section (cathode). V_D is the voltage across the diode in Fig. 2-1(a). It is positive when the anode is forward biased or positive with respect to the cathode and negative when the anode is reverse biased or negative with respect to the cathode. I_D and $-I_D$ are the forward (anode to cathode) and reverse (cathode to anode) diode currents. Note that different current and voltage scales are used so that the small forward voltages and reverse currents can be read clearly from the volt-ampere characteristic.

Forward Bias. When the anode is made positive with respect to the cathode, the diode is said to be *forward biased*. As the forward voltage V_D is increased, I_D increases very slowly until V_D sufficiently reduces the natural

34 / *Semiconductor Switching Devices*

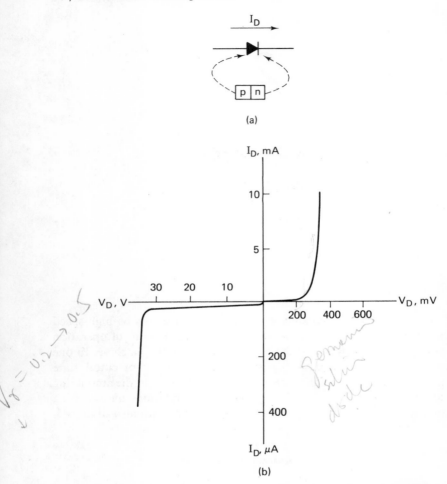

Fig. 2-1 (a) The schematic symbol for a junction diode. The arrow indicates the direction of conventional forward current. (b) The forward and reverse volt-ampere characteristics of a junction diode.

potential barrier of the *pn* junction. This forward voltage is known as the *cutin* voltage V_γ. (The greek letter *gamma* is used for the subscript because V_C is used to represent the collector-to-ground voltage in transistors and it is often used to represent one of the input voltages of logic gates.) As V_D is increased above V_γ, large increases in I_D take place. Figure 2-2 shows typical forward volt-ampere characteristics of a germanium diode and a silicon diode with comparable current ratings. The *cutin* voltage is in the order of 0.05 V to 0.2 V for germanium and is approximately 0.2 V to 0.5 V for most silicon diodes. It is sometimes called the *threshold* or *turn-on* voltage.

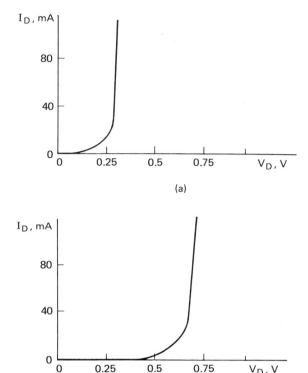

Fig. 2-2 Typical volt-ampere characteristics of (a) a germanium diode and (b) a silicon diode with comparable current ratings. The cutin voltage V_γ is approximately 0.2 V for germanium and is approximately 0.5 V for silicon.

Reverse Bias. When the anode is made negative with respect to the cathode, the diode is said to be *reverse biased*. As the reverse bias is increased, the majority carriers (holes in the *p* material and electrons in the *n* material) see a larger and larger barrier. However, the minority carriers (electrons in the *p* material and holes in the *n* material) see this voltage as forward bias. This accounts for the small reverse current which is alternatively called the *reverse saturation current, leakage current,* or *cutoff current* I_{co}. I_{co} is almost independent of voltage, but it is extremely sensitive to temperature. The slight increase in $-I_D$, as $-V_D$ is increased in Fig. 2-3(a), is due to shunt leakage resistance caused by surface effects and dirt at the junction. This resistance varies from several megohms for germanium to several hundred megohms for silicon. It may be assumed, with negligible error, that I_{co} is constant as $-V_D$ in increased. I_{co} is in the order of a few microamperes for

36 / *Semiconductor Switching Devices*

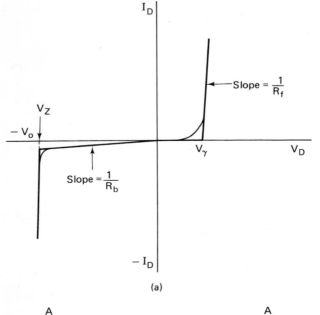

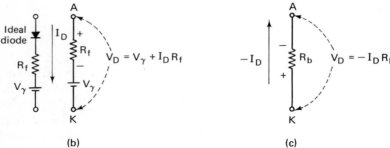

Fig. 2-3 (a) A diode volt-ampere curve with a piecewise linear approximation in dark lines. (b) The equivalent circuit of a forward biased diode. (c) The equivalent circuit of a reverse biased diode.

germanium diodes and is approximately 1,000 times less for silicon diodes. The variation in leakage current with temperature is given by

$$I_{co_f} = I_{co_i} \epsilon^{K(T_f - T_i)} \tag{2-1}$$

where

$K =$ a constant and is approximately equal to 0.075 in germanium and 0.13 in silicon;

T_f and $T_i =$ the final and initial temperatures in °C;

I_{co_f} and $I_{co_i} =$ the final and initial reverse saturation currents;

$\epsilon = 2.718$

The temperature dependance of germanium and silicon will be explained in the following examples.

Example 2-1: A germanium diode has a leakage current $I_{co} = 2$ μA at 25°C. What is the value of I_{co} at 34.3°C?
Solution:

$$I_{co_t} = 2 \text{ μA } [2.718^{(0.075)(34.3°C - 25°C)}]$$
$$= 2 \text{ μA } [\log^{-1}(0.698 \log 2.718)]$$
$$= 2 \text{ μA } [\log^{-1} 0.303]$$
$$= 2 \text{ μA } [2.01]$$
$$I_{co_t} \approx 4 \text{ μA}$$

Example 2-2: A silicon diode has a leakage current $I_{co} = 2$ nA at 25°C. What is the value of I_{co} at 30.4°C?
Solution:

$$I_{co_t} = 2 \text{ nA } [2.718^{(0.13)(30.4°C - 25°C)}]$$
$$= 2 \text{ nA } [\log^{-1}(0.701 \log 2.718)]$$
$$= 2 \text{ nA } [\log^{-1} 0.305]$$
$$= 2 \text{ nA } [2.02]$$
$$I_{co_t} \approx 4 \text{ nA}$$

Conclusion: I_{co} approximately doubles for every 9.3°C increase in temperature in germanium (Ex. 2-1), while it approximately doubles for every 5.4°C increase in temperature in silicon (Ex. 2-2). It should be remembered, however, that I_{co_t} is approximately 1,000 times smaller in silicon than in germanium. So even though I_{co} rises at a faster rate in silicon than germanium, silicon is still almost always preferred in high-temperature applications. Example 2-3 verifies the fact that at a given high temperature I_{co} is lower in silicon than in germanium.

Example 2-3: Find the leakage current at 125°C for the diodes in the above examples.
Solution: The germanium diode has a leakage current

$$I_{co_t} = 2 \text{ μA } [2.718^{(0.075)(125°C - 25°C)}]$$
$$= 2 \text{ μA } [\log^{-1}(7.5 \log 2.718)]$$
$$= 2 \text{ μA } [\log^{-1} 3.26]$$
$$= 2 \text{ μA } [1.82 \times 10^3]$$
$$I_{co_t} = 3.64 \text{ mA}$$

38 / Semiconductor Switching Devices

The silicon diode has a leakage current

$$I_{co_t} = 2 \text{ nA } [2.718^{(0.13)(125°C-25°C)}]$$
$$= 2 \text{ nA } [\log^{-1}(13 \log 2.718)]$$
$$= 2 \text{ nA } [\log^{-1} 5.64]$$
$$= 2 \text{ nA } [4.36 \times 10^5]$$
$$I_{co_t} = 0.872 \text{ mA}$$

Note that this current for silicon is less than one-fourth the current for germanium.

The Breakdown Voltage. When the reverse voltage exceeds a certain value, a large reverse current flows. An increase in reverse voltage of only one-tenth of a volt will cause the reverse current to increase from a few microamperes to several milliamperes. There should be an external resistance to limit this current, or else temperature will increase and ruin the junction. The abrupt increase in $-I_D$ is caused by either of the two following actions:

1. *Zener breakdown.* A reverse potential difference of only -6 V across a junction that is only 0.0001 inches wide means a field intensity of 60,000 V/in. The resulting high electric field intensity is so strong that it forces electrons to break covalent bonds, even at a low potential difference of -6 V.
2. *Avalanche breakdown.* At a sufficiently higher value of reverse voltage, minority carriers are accelerated across the junction with enough velocity to break up covalent bonds by impact or collision. The freed electrons are also accelerated by the high reverse potential and they break up other covalent bonds upon collision.

If a diode is *heavily* doped, the barrier zone is *thin* and Zener breakdown occurs first. However, if the diode is lightly doped, the barrier zone is wide and avalanche breakdown occurs first. The Zener effect is important only in diodes with low breakdown voltages (2 V to 6 V) while silicon diodes are available with avalanche breakdown voltages up to several hundred volts. However, the term *Zener diode* is commonly applied to all diodes that operate in the reverse breakdown region. The voltage at which breakdown occurs is alternatively called the *Zener voltage* V_z, the *avalanche voltage*, or the *peak inverse voltage*, PIV.

The Piecewise Linear Diode Equivalent. A useful large-signal equivalent circuit of a diode comes from approximating its volt-ampere curves. Figure 2-3(a) shows a piecewise linear approximation of a typical semiconductor diode. It is shown to have an infinite forward resistance until V_D equals the cutin voltage V_γ. At values of $V_D > V_\gamma$, the linearized diode has a *forward resistance* $R_f = \Delta V_D/\Delta I_D$ (typically 5–20 Ω).

The reverse biased diode is shown to have a *reverse* or *back resistance* R_b in the order of several megohms at $-V_D < V_z$ and a very low reverse resistance at $-V_D > V_z$. Figure 2-3(b) shows the equivalent circuit of a forward biased diode with $V_D > V_\gamma$ and $R_f = V_{Rf}/I_D = (V_D - V_\gamma)/I_D$. Figure 2-3(c) shows the equivalent circuit of a reverse biased diode. With $-V_D < V_z$, it is simply a large resistance $R_b = -V_D/-I_D$. The piecewise linear representation is used as a large-signal approximation of the diode in this text.

2-2 Signal Characteristics of Diodes

When a diode must be turned on and off with high-frequency signals, the ability of the diode to respond to these signals must be considered. The high-frequency limitations of the diode are discussed in this section.

Barrier Capacitance C_b. When a diode is reverse biased, a capacitance is formed by the barrier zone and the conducting p and n layers. This *barrier* or *depletion capacitance* C_b must be considered when high-frequency signals are applied to the diode. For example, a large barrier capacitance will act as a coupling capacitor causing the reverse biased diode to pass, rather than block, an undesired signal. The value of C_b is specified at 1 MHz and 0 V bias and includes lead and packaging capacitance. It varies from less than one to several picofarads in high-frequency diodes. Since the barrier zone becomes thicker as the reverse bias is increased, C_b is inversely proportional to $-V_D$.

Forward Recovery Time. When a diode has previously been reverse biased, it takes time for an applied forward bias to reduce the barrier zone. This time is known as the *turn-on* or *forward recovery time* of the diode. In digital circuits the diode is usually in series with the source permitting C_b to act as a coupling capacitor. Thus, the desired signal can be passed to the load even though the diode doesn't conduct instantaneously. As explained in Sec. 2-5, a series capacitor can provide an overdrive spike to actually speed up the load. In addition, the forward recovery time is usually much less than the reverse recovery time. Consequently, forward recovery time is usually not given in the manufacturer's specification sheet.

Reverse Recovery Time. When a diode is forward biased, it takes time for an applied reverse bias to remove minority carriers from the conducting junction. This time is known as the *turn-off* or *reverse recovery time* of the diode. While the diode remains on, the reverse bias is applied to the load and may cause an undesired switching of that circuit. Diodes are available with reverse recovery times of less than one nanosecond.

2-3 The Transistor Amplifier

Before the transistor as a switch is discussed, a review of the transistor as an amplifier is presented. In this section the volt-ampere characteristics, temperature effects, and signal relationships of the junction transistor are explained.

Biasing the Junction Transistor. The two types of junction transistors are shown in Fig. 2-4. Silicon (Si) or germanium (Ge) crystals are formed in either *pnp* or *npn* layers. The layers are called the *emitter E*, the *base B*, and the *collector C*. The emitter-base (EB) junction is forward biased, which reduces the EB barrier and permits the emitter to inject some of its majority carriers into the base. The base is thin (0.0001 in. or less) and lightly doped, so most of these carriers (95–99.5 percent) reach the collector-base (CB) junction where they can be attracted by the collector supply. In the case of a *pnp* transistor, a poistive V_{EB} will force holes to leave the emitter and flow into the base. Most of them reach the CB junction where a negative V_{CB} can collect them. The supplies are reversed for an *npn* transistor.

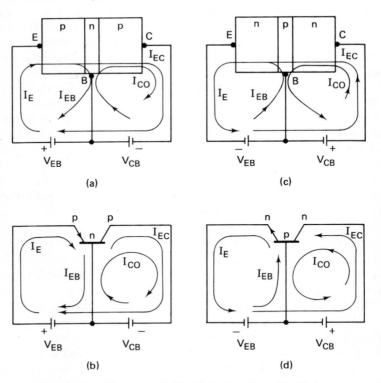

Fig. 2-4 (a)–(b) A properly biased *pnp* transistor. (c)–(d) A properly biased *npn* transistor.

Conclusion: *To properly bias a junction transistor as an amplifier, the emitter-base junction must be forward biased and the collector-base junction must be reverse biased.*

Transistor Currents. The currents mentioned above are designated I_E (the emitter current), I_{EB} (the current that goes from the emitter to the base), and I_{EC} (the current that goes from the emitter to the collector). In addition, there is a leakage current designated I_{CBO} or I_{CO}. I_{CBO} stands for the current between the collector and base with the emitter (input) open. This current is caused by the minority carriers on each side of the CB junction which see V_{CB} as forward bias. The direction of I_{CO} is shown in Fig. 2-4 from which the following relationships can be seen:

$$I_E = I_{EB} + I_{EC} \tag{2-2}$$

$$I_C = I_{EC} + I_{CO} \tag{2-3}$$

$$I_B = I_{EB} - I_{CO} \tag{2-4}$$

From Eq. (2-3)

$$I_{EC} = I_C - I_{CO} \tag{2-3a}$$

and from Eq. (2-4)

$$I_{EB} = I_B + I_{CO} \tag{2-4a}$$

Substituting (2-3a) and (2-4a) in Eq. (2-2) yields

$$I_E = I_B + I_C \tag{2-5}$$

The ratio of I_{EC} to the emitter current I_E is called α_{DC}, where

$$\alpha_{DC} = \frac{I_{EC}}{I_E} = \frac{I_C - I_{CO}}{I_E} \tag{2-6}$$

I_{CO} is only a few microamperes for germanium and is in the nanoampere range in silicon transistors so

$$\alpha_{DC} \approx \frac{I_C}{I_E} \tag{2-7}$$

The ac or small-signal current amplification factor is called α

$$\alpha = \left.\frac{\Delta I_{EC}}{\Delta I_E}\right|_{V_{CB}} \tag{2-8}$$

But since I_{CO} is constant at a given temperature,

$$\alpha = \left.\frac{\Delta I_C}{\Delta I_E}\right|_{V_{CB}} \tag{2-9}$$

Because I_{CO} is very small, α and α_{DC} are usually used interchangeably.

Transistor Amplifier Configurations. There are three types of transistor amplifier circuits. They are the *common base* (CB), the *common collector* (CC), and the *common emitter* (CE) amplifiers. Only the common emitter amplifier is widely used. The common base circuit has only a few applications. It may be used to match a low Z source to a high Z load or as a voltage amplifier. The common collector or *emitter follower* is widely used as an impedance matching device between a high Z source and a low Z load. The input and output signals are in phase in both the CB and CC amplifiers. Properly biased CB and CC amplifiers are shown in Fig. 2-5(a) and (b).

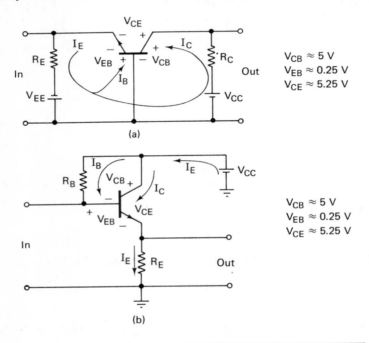

Fig. 2-5 (a) A common-base circuit properly biased in the amplifier region. (b) A common-collector circuit properly biased as an amplifier. (c) A comparison of the two circuits.

A comparison between these circuits is given in Fig. 2-5(c). Note that the CB has a current amplification A_i of less than one and that the CC has a voltage amplification A_v of less than one. This causes these circuits to have low values of power amplification A_p in comparison to the CE. Both circuits also have a great difference between their input and output impedances which makes them difficult to cascade. A knowledge of these CB and CC characteristics is sufficient for the study of digital circuits. If further information is desired, refer to transistor amplifier textbooks.

The Common Emitter Amplifier. The CE configuration is the most frequently used circuit. A simplified CE amplifier is shown in Fig. 2-6(a). Typical values of A_i, A_v, A_p, R_i, and R_o are given in Fig. 2-6(b). Notice the high values for both A_i and A_v. This results in very high values of power amplification.

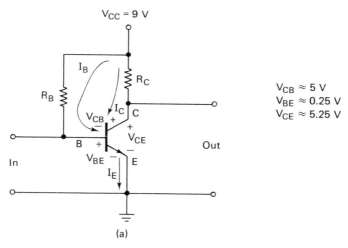

Quantity	Typical minimum	Typical maximum
A_i	20	100
A_v	50	500
A_p	2,000	10,000
R_i	1 kΩ	2 kΩ
R_o	20 kΩ	50 kΩ

(b)

Fig. 2-6 (a) A common emitter circuit properly biased as amplifier. (b) Typical values of A_i, A_v, A_p, R_i, and R_o for the CE amplifier.

Characteristic Curves. Figure 2-7 shows the various characteristic curves used in analyzing the CE circuit. The *output characteristics* [Fig. 2-7(a)] show how the output current I_C varies when the output voltage V_{CE} is changed and the input current I_B is held constant. A family of output curves is given in Fig. 2-7(a). Note that I_C is almost independent of V_{CE} but that it increases substantially as I_B is increased. This transistor has a maximum collector dissipation $P_{C\max}$ rating of 180 mW. By connecting all points at which I_C and V_{CE} combine to produce 180 mW a $P_{C\max}$ curve is produced.

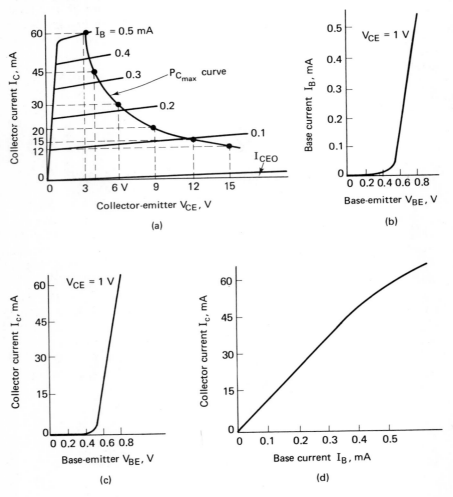

Fig. 2-7 Typical characteristic curves for a silicon *npn* transistor in the CE circuit. (a) Output, (b) input, (c) output current vs input voltage and (d) current transfer characteristics.

The $P_{C\max}$ rating is exceeded at all points to the right of this curve. The manufacturer specifies the maximum collector power at 25°C. For higher temperatures, P_C must be decreased. A typical P_C derating curve is shown in Fig. 2-8.

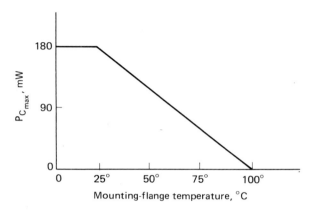

Fig. 2-8 A typical $P_{c\max}$ derating curve as supplied in a manufacturer's specification sheet.

The *input characteristics* in Fig. 2-7(b) show how the input current I_B varies when the input voltage V_{BE} is changed and the output voltage V_{CE} is held constant. As V_{CE} is increased, the effective base width decreases which causes a decrease in the recombination base current and accounts for the slight increase in I_C shown in Fig. 2-7(a). Because the decrease in I_B will be small as V_{CE} is increased above 1 V, only one input curve is given. This curve may be considered representative of all curves for values of V_{CE} greater than 1 V. Figure 2-7(c) shows how the output current I_C varies with input voltage V_{BE}, and Fig. 2-7(d) plots output current I_C vs input current I_B. This curve (Fig. 2-7d) is known as the *forward current transfer characteristic*.

Large-signal Current Amplification Factor β_{dc} or h_{FE}. The ratio of dc emitter-to-collector current I_{EC} to dc emitter-to-base current I_{EB} is called the dc or large-signal current amplification factor β_{dc} or h_{FE}.

$$\beta_{dc} = h_{FE} = \frac{I_{EC}}{I_{EB}} = \frac{I_C - I_{CO}}{I_B + I_{CO}} \qquad (2\text{-}10)$$

If $I_C \gg I_{CO}$ and $I_B \gg I_{CO}$,

$$\beta_{dc} = h_{FE} \approx \frac{I_C}{I_B} \qquad (2\text{-}11)$$

Small-signal Current Amplification Factor β or h_{fe}. The ac or small-signal current amplification factor of the CE circuit is called β or h_{fe}. The definition is given by

$$\beta = h_{fe} = \left.\frac{\Delta I_{EC}}{\Delta I_{EB}}\right|_{V_{CE}} \quad (2\text{-}12)$$

But since I_{co} is constant at a given temperature,

$$\beta = h_{fe} = \left.\frac{\Delta I_C}{\Delta I_B}\right|_{V_{CE}} \quad (2\text{-}13)$$

When the leakage current is small, β and β_{dc} are used interchageably.

The relationship between α and β is determined as follows: First, α is found in terms of β.

$$\alpha = \frac{\Delta I_C}{\Delta I_E} \quad (2\text{-}9)$$

$$= \frac{\Delta I_C}{\Delta I_C + \Delta I_B}$$

But from Eq. (2-13), $\Delta I_C = \beta(\Delta I_B)$. Hence,

$$\alpha = \frac{\beta(\Delta I_B)}{\beta(\Delta I_B) + (\Delta I_B)}$$

By dividing the numerator and denominator by ΔI_B, it is found that

$$\alpha = \frac{\beta}{\beta + 1} \quad (2\text{-}14)$$

Now, β is found in terms of α.

$$\beta = \frac{\Delta I_C}{\Delta I_B} \quad (2\text{-}13)$$

But $\Delta I_C = \alpha(\Delta I_E)$ and $\Delta I_B = (1 - \alpha)\Delta I_E$. Hence,

$$\beta = \frac{\alpha(\Delta I_E)}{(1 - \alpha)\Delta I_E}$$

By dividing the numerator and denominator by ΔI_E it is found that

$$\beta = \frac{\alpha}{1 - \alpha} \quad (2\text{-}15)$$

Signal Phase Relationships in the CE Amplifier. When the transistor is connected as a CE amplifier, the output voltage signal is 180° out of phase with the input voltage signal. A CE amplifier, with its waveforms, is shown in Fig. 2-9. A dc operating point is assumed with $V_{CE} = 5$ V and $V_{BE} = 0.6$ V.

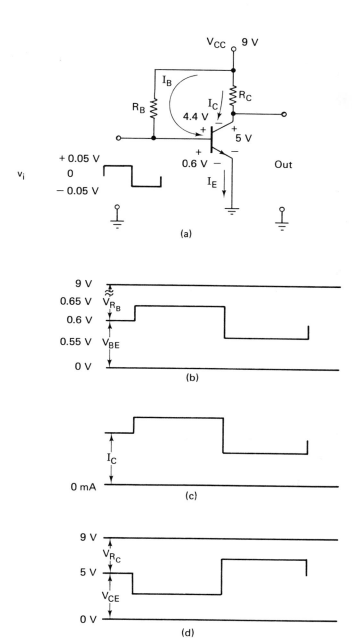

Fig. 2-9 (a) A common-emitter amplifier. (b) The instantaneous total (ac + dc) base-to-emitter voltage v_{BE}. (c) The instantaneous total collector current i_C. (d) The instantaneous total collector-to-emitter voltage v_{CE}.

When the input signal goes positive, it increases the EB forward bias v_{BE} to 0.65 V. This increases the collector current i_C, which increases the voltage drop on R_C. $v_{CE} = V_{CC} - v_{R_C}$ so when v_{R_C} increases, the output voltage v_{CE} decreases. During the second half cycle of input v_{BE} decreases, which causes i_C and v_{R_C} to decrease and v_{CE} to increase. This signal inversion characteristic of the CE amplifier is often used to advantage.

2-4 The Transistor Switch

In this section the *cutoff* and *saturation* regions are explained. Figure 2-10 shows the output characteristics of a CE circuit with the three regions of operation identified. Transistors of the *npn*-type are used because the higher mobility of the majority electrons permits faster switching. The *cutoff* region is the region below the $I_B = 0$ curve. The *saturation* region is the area to the left of the knee of the curves. The area between these two regions is the *active* or *amplifier* region. It is permissible for the load line to be to the right of the $P_{C\,max}$ curve because the transistor is switched so quickly through the active region.

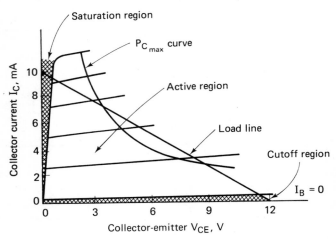

Fig. 2-10 Common emitter output characteristics showing the three regions of operation.

The Common Emitter Cutoff Region. When the input current is reduced to zero, the output current does not become zero. In fact, an appreciable leakage current may still flow in the output circuit. In the CB circuit it is the current that flows between the collector and base with the emitter (input) open. This value of I_C is called I_{CBO} or I_{CO}. In the CE circuit the leakage current is the current that flows between the collector and emitter with the base (input) open. It is designated I_{CEO}. The relationship between I_{CEO} and I_{CO} is found as follows:

In any transistor

$$I_C = I_{EC} + I_{CO} \qquad (2\text{-}3)$$

and from Eq. (2-6)

$$I_{EC} = \alpha I_E$$

so

$$I_C = \alpha I_E + I_{CO} \qquad (2\text{-}16)$$

Equation (2-16) is used with the common base amplifier. In the common emitter circuit I_C is found in terms of its input current I_B and β. Substituting Eq. (2-5) and Eq. (2-14) in Eq. (2-16) gives

$$I_C = \frac{\beta}{1+\beta}(I_B + I_C) + I_{CO}$$

from which

$$I_C = \beta I_B + (1 + \beta) I_{CO} \qquad (2\text{-}17)$$

Equation (2-17) shows that when the CE input is open-circuited so that $I_B = 0$,

$$I_C = (1 + \beta) I_{CO}$$

Therefore,

$$I_{CEO} = (1 + \beta) I_{CO} \qquad (2\text{-}18)$$

The leakage current in the CE circuit is greater than the leakage current in the CB circuit by the factor $1 + \beta$. The value of β is lower near cutoff, but it may still be as large as 10. Since β also increases with temperature, the common emitter circuit is often stabilized to prevent *thermal runaway*. That is, an increase in temperature increases β and the leakage current I_{CO}. The increase in I_{CO} is multiplied by $1 + \beta$ and causes an additional increase in temperature. If not controlled, I_{CO} and temperature continue to increase until the transistor is ruined. I_{CEO} appears in Fig. 2-7(a) as the collector current with $I_B = 0$. This value of I_{CEO} is measured at room temperature, but since I_{CO} approximately doubles for every increase of 9.3°C in germanium and 5.4°C in silicon (as shown in Exs. 2-1 and 2-2 of Sec. 2-1), the actual value of I_{CEO} may be much higher at the operating temperature. In switching circuits, a high value of I_{CO} can keep a circuit on when it should be off.

Temperature Stabilization. Figure 2-11 shows a modified CE circuit. V_{BB} and R_B are designed to keep the transistor cut off at the highest expected temperature. Cutoff will be maintained if $I_E = 0$. Then from Eq. (2-2), I_{EB}

50 / Semiconductor Switching Devices

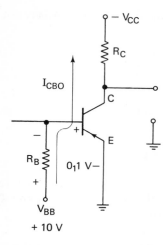

Fig. 2-11 A common emitter circuit with temperature stabilization. A reverse base current $-I_B = I_{CBO}$ flows as shown.

and I_{EC} equal zero; and Eqs. (2-3) and (2-4) reduce to $I_C = I_{CO}$ and $I_B = -I_{CO}$. I_E will equal zero if V_{BE} is a reverse voltage of about 0.1 V for germanium and is 0 V for silicon. This voltage is designated $V_{BE_{co}}$.

Conclusion: *A Ge transistor is at the edge of cutoff when both junctions are reverse biased. A Si transistor is at the edge of cutoff when the CB junction is reverse biased and $V_{BE} = 0$. Beyond cutoff both junctions are reverse biased in either type.*

Example 2-4: Using a GE transistor with an $I_{CO} = 2~\mu\text{A}$ and a $V_{BE_{co}} = 0.1$ V at 25°C, find the value of R_B that will just keep the transistor of Fig. 2-11 cut off at 80°C. Use $V_{BB} = 10$ V.
Solution: First determine I_{CO} at 80°C

$$I_{CO_f} = I_{CO}\epsilon^{K(T_f-T_i)} \tag{2-1}$$

(where $K = 0.075$ for GE).

$$I_{CO_f} = [2~\mu\text{A}][2.718^{(0.075)(80°-25°)}]$$
$$= [2~\mu\text{A}][\log^{-1}(4.13 \log 2.718)]$$
$$= [2~\mu\text{A}][\log^{-1} 1.8]$$
$$= [2~\mu\text{A}][63]$$
$$I_{CO_f} = 126~\mu\text{A}$$

Now determine R_B.

$$R_B = \frac{V_{BB} - V_{BE_{co}}}{I_{CO_{max}}} = \frac{10 - 0.1}{126 \times 10^{-6}}$$

$$R_B = 78.5~\text{k}\Omega$$

Resistor R_B will drop 9.9 V at 80°C and leave 0.1 V reverse bias across the EB junction in Fig. 2-11.

Transistor Breakdown Voltage. When the circuit of Fig. 2-11 is operated at *lower* temperatures, I_{CO} is less than 126 µA and V_{R_B} decreases. This causes the reverse bias across the emitter-base junction to *increase*. At room temperature $I_{CO} = 2$ µA, $V_{R_B} = (2$ µA$)(78.5$ k$) = 157$ mV, and $V_{BE} \approx V_{BB}$. The transistor must have an emitter-base *reverse breakdown voltage rating* BV_{EBO} that exceeds V_{BB}. BV_{EBO} varies from less than 1 V to as much as 100 V in some transistors. The third subscript O indicates that the collector is open-circuited.

In addition to BV_{EBO}, the manufacturer specifies the *reverse breakdown voltage rating of the CB junction* BV_{CBO}. The O means the emitter is open. This breakdown is caused either by *avalanche breakdown* or by *punch-through*. Avalanche breakdown is explained in Sec. 2-1. Punch-through occurs when the collector-base depletion region spreads through the base and reaches into the emitter. A large increase in current takes place and transistor action is lost. In thin base-alloy transistors punch-through occurs first and determines BV_{CBO}. Punch-through is also known as *reach-through*. A list of transistor breakdown voltage definitions is given in Table 2-1. The ratings do not always apply to all circuits.

Table 2-1 Transistor breakdown voltage symbols and definitions.

Symbol	Definition
BV_{EBO}	The dc reverse breakdown voltage of the EB junction (at a specified I_E) with the collector open.
BV_{CBO}	The dc reverse breakdown voltage of the CB junction (at a specified I_C) with the emitter open.
BV_{CEO}	The dc collector-to-emitter breakdown voltage (at a specified I_C) with the base open. $BV_{CEO} < BV_{CBO}$
BV_{CER}	The same as BV_{CEO} except that the base is returned to the emitter through a resistor. $BV_{CER} > BV_{CEO}$
BV_{CES}	The same as BV_{CEO} except that the base is shorted to the emitter. $BV_{CES} > BV_{CER}$
BV_{CEX}	The same as BV_{CEO} except that the EB junction is reverse biased through a specified circuit. $BV_{CEX} > BV_{CES}$

The Common Emitter Saturation Region. Figure 2-12 shows the input characteristics and the output current vs input voltage curve for a CE circuit. Note that these are similar to the diode curves of Fig. 2-2. When the emitter-base junction is forward biased, i_B and i_C remain near zero until v_{BE} sufficiently reduces the natural barrier potential of the *pn* junction. When v_{BE} exceeds the *cutin voltage* V_γ of the junction, i_B and i_C increase and the transistor enters the active region. The cutin voltage decreases with temperature but at room temperature on a linearized curve it is typically from 0.4 V to 0.6 V for silicon and from 0.1 V to 0.2 V for germanium.

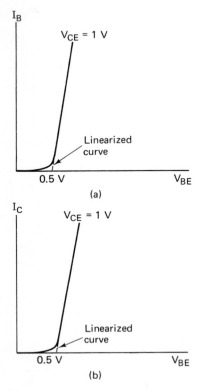

Fig. 2-12 (a) The input curves. (b) The I_C vs V_{BE} curves of a silicon transistor. I_B and I_C are both equal to zero until V_{BE} exceeds the cutin voltage $V_\gamma = +0.5$ V.

As v_{BE} is increased more and more, i_C increases until the transistor is nearly a short circuit. Resistance values of a few ohms are typical. Figure 2-13 shows an *npn* Si and an *npn* Ge transistor in saturation. As i_C increases, v_{R_C} increases and v_{CE} decreases to only a few tenths of a volt. When v_{CE} becomes less than v_{BE}, the collector-base junction becomes forward biased and the transistor is saturated.

$$I_{C_{sat}} = \frac{V_{CC} - V_{CE_{sat}}}{R_C} \approx \frac{V_{CC}}{R_C}$$

Thus, $I_{C_{sat}}$ is limited only by V_{CC} and R_C. The collector current becomes almost independent of I_B and the transistor's parameters. Figure 2-14 shows how it is possible to have less voltage across the entire transistor than there is across the EB junction.

$$V_{CE_{sat}} = V_{BE_{sat}} + V_{CB_{sat}} = (+0.7 \text{ V}) + (-0.4 \text{ V}) = 0.3 \text{ V}$$

Conclusion: *A transistor is in saturation when both junctions become forward biased.*

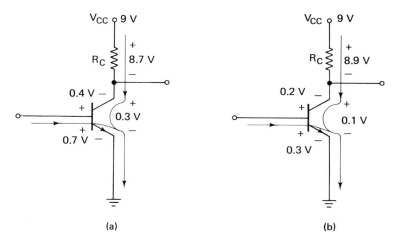

Fig. 2-13 A common emitter circuit in saturation. (a) Typical voltages for a silicon transistor. (b) typical voltages for a germanium transistor.

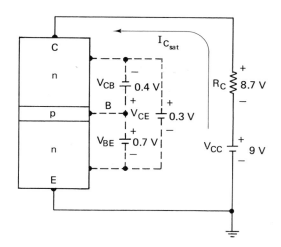

Fig. 2-14 The junction voltages are series opposing in saturation so that $V_{CE_{sat}}$ is less than $V_{BE_{sat}}$.

Although $I_{C_{sat}}$ is independent of the transistor's parameters, the parameter h_{FE} is important. $I_{C_{sat}}$ is easily determined for a given V_{CC} and R_C by V_{CC}/R_C. The minimum base current required to saturate the transistor is then given by

$$I_{B_{min}} = \frac{I_{C_{sat}}}{h_{FE}} \approx \frac{V_{CC}}{h_{FE}R_C} \tag{2-19}$$

Consequently, the manufacturer specifies the dc forward current transfer ratio h_{FE} of the transistor. h_{FE} is lower at the edge of saturation than it is in the active region and it varies over a wide range of values even for transistors of the same type. When calculating $I_{B_{min}}$, the minimum or worst-case h_{FE} should be used. This value is called $h_{FE_{min}}$.

A list of typical junction voltages is given in Table 2-2. These voltages

Table 2-2 Typical junction voltages for *npn* transistors at the edge of cutoff at the edge of turn-on (cutin), in the active and saturation regions. The polarities reverse for *pnp* transistors.

	Cutoff	Cutin	Active	Saturation
Ge	$V_{BE_{CO}} = -0.1$ V	$V_{BE} = V_\gamma = 0.2$ V	$V_{BE} = 0.25$ V	$V_{BE_{sat}} = 0.3$ V $V_{CE_{sat}} = 0.1$ V
Si	$V_{BE_{CO}} = 0$ V	$V_{BE} = V_\gamma = 0.5$ V	$V_{BE} = 0.6$ V	$V_{BE_{sat}} = 0.7$ V $V_{CE_{sat}} = 0.3$ V

vary with temperature and from transistor to transistor, but their use in analyzing digital circuits causes only a negligible error. $V_{CE_{sat}}$ increases and the EB cutin voltage, V_γ, decreases with temperature. Therefore, it is easier to turn on or saturate a transistor at higher temperatures.

2-5 Transient Response of the Transistor Switch

When the input current is varied between the turn-on and turn-off levels, the state of the transistor does not immediately change. It takes time for the transistor to turn on and turn off. Figure 2-15(a) shows a CE circuit and its collector current response to the input current waveform. $V_{CE_{sat}}$ is neglected in this discussion.

$$I_{C_{sat}} \approx \frac{V_{CC}}{R_C} = \frac{10 \text{ V}}{1 \text{ k}} = 10 \text{ mA}$$

$$I_{B_{min}} = \frac{I_{C_{sat}}}{h_{FE_{min}}} \approx \frac{10 \text{ mA}}{20} = 0.5 \text{ mA} \tag{2-19}$$

Delay Time t_d. Delay time t_d is the time that it takes a pulse to rise from its minimum value to 10 percent of its maximum value. In a transistor this delay is due, in part, to the time that it takes for the EB junction capacitance to discharge the reverse bias and the cutin potential to be reached. In addition, it takes time for the carriers to cross the base and begin entering the collector.

Rise Time t_r. Rise time t_r is the time that it takes a pulse to rise from 10 percent of its maximum value to 90 percent of its maximum value. Since the carriers travel along different paths, it takes time for i_C to pass through the active region and enter the saturation region. Circuit capacitance, the

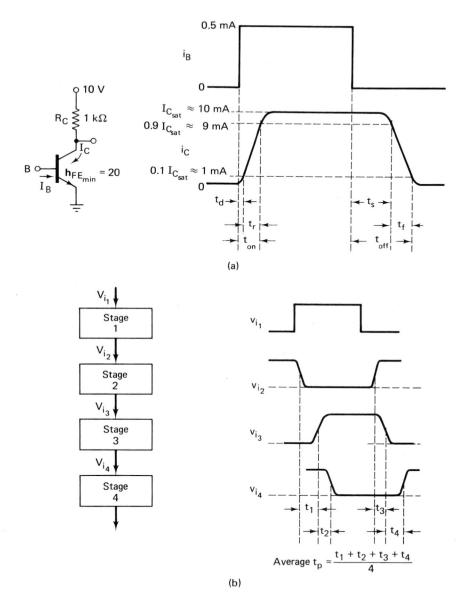

Fig. 2-15 (a) Input and output current waveforms showing delay time, rise time, storage time, and fall time. (b) The procedure for measuring propagation delay time.

frequency response of the transistor, and the amplitude of the input current affect collector current rise time.

Storage Time t_s. Storage time t_s is the time that it takes a pulse to fall from its maximum value to 90 percent of its maximum value. When a transistor is saturated, the CB junction becomes forward biased. The collector injects carriers into the base and before turn-off occurs these *stored* carriers must be returned to the collector. In the case of the *npn* transistor in Fig. 2-15 the collector emits electrons into the base. When the input current is removed, the transistor remains saturated until 10 percent of the electrons return to the collector. Storage time is also called *saturation delay time*.

Fall Time t_f. Fall time t_f is the time that it takes a pulse to fall from 90 percent of its maximum value to 10 percent of its maximum value. Since the carriers travel along different paths when returning to the collector, it takes time for i_C to pass through the active region and enter the cutoff region. Collector current fall time, t_f, is affected by the same factors as rise time, t_r.

Propagation Time t_p. Propagation time t_p is the time that it takes for a given logic level to pass through a circuit. It is standard practice to measure t_p as the time between the 50-percent levels of the input and output waveforms. Figure 2-15(b) illustrates the procedure used to measure propagation time. Because $t_d + t_r$ is not the same as $t_s + t_f$, the propagation time depends on whether the output is going from the low voltage level V_L to the high voltage level V_H, or vice-versa. Propagation time is usually measured over an even number of stages and the *average propagation time* $t_{p_{av}}$ is calculated [Fig. 2-15(b)]. This average time is sometimes called a *unit propagation delay*.

Turn-on Time $t_{on} = t_d + t_r$ and Turn-off Time $t_{off} = t_s + t_f$. High-speed transistors have turn-on and turn-off times in the nanosecond range. When switching speed is a primary consideration, however, special circuit techniques are used. One of these techniques is called *overdrive* in which the input current exceeds the minimum turn-on and turn-off values. The other utilizes clamping diodes to prevent saturation and eliminate the storage effects.

2-6 Overdrive

Applying Eq. (2-19) to Fig. 2-16(a) yields

$$I_{B_{min}} = \frac{V_{CC}}{R_C \mathbf{h}_{FE_{min}}} = \frac{12}{(2)(20)} = 0.3 \text{ mA}$$

The base current signal in Fig. 2-16(b) will just saturate Q_1 with 0.3 mA and it will just cut it off with 0 mA. However, the transistor will turn on and turn off much faster with the *overdrive* base current waveform shown in Fig. 2-16(c). The collector current response to both input signals is given in Fig. 2-17. Without overdrive, i_C climbs to $I_{C_{sat}} = \mathbf{h}_{FE}I_B = (20)(0.3) = 6$ mA

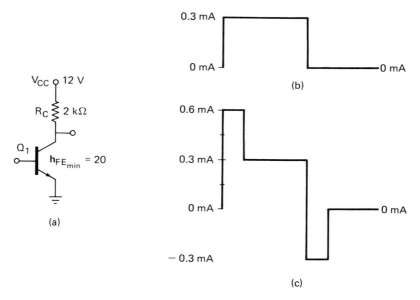

Fig. 2-16 (a) A transistor with an $I_{C_{sat}} \approx V_{CC}/R_C = 6$ mA and an $I_{B_{min}} = I_{C_{sat}}/h_{EE_{min}} \approx 0.3$ mA. (b) Base current waveform to just saturate and cut off the transistor. (c) Ideal base current waveform for an overdrive factor **od** = 2.

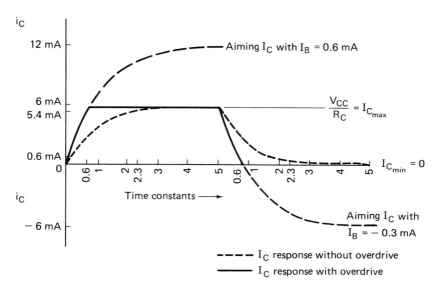

Fig. 2-17 Collector current response for the circuit of Fig. 2-16(a) with and without overdrive.

58 / *Semiconductor Switching Devices*

in 5 time constants ($5\,\tau$) and reaches $0.9\,I_{C_{sat}} = 5.4$ mA in $2.3\,\tau$. With the overdrive applied by Fig. 2-16(c), i_C exponentially seeks a maximum of $\mathbf{h}_{FE}I_B = (20)(0.6) = 12$ mA; however, collector current is still limited to $V_{CC}/R_C = 6$ mA. Since i_C now only has to climb to 45 percent (5.4 mA/12 mA) of its exponential maximum, turn-on is completed in $0.6\,\tau$.

Turn-off time is affected in the same way, as shown in Fig. 2-17. A base current of 0 mA will turn off the transistor in $2.3\,\tau$, but a reverse base current of -0.3 mA results in a faster turn-off. The collector current now exponentially seeks $\mathbf{h}_{FE}I_B = (20)(-0.3) = -6$ mA, but it stops at 0 mA and reaches $0.1\,I_{C_{sat}} = 0.6$ mA in $0.6\,\tau$. Thus, both turn-on and turn-off times are reduced from $2.3\,\tau$ to $0.6\,\tau$ as shown in Fig. 2-17.

The time constant τ increases with R_C, the collector-junction capacitance C_{cb}, and $\mathbf{h}_{FE}$, and it decreases with the common emitter *gain-bandwidth product* f_t (the frequency at which $\mathbf{h}_{fe}$ is unity). It should be noted, however, that while the time constant τ increases with $\mathbf{h}_{FE}$, so does the exponential maximum aiming collector current I_C. If heavy overdrive is used, however, rise time and fall time are almost independent of $\mathbf{h}_{FE}$.

Overdrive Factor od. The ratio of the actual base current I_B to the base current that will just saturate the transistor $I_{B_{min}}$ is called the *overdrive factor* **od**.

$$\mathbf{od} = \frac{I_B}{I_{B_{min}}} \tag{2-20}$$

The **od** ratio must be at least unity for the transistor to be in saturation.

Commutating (Speed-up) Capacitors. The input waveform of Fig. 2-16(b) can be approximated by adding a capacitor across the input biasing resistor. A circuit with a *commutating* or *speed-up* capacitor is shown in Fig. 2-18(a) and its waveforms are given in Fig. 2-18(b). When the input voltage v_i rises from 0 to 5 V, the voltage across C_1 remains zero. Since it takes time for C_1 to charge, all of the 5 V is dropped across the source resistance R_s and the emitter-base junction. The additional forward bias causes a high base current, part of which charges C_1.

When C_1 is fully charged to $v_i - (I_{B_{min}}R_s + V_{BE_{min}})$, $i_{C_1} = 0$. $V_{BE_{min}}$ is the base-to-emitter forward bias that will produce $I_{B_{min}}$. R_1 is designed to maintain saturation after C_1 is fully charged. For the circuit of Fig. 2-18(a)

$$I_{B_{min}} = \frac{V_{CC}}{R_C \mathbf{h}_{FE_{min}}} \tag{2-19}$$

$$= \frac{12\text{ V}}{2\text{ k} \times 20}$$

$$I_{B_{min}} = 0.3\text{ mA}$$

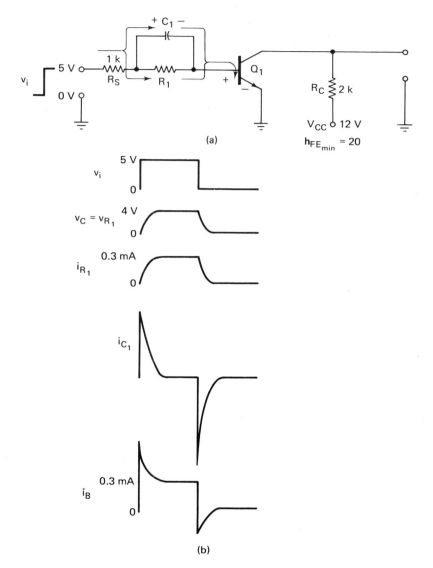

Fig. 2-18 (a) A transistor switch including a speed-up capacitor. (b) Typical waveforms for the circuit.

From the input curves in Fig. 2-19(a), a $V_{BE} = 0.7$ V will produce 0.3 mA of base current and keep the transistor in saturation in the steady-state condition. From Fig. 2-19(b),

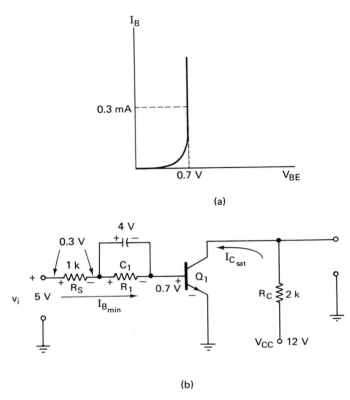

Fig. 2-19 (a) The input curves for the transistor in Fig. 2-18. (b) The circuit in the steady-state *on* condition.

$$R_1 = \frac{v_i - (I_{B_{min}} R_s + V_{BE_{min}})}{I_{B_{min}}}$$

$$= \frac{5 - [(0.3 \text{ mA})(1 \text{ k}) + (0.7 \text{ V})]}{0.3 \text{ mA}}$$

$$= \frac{5 - 1 \text{ V}}{0.3 \text{ mA}}$$

$$R_1 = 13.33 \text{ k}\Omega$$

An exact calculation of C_1 is difficult because of the number of parameters involved. It is common practice for the designer to connect a trimmer across R_1 and adjust it for best turn-on and turn-off times. A "ballpark" or starting-point value may be calculated as follows:

Assume:
    1. A desired turn-on time of 0.4 μs.

2. An overdrive factor **od** = 2 will provide this turn-on time.
3. An average current of 0.3 mA through R_1.

From

$$\mathbf{od} = \frac{I_B}{I_{B_{min}}} \tag{2-20}$$

$$I_B = (\mathbf{od})(I_{B_{min}})$$
$$= (2)(0.3 \text{ mA})$$
$$I_B = 0.6 \text{ mA}$$

C_1 must provide 0.3 mA average overdrive current. The equation for the current through a capacitor is

$$I_C = C\frac{\Delta V_C}{\Delta t}$$

From which

$$C = I_C \frac{\Delta t}{\Delta V_C}$$

The speed-up capacitor C_1 must provide 0.3 mA while the voltage across it changes from 0 V to 4 V in 0.4 μs. Therefore,

$$C_1 = (3 \times 10^{-4})\left(\frac{4 \times 10^{-7}}{4}\right)$$
$$C_1 = 30 \text{ pF}$$

When the input voltage v_i drops to 0 V, the capacitor remains charged to 4 V to momentarily apply a large reverse bias to the EB junction. Figure 2-20 shows the circuit at this instant. The negative voltage on the base forces the stored electrons in the base to quickly move back to the collector. As illustrated in Fig. 2-18(b), the reverse base current drops to zero when C_1

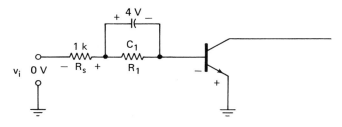

Fig. 2-20 The input circuit at the instant that v_i in Fig. 2-18(a) drops to 0 V.

finishes discharging. (Another way to analyze the action of the speed-up capacitor is in terms of the overcompensated attenuator of Sec. 1-9.)

Despite the advantages shown in Fig. 2-18(b), the speed-up capacitor has several disadvantages: It slows down the input source because it increases the load capacitance of that circuit; it couples noise pulses to the base without attenuation; and it increases the cost of the circuit. Consequently, the capacitor is frequently omitted. Overdrive is still possible by designing R_1 to permit a continuous $I_B > I_{B_{min}}$. However, this increases storage time and thus slows down turn-off. Obviously, other techniques must be used to prevent a slow-down of turn-off time.

2-7 Nonsaturating Techniques

Turn-off time can be greatly reduced if saturation is prevented. This is accomplished by keeping the collector-base junction always reverse biased even with $I_B > I_{B_{min}}$. The storage effects are thereby eliminated.

Collector Clamp. There are several nonsaturating techniques, the simplest of which is the *collector clamp*. Figure 2-21 shows a transistor switch with a collector clamping diode D_1. R_2 is designed to keep the transistor cut off at the highest expected temperature and $v_i = 0$ V. R_1 is designed to keep the transistor in saturation with $v_i = 5$ V and C_1 provides overdrive. The diode D_1 will prevent saturation by clamping v_{CE} at $V_{CL} - V_D$. The clamping voltage V_{CL} may be any voltage less than V_{CC} as long as $V_{CL} - V_D > V_{BE_{sat}}$. The clamping diode must have a recovery time substantially less than the recovery time of the transistor. When $v_i = 0$ V, the transistor is cut off, $v_{CE} = V_{CC}$, and D_1 is reverse biased. When $v_i = 5$ V, $i_B > I_{B_{min}}$ and i_C rises toward

$$I_{C_{sat}} = \frac{V_{CC} - V_{CE_{sat}}}{R_C} = \frac{10 - 0.4}{1 \text{ k}} = 9.6 \text{ mA}$$

However, when v_{CE} drops to less than V_{CL}, the diode conducts and clamps v_{CE} at $V_{CL} - V_D$. Assuming a germanium diode with a forward voltage drop equal to 0.2 V, v_{CE} is clamped at 1.5 V $-$ 0.2 V $= 1.3$ V. $v_{CB} = v_{CE} - V_{BE_{sat}}$ $= 1.3$ V $- 0.7$ V $= 0.6$ V to keep the CB junction reverse biased.

The disadvantage of the simple collector clamp is that it doesn't prevent i_C from increasing; it simply prevents i_{R_C} from increasing. When i_B increases above $I_{B_{min}}$, the additional collector current is supplied by the V_{CL} source. If temperature or noise increases i_B, both the collector and diode power dissipations increase and both components may be ruined.

Single-diode Back Clamp. An improved nonsaturating technique is shown in Fig. 2-22. The collector current is controlled by negative feedback. As soon as the collector voltage drops below a certain value, the base current

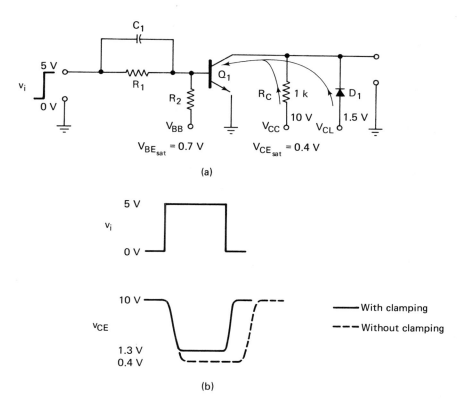

Fig. 2-21 (a) A switch with a collector clamp to prevent saturation and to eliminate storage effects. (b) The waveforms for the circuit.

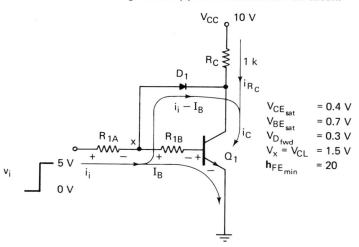

Fig. 2-22 A single-diode back clamping circuit.

63

will be limited. $R_{1A} + R_{1B} = R_1$ and are chosen so that $V_{R_{1B}} = 0.8$ V and $V_x = V_{R_{1B}} + V_{B_{sat}} = 1.5$ V. V_x is the clamping voltage V_{CL} in this circuit. When $v_i = 0$ V, the transistor is cut off, $v_{CE} = V_{CC} = 10$ V, and D_1 is reverse biased. When $v_i = 5$ V, the input current i_i increases. At first it flows entirely into the base and, as i_C increases, v_{R_C} increases and v_{CE} decreases. When $v_{CE} < V_x = 1.5$ V, D_1 comes on and shunts some of the input current away from the base. If the diode has a forward voltage drop of 0.3 V, v_{CE} is clamped at 1.5 V − 0.3 V = 1.2 V and $v_{CB} = 0.5$ V to prevent saturation. The diode handles only the extra input current before it is amplified by the transistor. The extra input current also flows through the collector unamplified.

The following procedure is used to calculate R_{1A} and R_{1B}:

Example 2-5: Design the circuit of Fig. 2-22 so that Q_1 *just* stays out of saturation when $v_i = 5$ V.

Solution: In order for i_C to be just below $I_{C_{sat}}$, the CB junction must be slightly reverse biased. Let $v_{CB} = 0.5$ V. Then

$$v_{CE} = v_{CB} + V_{BE_{sat}}$$
$$= 0.5 + 0.7$$
$$v_{CE} = 1.2 \text{ V}$$

Since $V_D = 0.3$ V, the voltage at point X must be

$$V_x = V_{CL} = V_D + v_{CE}$$
$$= 0.3 + 1.2$$
$$V_{CL} = 1.5 \text{ V}$$

At the point where D_1 *just* comes on, $i_D \approx 0$; thus, $i_C \approx i_{R_C}$ and $i_B \approx i_i$.

$$i_C \approx i_{R_C} = \frac{V_{CC} - v_{CE}}{R_C}$$
$$= \frac{10 - 1.2}{1 \text{ k}}$$
$$i_C \approx 8.8 \text{ mA}$$

and

$$i_B = \frac{i_C}{h_{FE_{min}}}$$
$$\approx \frac{8.8 \text{ mA}}{20}$$
$$i_B \approx 0.44 \text{ mA}$$

When $i_B = 0.44$ mA, $i_C = 8.8$ mA, $v_{CE} = 1.2$ V, and D_1 *just* comes on. For $V_{CL} = 1.5$ V, $V_{R_{1B}}$ must be 0.8 V. Therefore,

$$R_{1B} = \frac{V_{CL} - V_{BE_{sat}}}{i_B}$$

$$\approx \frac{0.8 \text{ V}}{0.44 \text{ mA}}$$

$$R_{1B} \approx 1.81 \text{ k}\Omega$$

and

$$R_{1A} = \frac{v_i - V_{CL}}{i_i}$$

$$\approx \frac{5 - 1.5}{0.44 \text{ mA}}$$

$$R_{1A} \approx 7.96 \text{ k}\Omega$$

If v_i increases above 5 V, i_i and $v_{R_{1A}}$ increase and D_1 conducts harder to hold V_{CL} and i_B relatively constant.

Double-diode Back Clamp. Figure 2-23(a) shows an improvement over the circuit of Fig. 2-22. A germanium diode is used to shunt the excess input current away from the base and a silicon diode is used to supply the voltage $V_{R_{1B}}$ of Fig. 2-22. As shown in Fig. 2-23(b), D_2 has an almost constant voltage drop equal to 0.8 V. This makes the clamping voltage $V_{CL} = V_x = V_{D_2} + V_{BE}$ a regulated 1.5 V. When v_{CE} drops to less than 1.5 V, D_1 conducts and clamps v_{CE} at $V_x - V_{D_1} = 1.5 \text{ V} - 0.3 \text{ V} = 1.2$ V. $v_{CB} = v_{CE} - V_{BE_{sat}} = 1.2 \text{ V} - 0.7 \text{ V} = 0.5$ V. R_1 is the same as R_{1A} in Fig. 2-22. The double-diode back clamp circuit was developed by R. H. Baker and it is frequently called the *Baker clamp*.

Nonsaturation by Circuit Design. Saturation can also be prevented by proper circuit design. Figure 2-24(a) shows a transistor switch with the emitter supplied by a constant-*current* source and a base input of ± 1 V. The collector current is limited by restricting the emitter current; then R_C is designed to keep V_{CB} positive.

The circuit is best explained by using Thévenin's theorem. In Fig. 2-24(b) open-circuiting the emitter causes diode D_1 to conduct. If it is assumed that the diode has a forward voltage drop equal to 0.3 V, $V_{E_{TH}} = -V_D = -0.3$ V and

$$I_{R_E} = \frac{-V_D - V_{EE}}{R_E} = \frac{-0.3 \text{ V} - -12 \text{ V}}{1.2 \text{ k}} = 9.75 \text{ mA}$$

If the emitter circuit is then closed with -1 V at the base, as in Fig. 2-24(c), $v_{BE} = -0.7$ V and Q_1 is cut off. Hence, D_1 stays on, v_E remains at -0.3 V

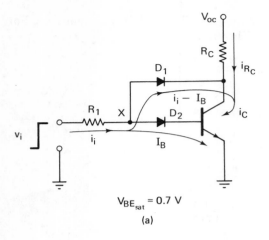

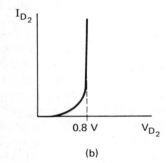

Fig. 2-23 (a) A double-diode back clamping circuit. (b) The volt-ampere characteristics of D_2.

and $i_{R_E} = 9.75$ mA as previously computed. When $+1$ V is applied, as in Fig. 2-24(d), $v_{BE} = +1.3$ V for an instant. Q_1 conducts and, if $V_{BE_{sat}} = 0.7$ V, v_E rises to $+0.3$ V to cut off D_1. The current paths for this condition are shown in Fig. 2-24(e).

$$i_E = \frac{V_{EE} - v_E}{R_E}$$

$$= \frac{12 - 0.3 \text{ V}}{1.2 \text{ k}\Omega}$$

$$i_E = 10.25 \text{ mA}$$

Regardless of $\mathbf{h}_{FE}$,

$$i_C \approx i_E \approx 10 \text{ mA}$$

To prevent saturation, V_{CB} must remain positive; thus, the collector voltage v_C must *not* drop below $v_i = +1$ V. R_C is designed to maintain a small

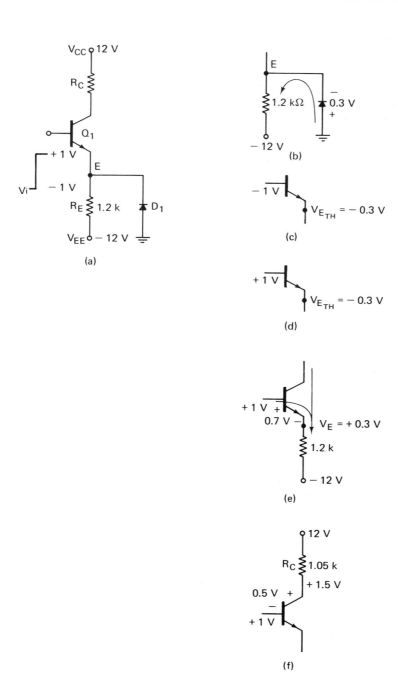

Fig. 2-24 (a) Nonsaturation by emitter current limiting and proper circuit design. (b) The current path with the emitter open-circuited. (c) The EB junction with $v_i = -1$ V. (d) The EB junction at the instant that $+1$ V is applied. (e) The EB circuit after Q_1 comes on. (f) The CB circuit with $+1$ V input and a properly designed R_C.

reverse bias of say 0.5 V across the CB junction. As shown in Fig. 2-24(f), a collector voltage of +1.5 V will maintain a 0.5 V reverse bias for the CB junction. Hence,

$$R_C = \frac{V_{CC} - v_C}{I_C}$$

$$\approx \frac{12 - 1.5 \text{ V}}{10 \text{ mA}}$$

$$R_C \approx 1.05 \text{ k}\Omega$$

2-8 Reducing Delay Time t_d

Figure 2-25 shows a transistor switch with a clamping diode D_1 in the base circuit. When Q_1 is cut off at temperatures below the maximum operating temperature, V_{BE} approaches V_{BB}. With this arrangement, D_1 conducts to clamp the base at $-V_D$. By keeping the reverse bias small, the delay time t_d is reduced. In addition, the EB breakdown voltage will not be exceeded.

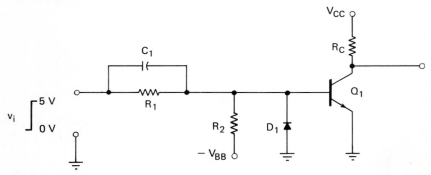

Fig. 2-25 A clamped base to reduce delay time, t_d.

2-9 Field Effect Transistors (FETs)

The *field effect transistor* (FET) is a device in which current is entirely due to majority carriers and its flow is controlled by an electric field. Since current is due to only one type of carrier, the transistors are called *unipolar* transistors. The conventional transistor is a *bipolar* device because both holes and electrons flow. There are two types of FETs:

1. The junction FET (JFET or FET).
2. The insulated gate FET (IGFET) which is also called the metal oxide semiconductor FET (MOSFET).

The Junction Field Effect Transistor. If a voltage is applied to a slab of *p* or *n* material, majority carriers will flow as shown in Fig. 2-26. Since current flow will reverse if the supplies are reversed, the *source* and *drain* are interchangeable. The majority carriers always enter the slab from the source *S* terminal and leave through the drain *D* terminal. The amount of current depends on the degree of doping and the dimensions of the slab. I_D is directly proportional to the width of the slab. By diffusing *p*-type regions into the *n* slab, or *n*-type regions into the *p* slab, the effective width may be controlled. Figure 2-27(a) shows the structure of an *n*-channel JFET. The heavily doped *p* regions form the *gate G* of the transistor. The *channel* is the area between the gate regions through which majority carriers flow. The *pn* junctions are reverse biased by a V_{GS} supply.

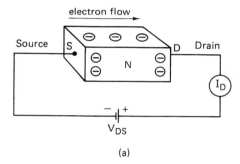

(a)

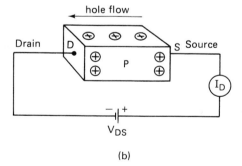

Fig. 2-26 (a) A slab of *n*-type material with an external supply. (b) A slab of *p*-type material with an external supply.

(b)

Operation. As in any reverse biased *pn* junction, a depletion region is formed. Positive ions on the *n* side and negative ions on the *p* side produce an electric field through which current is zero. As the reverse bias V_{GS} is increased, the depletion region grows wider, the effective width of the channel becomes smaller, and I_D decreases. Eventually, the reverse voltage removes all the free charge from the channel and *pinches off* the current flow. The

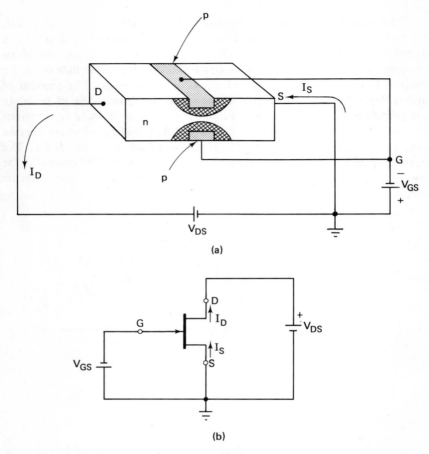

Fig. 2-27 (a) The structure of *n*-channel FET. (b) The schematic symbol for an *n*-channel FET. For a *p*-channel FET, reverse the supplies and arrows.

reverse voltage V_{GS} that pinches off the channel is called the *pinch-off voltage* V_p.

When the JFET is operated with the gate-to-channel junction reverse biased, it is said to be in the *depletion mode* and it has a very high input resistance. Values of more than 100 MΩ are typical. Forward biasing the gate-to-channel junction produces more carriers in the channel and increases current. This is known as the *enhancement mode* of operation. The JFET is normally operated in the depletion mode because in the enhancement mode the forward biased gate-source diode presents a low input resistance. However, it can be operated with a small forward bias and still present a high input impedance if the cutin voltage of the silicon junction is not exceeded.

FET Volt-Ampere Characteristics. Assume that V_{DS} is increased from zero as shown in Fig. 2-28(a). At first the FET acts as a low resistance, the value of which depends on V_{GS}. Consider the curve for $V_{GS} = 0$. With $V_{DS} = 0$, the entire channel is open and as V_{DS} is increased, I_D increases. With higher current, the *IR* drop across the channel increases and an increasing reverse bias is developed across the *pn* junction. Since the *IR* drop is greater near the drain than the source, the reverse bias is not uniform along the channel. The result, as shown in Fig. 2-29, is a depletion zone that is

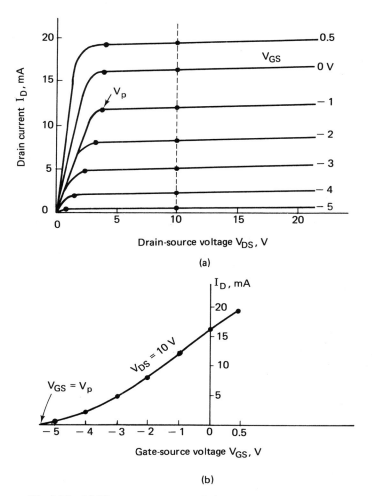

Fig. 2-28 (a) The common-source drain or output characteristics. Note that pinch-off occurs at lower values of V_{DS} as V_{GS} is made more negative. (b) A static transfer curve for $V_{DS} = 10$ V.

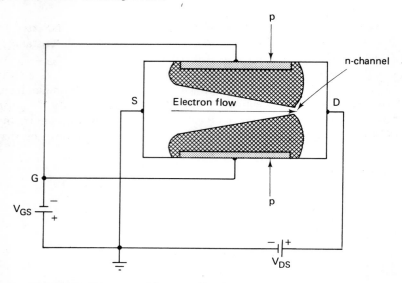

Fig. 2-29 The reverse bias caused by the *IR* drop along the channel is not uniform which results in a wider depletion region near the drain.

wider near the drain. The output curve begins to flatten out indicating the higher resistance of the narrower channel. Eventually, the ohmic drop produces enough reverse bias to pinch off the channel. It is not possible to completely pinch off the channel because if I_D is made zero, the *IR* drop that causes the depletion zone will not exist.* Instead, I_D increases at first and then levels off as pinch-off occurs. The drain-source pinch-off voltage is the voltage V_{DS} at which the FET enters its constant current region. As V_{GS} is made negative, pinch off occurs at lower values of V_{DS}.

A static transfer curve for $V_{DS} = 10$ V is shown in Fig. 2-28(b). Because the output curves are almost parallel beyond pinch-off, the transfer curve for $V_{DS} = 10$ V is representative of all curves for values of V_{DS} in the constant current region.

Cutoff. As shown in Fig. 2-28(b), with a constant V_{DS}, I_D decreases as V_{GS} increases toward V_p. When $V_{GS} = V_p$, the channel closes and $I_D = 0$. In practice, however, a small leakage current still flows. The value of this leakage current is specified by the manufacturer as $I_{D_{off}}$ at some value of V_{DS} and V_{GS}.

The *gate cutoff current* is specified as the gate-to-source current with the drain shorted to the source I_{GSS} and $V_{GS} = V_p$. $I_{D_{off}}$ and I_{GSS} are both in the nanoampere range for silicon transistors.

---

*This is similar to trying to completely cut off a vacuum tube with cathode bias; it is impossible because it is the current that produces the bias.

The construction of the JFET shown in Fig. 2-27(a) is not practical because of the difficulties encountered when trying to diffuse impurities on each side of the bar. A more practical structure is shown in Fig. 2-30. The n channel is epitaxially grown onto a p-type substrate and a p-type gate is diffused into the channel. The substrate may be used as a second gate.

The Insulated Gate FET IGFET or Metal Oxide Semiconductor (FET MOSFET). The metal oxide semiconductor FET can operate in the enhancement mode and still present higher impedances than the JFET. As illustrated in Fig. 2-31, two n-type regions are diffused in a p-type substrate. One acts as the source, the other as the drain. A layer of silicon dioxide (SiO_2) 10^{-5}-cm

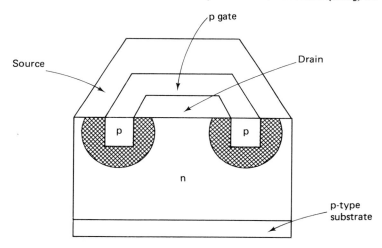

Fig. 2-30 A practical n-channel JEFT structure.

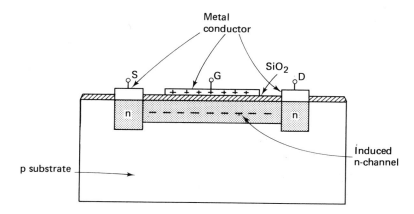

Fig. 2-31 Construction of an enhancement mode n-channel MOSFET.

thick is grown on the top of the n bar. The gate is a layer of aluminum on top of the oxide. The thin SiO_2 layer is broken, and aluminum contacts are made with the source and drain. The aluminum gate and the n channel are two conductors and, separated by the oxide insulating layer, they form a capacitor.

The *metal* gate, *oxide* layer, and *semiconductor* bar are the reason that it is called a MOSFET. The insulating layer between the gate and channel leads to the name *insulated gate* FET IGFET. The MOSFET can be made to operate in either the enhancement mode or the depletion mode. Its input resistance is one hundred times higher than the JFET with values of more than 10^{12} Ω being typical.

The Enhancement MOSFET. If a voltage V_{DS} is applied between the drain and source while V_{GS} is held at zero, one of the two pn junctions is reverse biased. Therefore, I_D at $V_{GS} = 0$ is only the small leakage current of a silicon junction. The drain current with $V_{GS} = 0$ is designated I_{DSS}. As V_{GS} is made positive, an n channel is *induced* just below the gate. The positive gate pulls minority electrons into this region and holes are filled. The negative charge that results cannot be used to produce a current because of a strong covalent bond.

As V_{GS} is made more positive, the charge required by the gate-to-semiconductor capacitor becomes greater, so more electrons are forced to move into the region below the gate. Since the other atoms in this region have formed covalent bonds, these electrons are free to act as majority carriers. The source-to-drain current is *enhanced* by a positive gate-to-source voltage; hence, the name enhancement FET. The value of V_{GS} at which I_D reaches a specified small value is called the *gate-source threshold voltage* V_{GST}. The value of V_{GS} required to produce a current $I_{D_{on}}$ (approximately the maximum value on the output characteristics) is also usually specified by the manufacturer. The output characteristics, transfer curve, and schematic symbols for an n-channel enhancement type MOSFET are shown in Fig. 2-32. The substrate connection may be brought out to form a tetrode. The p-type substrate and the induced n channel form a pn junction. If this junction is reverse biased, it operates as a junction FET. However, it is usually internally connected to the source to form a triode. In some cases the symbol is the same as the one used for the JFET with the substrate understood to be connected to the source.

The Depletion MOSFET. The MOSFET is also made to operate in the depletion mode. A thin n channel is diffused between the source and drain to permit a large drain-to-source current I_{DSS} with $V_{GS} = 0$. The structure is the same as shown in Fig. 2-31 except that the channel is diffused. If the gate-source voltage V_{GS} is made negative, electrons are forced out of the channel leaving behind a depletion zone. The depletion region is wider near the drain

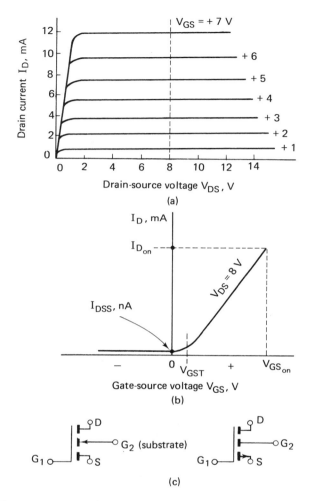

Fig. 2-32 (a) Drain characteristics. (b) A transfer curve for $V_{DS} = 8$ V. (c) The schematic symbol of an n-channel enhancement type MOSFET.

because of the ohmic drop caused by I_D. The output characteristics and a transfer curve for $V_{DS} = 8$ V are shown in Fig. 2-33. The depletion type MOSFET can also be operated in the enhancement mode. A positive voltage between the gate and source induces additional electrons into the n channel. The result, as shown in Fig. 2-33(a), is higher values of drain current.

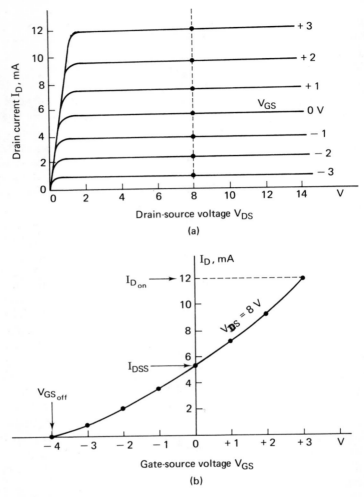

Fig. 2-33 (a) Output characteristics. (b) A transfer curve for $V_{DS} = 8$ V of an *n*-channel depletion type MOSFET. It may be operated in the enhancement mode by making V_{GS} positive.

2-10 The FET as a Switch

The FET has both advantages and disadvantages as a switch. The two voltage levels used to represent a logical 1 and a logical 0 can switch it between two vastly different drain-to-source resistances. When on, values as low as 100 Ω are possible. The output current swings from a few nanoamperes to several milliamperes; the output voltage swings from about 2 V to V_{DD}. The very high input impedance (more than 10^8 Ω for JFET and more than 10^{12} Ω for IGFET are typical) means that it will not load down its driver stage. The number of stages that a circuit can drive is called its *fan-out*. Because of the low current requirements of the gate-source input, there is practically no dc fan-out limitation. Capacitance, however, does affect the switching times. The input capacitance C_{gs} must charge and discharge and thus it limits switching speed. Values as high as 10 pF are typical and propagation delays are in the order of 100 ns. The input capacitance of the MOSFET may be used as a short-duration storage device. The small size, low power requirements and simple fabrication procedure of the MOSFET make it very practical in integrated circuitry.

2-11 Summary

Digital computers use semiconductor diodes and transistors as high-speed switches. Diodes act as closed switches when the anode is made positive with respect to the cathode (forward bias) and as open switches when the anode is negative with respect to the cathode (reverse bias). The conventional bipolar transistor acts as a closed switch (saturation) when *both* junctions are forward biased, and it acts as an open circuit (cutoff) when both junctions are reverse biased (the silicon type is at the edge of cutoff when $V_{BE} = 0$). When the emitter-base junction is forward biased and the collector-base junction is reverse biased, the transistor is in its active or amplifier region of operation. The common emitter is the most frequently used circuit configuration.

Silicon is more sensitive to temperature than germanium, but because its leakage current is initially 1,000 times lower (nA compared to μA at 25°C), it is still used in all high-temperature applications. Stabilization circuits may be added to assure cutoff at very high temperatures.

The *npn*-type transistor is preferred to the *pnp*-type transistor because the higher mobility of the majority electrons permits faster switching. Overdrive and nonsaturating techniques may be employed to reduce switching times.

The field effect transistor FET is a unipolar device that comes in two varieties: the junction FET (JFET) and the metal oxide semiconductor FET (MOSFET). They, too, can be switched between two widely different conduction states. Although the bipolar transistor is a faster switch and has lower values of output resistance and voltage when on, the FET has no dc fan-out limitations, no offset voltage to overcome when turning on, and it is ideal in large-scale integration (LSI).

Problems

2-1 The diode of Fig. 2-34 has a reverse saturation current $I_{co} = 10\ \mu A$ and a forward voltage drop $V_D = 0.2$ V. Calculate
  (a) I_D with the switch in position A.
  (b) V_D with the switch in position B.

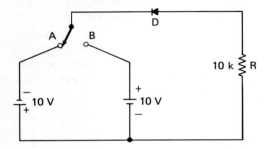

Fig. 2-34

2-2 Draw the piecewise linear characteristics of the diode curves shown in Fig. 2-35. Determine

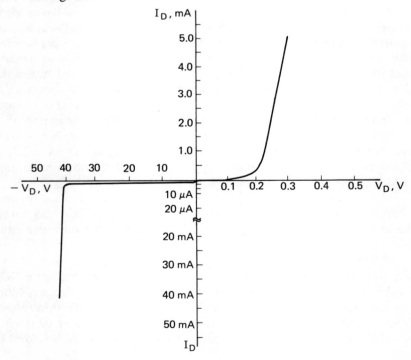

Fig. 2-35

(a) The cutin voltage V_γ and the forward resistance R_f.
(b) The back resistance R_b at $-V_D < V_Z$.
(c) The back resistance R_b at $-V_D > V_Z$.

2-3 Using the answers to Prob. 2-2 for the diode of Fig. 2-34, calculate
(a) V_D and V_R with the switch in position A.
(b) V_D and V_R with the switch in position B.

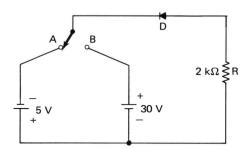

Fig. 2-36

2-4 A silicon diode has a reverse saturation current $I_{co} = 5$ nA at 25°C. Calculate
(a) I_{co} at 100°C.
(b) I_{co} at 105.4°C.
(c) What conclusion can be made from the answers to (a) and (b)?

2-5 A germanium diode has a reverse saturation current $I_{co} = 5$ μA at 25°C. Calculate
(a) I_{co} at 100°C.
(b) I_{co} at 109.3°C.
(c) What conclusion can be made from the answers to (a) and (b)?
(d) What conclusion can be made from the answers to part (a) of Probs. 2-4 and 2-5?

2-6 The circuit of Fig. 2-9(a) has a base current $I_B = 0.04$ mA and a dc current amplification factor $h_{FE} = 50$. Determine the values of R_C and R_B for the voltages shown on the diagram.

2-7 A transistor has a leakage current $I_{co} = 10$ μA, a current amplification factor $\alpha = 0.975$, and a base current $I_B = 0.02$ mA. Calculate the leakage current I_{CEO} that will flow if this transistor is connected as a a common emitter amplifier.

2-8 If the circuit of Fig. 2-11 has a base-to-emitter cutoff voltage $V_{BEco} = 0.1$ V, $R_B = 100$ kΩ, and $V_{BB} = 18$ V, calculate
(a) The maximum leakage current I_{CBO} that can flow and still have the transistor remain off.
(b) The maximum temperature at which this circuit remains cut off if $I_{co} = 10$ μA at 25°C. (The V_{BEco} and I_{co} indicate that the transistor is of the germanium type.)

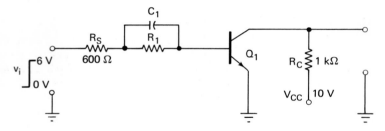

Fig. 2-37

2-9 The transistor of Fig. 2-37 has a minimum $h_{FE} = 25$. Using $V_{BE_{sat}} = 0.8$ V, $V_{CE_{sat}} \approx 0$ V, and an overdrive ratio od $= 2$ for a desired turn-on time $t_{on} = 0.3$ μs, calculate
(a) The value of R_1.
(b) The value of C_1.

2-10 Repeat Prob. 2-9 with $h_{FE_{min}} = 10$.

2-11 The diode of Fig. 2-38 has a forward voltage drop $V_D = 0.2$ V. $V_{BE_{sat}} = 0.3$ V and $V_{BE_{co}} = -0.1$ V. Calculate
(a) The maximum leakage current I_{CBO} that can flow and still have Q_1 remain off with $v_i = 1$ V.
(b) The value of V_{CB} when $V_{BE} = V_{BE_{sat}}$.
(c) The value of V_D when Q_1 is off.

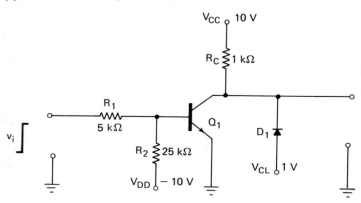

Fig. 2-38

2-12 If the transistor in Prob. 2-11 has an $h_{FE} = 20$, calculate
(a) Current flow through the diode if $I_B = 0.7$ mA.
(b) Current flow through the diode if $I_B = 1.4$ mA.

2-13 The transistor of Fig. 2-39 is of the Ge type and has an $h_{FE_{min}} = 20$. Use data from Table 2-2.

(a) Verify that Q_1 is in saturation with $v_i = -5$ V.
(b) Determine the value of v_i that will *just* permit Q_1 to saturate.
(c) If $I_{CO} = 2$ μA at 25°C, what is the highest temperature at which Q_1 will remain cut off with $v_i = 0$ V?

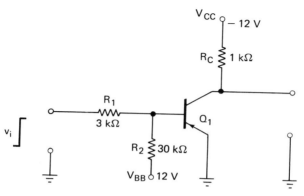

Fig. 2-39

2-14 The diode of Fig. 2-40 has a forward voltage drop $V_D = 0.2$ V and Q_1 has a $V_{BE_{sat}} = -0.3$ V.
(a) Calculate V_{BE}, V_{CE}, V_{CB}, and I_D when $v_i = +0.5$ V.
(b) Using $I_B = 0.2$ mA, calculate the value of R_C that will maintain a $V_{CB} = -0.5$ V to prevent saturation when $v_i = -0.5$ V.

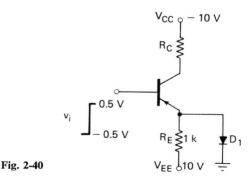

Fig. 2-40

2-15 Given $V_{BE} = 0.3$ V, $V_D = 0.2$ V, and $h_{FE_{min}} = 20$, design the circuit of Fig. 2-41 so that Q_1 is out of saturation, with a $V_{CB} = 0.3$ V, when $v_i = 3$ V.

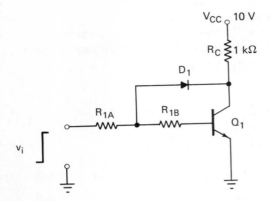

Fig. 2-41

2-16 The circuit of Fig. 2-42 has a drain supply $V_{DD} = 20$ V and uses source self-bias. Calculate the values of R_D and R_S that will provide quiescent values of $I_D = 5$ mA, $V_{DS} = 10$ V, and $V_{GS} = -2$ V.

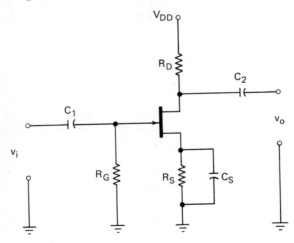

Fig. 2-42

3

COMPUTER MATH AND LOGIC

Two basic requirements for computer design are a system of numbers that can easily be used by the computer and a system of mathematics that describes every operation to be performed. The decimal system may be used, but it has disadvantages. For example, the decimal digits $0, 1, \ldots, 9$ may be represented by 0 V, 1 V, $\ldots$, 9 V; with these levels, however, errors may occur. As explained in Chap. 2, diodes and transistors are not perfect switches. When a semiconductor switch is closed, the output voltage will differ from the desired output by either V_D or $V_{CE\,\text{sat}}$. Thus, the output may be as high as 0.7 V instead of 0 V, and the computer may not distinguish one numeral level from another.

This disadvantage is overcome by letting 0 V represent 0, 5 V represent 1, and so on to where 45 V represents 9. This, of course, greatly increases the power requirements and the overall physical size of the system. The circuits in a pure decimal system are also more difficult to design because they require ten distinct levels or states. When the decimal system is used, it is normally in some form of binary code. Several of these codes are discussed in Sec. 3-4.

With the binary (base 2) system, using only the (two) digits 0 and 1, the above disadvantages are eliminated. The binary digit (*bit*) 0 may be represented by 0 V and the digit 1 may be represented by 5 V. This leaves enough separation between the two voltages to allow for the level shifts introduced by imperfect switches with much smaller input voltages. Also, with only two digits and two voltage levels, the binary system is easily adapted to bistable switching circuits.

84 / Computer Math and Logic

The binary system is discussed in detail in this chapter and general rules are given for any base. The system of math known as Boolean algebra, which is used in the design of digital computers, is then discussed.

3-1 Number Systems

The *base* or *radix* of a number system refers to the number of different digits that can appear in each position of a number. Each digit has two values: an *intrinsic* value and a *place* or *position* value. For example, in the decimal number 222, each digit has an intrinsic value of 2. However, their place or position values are different. They are

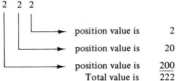

```
                position value is      2
                position value is     20
                position value is    200
                Total value is       222
```

The weight or position value of any digit in any system is determined in the following manner:

$Base^3 \quad Base^2 \quad Base^1 \quad Base^0 \quad \bullet \quad Base^{-1} \quad Base^{-2} \quad Base^{-3}$

Radix point

Any digit just to the left of the radix point has a total weight equal to that digit times the base raised to the zero power. Any digit two places to the left of the radix point has a total weight equal to that digit times the base raised to the first power. The weight of a digit just to the right of the radix point equals that digit times the base raised to a negative one exponent. This will be clarified by applying these rules to the familiar decimal (base 10) system.

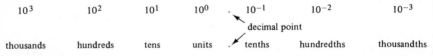

$10^3 \qquad 10^2 \qquad 10^1 \qquad 10^0 \qquad 10^{-1} \qquad 10^{-2} \qquad 10^{-3}$

decimal point

thousands   hundreds   tens   units   tenths   hundredths   thousandths

Example 3-1: Determine the place value of each digit in the decimal number 356.7.

Solution:

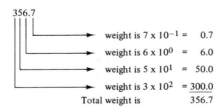

```
356.7
              weight is 7 x 10^-1 =    0.7
              weight is 6 x 10^0  =    6.0
              weight is 5 x 10^1  =   50.0
              weight is 3 x 10^2  =  300.0
              Total weight is        356.7
```

Example 3-2: Determine the weight or place value of each digit in the base six number 243.3.
Solution:

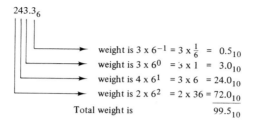

The Binary System. The place value of each binary digit or *bit* is determined in the same way. The radix point is now called a binary point.

$$2^6 \quad 2^4 \quad 2^3 \quad 2^2 \quad 2^1 \quad 2^0 \quad \bullet \quad 2^{-1} \quad 2^{-2} \quad 2^{-3} \quad 2^{-4} \quad 2^{-5}$$
$$\text{Binary point}$$
$$32s \quad 16s \quad 8s \quad 4s \quad 2s \quad 1s \quad \bullet \quad \tfrac{1}{2}s \quad \tfrac{1}{4}s \quad \tfrac{1}{8}s \quad \tfrac{1}{16}s \quad \tfrac{1}{32}s$$

Example 3-3: Determine the weight or place value of each bit in the binary number 10111.101.
Solution:

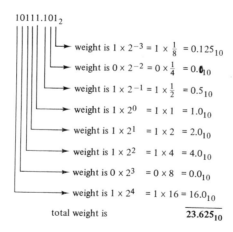

Counting in Binary. Logic design problems in this text are limited to a maximum of four variables, but the principles are the same no matter how many variables appear. Binary numbers of up to four bits should be recognized immediately without going through conversion procedures. Table 3-1 gives a list of important binary numbers that should be quickly recognized.

Table 3-1 Binary-decimal equivalents.

Decimal	Binary
1/16	0.0001
1/8	0.001
1/4	0.01
1/2	0.1
0	0.0
1	1.0
2	10.0
3	11.0
4	100.0
5	101.0
6	110.0
7	111.0
8	1000.0
9	1001.0
10	1010.0
11	1011.0
12	1100.0
13	1101.0
14	1110.0
15	1111.0
16	10000.0
32	100000.0
64	1000000.0
128	10000000.0
256	100000000.0
512	1000000000.0
1024	10000000000.0

3-2 Conversion Techniques

The weights or place value system of converting from one base to another is sometimes tedious. Therefore, other systems have been devised. In this section techniques used to convert whole numbers, fractions, and mixed numbers from any base to any other base will be explained. Emphasis is placed on the binary (base 2), octal (base 8), and hexadecimal (base 16) systems, all extensively used in digital computers.

Converting Decimal Whole Numbers to Any Other Radix. To convert a decimal whole number to any other base, follow these rules:

1. Divide the decimal number by the new radix.
2. Any remainder becomes the lowest-order digit of the new radix number.
3. The quotient is then divided by the new radix.

4. Any remainder becomes the next digit of the new radix number.
5. Repeat (3) and (4) until the quotient is 0.

Example 3-4: Convert 23_{10} to binary.
Solution:

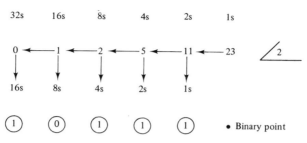

First divide 23 by 2; the quotient is 11 with a remainder of ①. Then divide 11 by 2; the quotient is 5 with a remainder of ①. Then, 2 into 5 equals 2 with a remainder of ①; 2 into 2 equals 1 with a remainder of ⓪; and, finally, 2 into 1 equals 0 with a remainder of ①. Thus, $23_{10} = 10111_2$.

In Ex. 3-4, when dividing 23_{10} by 2, the number of twos in 23 ones is being determined. There are 11 twos and ① one remaining. This ① is placed in the units position of the binary equivalent. Then the number of fours in 11 twos is determined by dividing 11 by 2. There are 5 fours and ① two remaining, which goes in the twos position of the binary equivalent. This procedure is continued until it is determined that there are 0 thirty-twos with ① sixteen remaining.

Example 3-5: Convert 120_{10} to binary.
Solution:

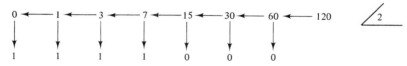

Thus, $1111000_2 = 120_{10}$.

Example 3-6: Convert 127_{10} to octal.
Solution:

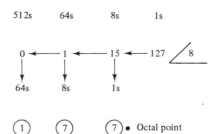

In 127 ones there are 15 eights and ⑦ ones remaining. In 15 eights there is 1 sixty-four with ⑦ eights remaining. And in 1 sixty-four there are 0 five hundred and twelves with ① sixty-four remaining. Thus, $177_8 = 127_{10}$. This may be checked by adding weights.

```
        177
         |  |  |
         |  |  └──► weight is 7 x 8⁰ =   7₁₀
         |  └─────► weight is 7 x 8¹ =  56₁₀
         └────────► weight is 1 x 8² =  64₁₀
                    total weight is    127₁₀
```

Converting Whole Numbers from Any Radix to Decimal. To convert a whole number from any base to decimal, follow this rule: Multiply by the other base, add; multiply by the other base, add, starting from the left.

Example 3-7: Convert 10101_2 to decimal.
Solution:

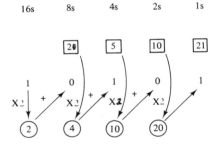

This procedure says that there are ② eights in 1 sixteen. With 0 eights in the binary number itself, there is a total of ⎡2⎤ eights. Then, in ⎡2⎤ eights there are ④ fours plus the 1 four in the binary number for a total of ⎡5⎤ fours. There are ⑩ twos in ⎡5⎤ fours plus the 0 twos in the binary number for a total of ⎡10⎤ twos. And finally, there are ⑳ ones in ⎡10⎤ twos plus the 1 one in the binary number for a total of ⎡21⎤ ones. Thus, $10101_2 = 21_{10}$.

Example 3-8: Convert 123_8 to decimal.
Solution:

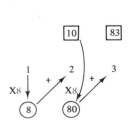

Thus, $123_8 = 83_{10}$. This answer may be checked by adding weights.

Conversion Techniques / 89

```
  1 2 3
  │ │ └─► weight is 3 x 8⁰ =  3₁₀
  │ └───► weight is 2 x 8¹ = 16₁₀
  └─────► weight is 1 x 8² = 64₁₀
          total weight is    83₁₀
```

Converting Decimal Fractions to Any Other Radix. To convert decimal fractions to any other base, follow these rules:

1. Multiply the decimal fraction by the new radix.
2. The digit to the left of the decimal point in the product is the highest-order digit in the new-radix fraction.
3. The fraction part of the product is multiplied by the new radix.
4. The digit to the left of the decimal point in this product is the next digit in the new-radix fraction.
5. Repeat (3) and (4) until the product is zero.

Example 3-9: Convert 0.625_{10} to binary.
Solution:

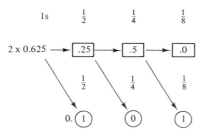

Thus, $0.625_{10} = 0.101_2$.

In Ex. 3-9, multiplying by 2 indicates that there are one and a quarter ①.|25| halves in 0.625 ones. The ① is placed in the halves position of the binary equivalent. Multiplying the |25| halves by 2 indicates that there are ⓪|.5| quarters in |25| halves. The ⓪ is placed in the quarters position of the answer. Finally, multiplying the |.5| quarters by 2 indicates that there is ①|.0| eighth in |.5| quarters. The ① is placed in the eighths position of the binary equivalent. With |.0| eighths remaining there are no sixteenths.

Example 3-10: Convert 0.390625_{10} to octal.
Solution:

```
                 1s           ⅛s          1/64 s

     8 x 0.390625  ──►  .125  ──►  .0

                        ⅛s         1/64 s

                        0.3         .1
```

Thus, $0.390625_{10} = 0.31_8$. This answer may be checked by adding weights.

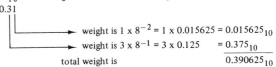

When decimal fractions are converted to another radix, they may not come out even. The answer will automatically be carried to the capacity of the machine, which may be 36 digits or more. However, a lower limit must be set on solutions to text problems. For a binary fraction to accurately represent a decimal fraction, the binary fraction should be carried at least four places for each decimal digit. Since octal is close to decimal, carry the octal fraction just one place more than there are digits in the decimal number.

Example 3-11: Convert 0.27_{10} to binary.
Solution:

$2 \times 0.27 \rightarrow .54 \rightarrow .08 \rightarrow .16 \rightarrow .32 \rightarrow .64 \rightarrow .28 \rightarrow .56 \rightarrow .12$

$0.0 \quad 1 \quad 0 \quad 0 \quad 0 \quad 1 \quad 0 \quad 1$

Thus, $0.27_{10} \approx 0.01000101_2 = 0.26953125_{10}$.

Example 3-12: Convert 0.26_{10} to octal.
Solution:

$8 \times 0.26 \rightarrow .08 \rightarrow .64 \rightarrow .12$

$0.2 \quad 0 \quad 5$

Thus, $0.26_{10} \approx 0.205_8 = 0.259765625_{10}$.

Converting Fractions from Any Radix to Decimal. To convert fractions from any base to decimal, follow this rule: Divide by the other base, add; divide by the other base, add, starting from the right.

Example 3-13: Convert 0.101_2 to decimal.
Solution:

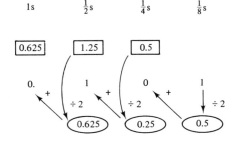

Dividing the rightmost digit by 2 indicates that there are 0.5 quarters in 1 eighth. With 0 quarters in the binary number itself, there is a total of 0.5 quarters. Then, in 0.5 quarters there are 0.25 halves. With the 1 half in the binary number, there is a total of 1.25 halves. And, finally, there are 0.625 units in 1.25 halves. Thus, $0.101_2 = 0.625_{10}$.

Example 3-14: Convert 0.12_8 to decimal.
Solution:

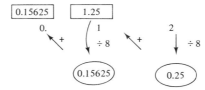

Thus, $0.12_8 = 0.15625_{10}$. This answer may be checked by weights.

```
0.12
 ││
 │└──→ weight is 2 x 8⁻² = 2 x 0.015625 = 0.03125₁₀
 └───→ weight is 1 x 8⁻¹ = 1 x 0.125   = 0.125₁₀
       total weight is                   0.15625₁₀
```

Converting from Any Base to Any Other Base. It is possible to convert from any base to any other base by going through base ten. For example, if a base 6 number is to be expressed in base 5, first convert the base 6 number to decimal and then convert the decimal number to base 5.

Example 3-15: Convert 35_6 to base 5.
Solution: First convert 35_6 to decimal by the $X6, +; X6, +$ method.

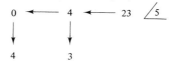

Now convert 23_{10} to base five by dividing by 5.

```
0 ←──── 4 ←──── 23 ╱5
│       │
↓       ↓
4       3
```

Thus, $35_6 = 43_5$.

Special-case Conversions. There are certain conversions that can be done directly. Examples of these are base 3-base 9, base 2-base 8, and base 2-base 16 (hexadecimal). The direct conversion between binary-octal and binary-hexadecimal is the reason that the octal and hexadecimal systems are used in computers.

The hexadecimal or radix sixteen system needs explaining. There are sixteen different symbols that can appear in each position of the number. The ten decimal numerals are used, and any other convenient symbols may be employed to represent the six numerals, 10_{16}, 11_{16}, 12_{16}, 13_{16}, 14_{16}, and 15_{16}. It is common practice to use the letters A, B, C, D, E, and F. Table 3-2 gives a list of decimal-hexadecimal conversions. The conversion techniques previously explained are still valid.

Table 3-2 Hexadecimal-decimal conversion.

Decimal	Hexadecimal	Decimal	Hexadecimal
0	0		
1	1	12	C
2	2	13	D
3	3	14	E
4	4	15	F
5	5	16	10
6	6	256	100
7	7	4096	1000
8	8	0.0625	0.1
9	9	0.00390625	0.01
10	A	5000	1388
11	B	6700	1A2C

As noted earlier, the reason for using the octal and hexadecimal systems is the direct conversion between these and the binary system. The methods will be explained in the following examples.

Example 3-16: Convert 11110.10101_2 to octal.
Solution: Starting at the binary point, divide the binary number into groups of three. Then express each three-bit number as an octal digit.

$$\begin{array}{cccc} 011 & 110. & 101 & 010 \\ 3 & 6\,. & 5 & 2 \end{array}$$

Note that the addition of the 0 on the extreme left of the whole number has no effect on the octal digit. However, the 0 added to the extreme right of the fraction yields a three-bit number the value of which is 2. Hence, $11110.10101_2 = 36.52_8$.

Example 3-17: Convert 27.64_8 to binary.
Solution: Simply express each octal digit as a three-digit binary number.

$$\begin{array}{cccc} 2 & 7\,. & 6 & 4 \\ 010 & 111\,. & 110 & 100 \end{array}$$

Any zeros on the extreme left of the whole number and on the extreme right of the fraction may be dropped. Hence, $27.64_8 = 10111.1101_2$.

Example 3-18: Convert 10110.110101_2 to hexadecimal.
Solution: Starting at the binary point, divide the binary number into groups of 4.

$$\begin{array}{cccc} 0001 & 0110 & .\ 1010 & 1000 \\ 1 & 6 & .\ \ A & 8 \end{array}$$

Note that the addition of the three zeros on the extreme right of the fraction results in a four-bit number the value of which is 8. Hence, $10110.10101_2 = 16.A8_{16}$.

Example 3-19: Convert $46.C2_{16}$ to binary.
Solution: Simply express each hexadecimal digit as a four-digit binary number.

$$\begin{array}{cccc} 4 & 6 & .\ \ C & 2 \\ 0100 & 0110 & .\ 1100 & 0010 \end{array}$$

Any zeros on the extreme left of the whole number and on the extreme right of the fraction may be dropped. Hence, $46.C2_{16} = 1000110.1100001_2$.

As indicated earlier, only the binary system is actually used by the computer, but the large number of digits required to express high numbers makes the binary system cumbersome for humans. They must take great care to prevent errors when they read large binary numbers. This problem is greatly reduced if the binary number is broken into groups of three or four and if these three-bit or four-bit groups are expressed by their octal or hexadecimal equivalents.

3-3 Binary Arithmetic

There are a number of ways that a computer performs arithmetic operations. In this section direct binary addition, subtraction, multiplication, and division are explained. In later sections subtraction with *complements* and addition in decimal codes are explained.

Binary Addition. The two numbers to be added are called the *addend* and the *augend*. With the addition of a *carry-in*, there are eight possible combinations that may have to be added. They are listed in Table 3-3.

If there is only one 1, the sum is 1 with a 0 to be carried (the *carry-out*). If there are two ones, the sum is 0 with the two ones being represented by a carry-out of 1. If there are three ones, the sum is 1 with the remaining two ones being represented by a carry-out of one. The carry-out from one column becomes the carry-in to the next column. Recall that the column one place to

Table 3-3 Binary addition of three digits. The addition of an addend, an augend, and a carry-in results in a sum and a carry-out.

Addend	Augend	Carry-in	Sum	Carry-out
0	0	0	0	0
0	0	1	1	0
0	1	0	1	0
1	0	0	1	0
0	1	1	0	1
1	0	1	0	1
1	1	0	0	1
1	1	1	1	1

the left has a position value twice that of the column involved in the addition. Thus, the sum of two ones is represented by carrying a one into the next higher-order column.

Example 3-20: Add 1111_2 and 110_2.
Solution:

16s	8s	4s	2s	1s
	1	1	1	1
		1	1	0
1	1	1	0	
1	0	1	0	1

Starting with the lowest-order column in Ex. 3-20, $1 + 0$ ones is one 1 with a carry-out of 0. This carry-out is the carry-in of the 2s column. Adding the digits $1 + 1 + 0$ in the 2s column results in a sum of 0 and a carry-out of 1. That is, the two 2s are represented as zero 2s and one 4. The sum of $1 + 1 + 1$ in the 4s column is 1 with a carry-out of 1. That is, the three 4s are represented by one 4 and one 8. Finally, the sum of $1 + 1$ in the 8s column is represented as zero 8s and one 16.

The solution may be checked by converting each binary number to decimal. The addend $1111_2 = 15_{10}$, the augend $110_2 = 6_{10}$, and the sum $10101_2 = 21_{10}$.

Binary Subtraction. The two numbers involved in the problem are called the *minuend* and the *subtrahend*. With the inclusion of a *borrow-in*, there are eight possible combinations that must be considered when doing subtraction. They are listed in Table 3-4.

There are four times when it is necessary to borrow. If the subtrahend or the borrow-in is 1 and the minuend is 0, it is necessary to borrow a 1 from the minuend *to the left*. This borrowed 1 has a value equal to two 1s when

Binary Arithmetic / 95

Table 3-4 Binary subtraction of three digits. The subtraction of a subtrahend and a borrow-in from the minuend results in a difference and a borrow-out.

Minuend	Subtrahend	Borrow-in	Difference	Borrow-out
0	0	0	0	0
0	0	1	1	1
0	1	0	1	1
1	0	0	1	0
0	1	1	0	1
1	0	1	0	0
1	1	0	0	0
1	1	1	1	1

moved one place to the right, so the subtraction of either a subtrahend or a borrow-in from a minuend of 0 results in a difference of 1 and a *borrow-out* of 1. This borrow-out represents the 1 borrowed from the minuend to the left.

If both the subtrahend and the borrow-in are 1s and the minuend is 0, it is again necessary to borrow a 1 *from the minuend to the left.* This adds two 1s to the minuend of the column in which subtraction is being performed. The result is a difference of 0 and a borrow-out of 1.

If the minuend is 1 and the subtrahend and borrow-in are both 1s, it is again necessary to borrow. The two 1s subtracted from the borrow (two 1s) plus the minuend of 1 means that two 1s are being subtracted from three 1s. The result is a difference of 1 and a borrow-out of 1. The borrow-out of one column becomes the borrow-in to the next higher-order column. This will become clearer after the following examples are explained.

Example 3-21: Subtract 1100 from 10110.
Solution:

```
        16s   8s   4s   2s   1s
         0    1
              1
         ✗    0    1    1    0
              1    1    0    0
        ─────────────────────────
         0    1    0    1    0
```

The solution begins with the lowest-order column where 0 from 0 is 0 with a borrow-out of 0. In the 2s column, $1 - 0 = 1$ with a borrow-out of 0. In the 4s column, $1 - 1 = 0$ with a borrow-out of 0. In the 8s column, $0 - 1$ requires a borrow of 1 from the 16s column. This results in a zero in the minuend of the 16s column and two 1s in the minuend of the 8s column. Now, subtracting one 8 from the two 8s results in a difference of 1.

The solution may be checked by converting each binary number to decimal or by adding the difference to the subtrahend; the sum should equal the minuend. The minuend $10110_2 = 22_{10}$, the subtrahend $1100_2 = 12_{10}$, and the difference $1010_2 = 10_{10}$.

Example 3-22: Subtract 1001_2 from 10100_2.
Solution:

	16s	8s	4s	2s	1s
		1		~~1~~	1
	0	1	0	1	1
	~~1~~	0	~~1~~	0	0
		1	0	0	1
		1	0	1	1

The new point brought out in Ex. 3-22 is the procedure to be followed when borrowing from a minuend of 0. In the units column, $0 - 1$ requires a borrow of 1. Since the minuend in the 2s column is 0, a 1 is borrowed from the 4s column. This 1 is represented as two 1s in the 2s column. One of these two 1s is now borrowed and represented as two 1s in the minuend of the units column. Subtraction of the units now results in a difference of 1. In the 2s column, $1 - 0 = 1$, and in the 4s column, $0 - 0 = 0$. It is again necessary to borrow in the 8s column. The 1 in the minuend of the 16s column is brought into the 8s column as two 1s which results in a difference of 1.

The solution is checked by converting the binary numbers to decimal. The minuend $10100_2 = 20$, the subtrahend $1001_2 = 9$, and the difference $1011_2 = 11_{10}$.

Binary Multiplication. Multiplication in the binary system is simple because the *multiplicand* can only be multiplied by a *multiplier* of 1 or 0. The result is called the *product*.

When the multiplier is 1, simply rewrite the multiplicand as the partial product. When the multiplier is 0, the partial product is 0. The only problem is the addition of the *partial products*.

Example 3-23: Multiply 1011_2 by 101_2.
Solution:

32s	16s	8s	4s	2s	1s
			1	0	1	1
				1	0	1
		1	0	1	1
	0	0	0	0	
1	0	1	1		
1	1	0	1	1	1

If the multiplier contains several ones, there may be more than three 1s to be added. Simply carry one 1 for each pair of ones.

Example 3-24: Multiply 1011.1_2 by 11.11_2.
Solution:

```
              1 0 1 1 . 1
                1 1 . 1 1
              ─────────────
              1 0 1 1 1        ½s column of partial products
            1 0 1 1 1
          1 0 1 1 1
        1₁ 0₁ 1₁ 1₁ 1₁  ₁
        ───────────────────
        1 0 1 0 1 1. 0 0 1
```

Adding the four 1s as in Ex. 3-24 in the halves column of the partial products results in a sum of 0 with a carry-out of two 1s. That is, the four $\frac{1}{2}$s are represented as zero $\frac{1}{2}$s and two 1s. These two carries produce five 1s in the 1s column. The sum of five 1s is 1 with a carry-out of two 1s into the next column. Thus, the five 1s are represented as one 1 and two 2s. This procedure is continued until all partial products have been added. Since there is a total of three binary places in the multiplier and multiplicand, a binary point is placed three digits in from the right in the answer.

Binary Division. The numbers involved in the problem are called (1) the *dividend* which is divided by (2) the *divisor* and (3) the *quotient* which is the resultant.

Example 3-25: Divide 11001_2 by 101_2.
Solution:

```
              101
         ─────────
      101 | 11001
            101
            ───
             101
             101
             ───
             000
```

Since the quotients of each step of a long-division problem can only be 1 or 0, there is no need for trial-and-error solutions, as is sometimes necessary in decimal division. First divide 101 into the first three digits of the dividend 110; the quotient is 1. The procedure then continues as in decimal, resulting in an answer of 101. The solution may be checked by converting each number to decimal or by multiplying the quotient by the divisor; the product should equal the dividend. The divisor $101_2 = 5_{10}$ and the dividend $11001_2 = 25_{10}$; thus, the quotient 101_2 is correct.

3-4 Complements

Negative numbers are often stored in *complement* form. This enables a computer to subtract using its adder circuitry. The two systems used are the radix (Rs) system which is the *tens* complement system in decimal, the *twos* complement system in binary, etc., and the *radix minus one* (R − 1) system which is the *nines* complement system in decimal, the ones complement system in binary, etc. Either, but not both, is used. Which system is chosen is at the discretion of the designer. Both systems, including the procedures for arithmetic using complements, are explained in this section.

The Radix Minus One Complement. The R − 1 complement of a number is formed by subtracting each digit from one less than the radix. This means that each digit is subtracted from the highest digit in the system. For example,

The R − 1 complement of 3_{10} $= 9 - 3$ $= 6_{10}$
The R − 1 complement of $107_{10} = 999 - 107$ $= 892_{10}$
The R − 1 complement of $1011_2 = 1111 - 1011 = 0100_2$
The R − 1 complement of 64_8 $= 77 - 64$ $= 13_8$

A very important simplification should be noted in the binary system. Since subtracting 1 from 1 leaves 0 and since 0 from 1 leaves 1, the 1s complement of any binary number may be formed by simply changing each 0 to 1 and each 1 to 0. This is known as complementing each bit. Hence, the 1s complement of $1011011101_2 = 0100100010_2$.

The Radix Complement. The Rs complement is formed by first forming the R − 1 complement and then adding a 1 to the lowest-order digit. For example,

The Rs complement of 3_{10} $= (9 - 3) + 1$ $= 6 + 1$ $= 7_{10}$
The Rs complement of $107_{10} = (999 - 107) + 1$ $= 892 + 1$ $= 893_{10}$
The Rs complement of $1011_2 = (1111 - 1011) + 1 = 0100 + 1 = 0101_2$
The Rs complement of 64_8 $= (77 - 64) + 1$ $= 13 + 1$ $= 14_8$

The Rs complement of the above numbers may also be formed by subtracting 3_{10} from 10_{10}, 107_{10} from 1000_{10}, 1011_2 from 10000_2, and 64_8 from 100_8. However, this requires borrowing and is more difficult to do mentally in systems other than decimal.

A simplified procedure is also available for the Rs system in radix 2. To form the 2s complement of any binary number, simply complement each bit to the left of the lowest-order 1. Hence, the 2s complement of $1010111000_2 = 0101001000$.

Two special cases should be noted: The 2s complement of $0000_2 = 0000_2$ because there is no *lowest-order 1* and the 2s complement of $1000_2 = 1000_2$ because there are no bits *to the left of* the lowest-order 1.

Subtraction Using the Rs Complement System. The difference between two numbers may be obtained by adding the Rs complement of the subtrahend to the minuend and dropping the highest-order carry. The subtrahend must contain the same number of digits as the minuend. This poses no difficulties in a computer. Since there are only 1s and 0s and no blanks in a computer, a thirty-digit machine expresses every number using *all* thirty digits.

Example 3-26: Subtract 63_{10} from 175_{10} by using the 10s complement system.
Solution:

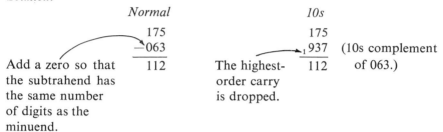

	Normal		10s	
	175		175	
	−063		₁937	(10s complement
Add a zero so that	112	The highest-	112	of 063.)
the subtrahend has		order carry		
the same number		is dropped.		
of digits as the				
minuend.				

Example 3-27: Subtract 1010_2 from 1011_2 by using the 2s complement system.
Solution:

	Normal		2s	
	1011		1011	
	−1010		₁0110	(2s complement
	0001	The highest-	0001	of 1010.)
		order carry		
		is dropped.		

All negative numbers are stored in complement form. If the subtrahend is larger than the minuend, the negative answer is automatically in complement from.

Example 3-28: Subtract 1111_2 from 1000_2 by using the 2s complement system.
Solution:

Normal		2s	
1000		1000	
−1111		0001	(2s complement of 1111)
−0111	The absence of a	1001	(2s complement of 0111)
	carry indicates a		
	negative answer.		

Subtraction Using the R − 1 Complement System. The difference between two numbers may also be obtained by adding the R − 1 complement of the subtrahend to the minuend. The highest-order carry is brought around and added to the lowest-order digit of the sum. This is known as *adding the end-around carry*. As in the Rs system, the subtrahend must have the same number of digits as the minuend.

Example 3-29: Subtract 42_{10} from 6255_{10} by using the 9s complement system.
Solution:

```
    Normal        9s
     6255         6255
    −0042         9957   (9s complement of 0042.)
     ────         ────
     6213         6212
                  ↳  1   Add the end-around carry.
                  ────
                  6213
```

Example 3-30: Subtract 101_2 from 1011_2 by using the 1s complement system.
Solution:

```
    Normal        1s
     1011         1011
    −0101         1010   (1s complement of 0101.)
     ────         ────
     0110         0101
                  ↳  1   Add the end-around carry.
                  ────
                  0110
```

If the answer is negative, it is automatically produced in complement form.

Example 3-31: Subtract -110_2 from -1111 by using the 1s complement system.
Solution:

```
    Normal        1s
    −1111         0000   (1s complement of 1111.)
    −⊕0110        0110
    ─────         ────
    −1001         0110   (1s complement of 1001.)
```

In the normal procedure the minus sign of the subtrahend is changed to a plus sign. Thus, the subtrahend is stored in true form by the machine. The absence of an end-carry indicates that the answer is negative.

Arithmetic with a Sign Bit. The negative and positive signs are represented by the bits 1 and 0, respectively. The *sign bit* may be handled in several different ways: It can be separated from the number by a radix point which is either always to the extreme left or the extreme right of the number. Thus, $+15 = 0.1111 \times 2^4$ and $-15 = 1.1111 \times 2^4$, or $+15 = 1111.0 \times 2^0$ and $-15 = 1111.1 \times 2^0$. Other computers actually have a floating radix point. Handling the exponent is not discussed in this text because it is actually a systems problem. However, examples are given below to show how addition is performed with a sign bit. The radix point is to the extreme left in these examples and the 1s complement system is used.

Example 3-32: Add $+.1000$ to $+.0100$ (both positive).
Solution:

```
         Normal        1s
         +.1000        0.1000
         +.0100        0.0100
         +.1100        0.1100
```

Example 3-33: Add $+.1010$ to $-.0100$ (larger positive).
Solution:

```
         Normal         1s
         +.1010        0.1010
         -.0100       ₁1.1011    (1s complement of .0100.)
         +.0110        0.0101
                           1    Add the end-around carry.
                       0.0110
```

Example 3-24: Add $+.0110$ to $-.1110$ (larger negative).
Solution:

```
         Normal         1s
         +.0110        0.0110
         -.1110        1.0001    (1s complement of .1110.)
         -.1000        1.0111    (1s complement of .1000.)
```

Example 3-35: Add $-.0101$ to $-.1010$ (both negative).
Solution:

```
         Normal         1s
         -.0101        1.1010    (1s complement of .0101.)
         -.1010       ₁1.0101    (1s complement of .1010.)
         -.1111        0.1111
                           1    Add the end-around carry.
                       1.0000    (1s complement of .1111)
```

3-5 Decimal Codes

The disadvantages of the decimal system, pointed out in the introduction of this chapter, are overcome by using *binary-coded decimal* BCD numbers. With these codes, the binary-decimal conversion of large numbers also becomes simple. Although many codes could be used, only a few find wide use. These are now discussed.

The Natural Binary-coded Decimal NBCD System. The *natural binary-coded decimal* NBCD system is also called the 8421 code. In this system every decimal digit is represented by a four-digit binary number having the natural weights 8421. The code numbers for each of the ten decimal digits are given in Table 3-5. Decimal numbers containing more than one digit are represented by NBCD numbers containing a four-bit group for each decimal digit. Thus, as shown in Table 3-5, decimal number 10 is 0001 0000 in NBCD. The first four-bit group represents the 1 in the tens position of the decimal number and the second four-bit group represents the 0 in the units position of the decimal number.

Table 3-5 Decimal-NBCD conversion.

Decimal	NBCD system (8421)		
0			0000
1			0001
2			0010
3			0011
4			0100
5			0101
6			0110
7			0111
8			1000
9			1001
10		0001	0000
99		1001	1001
256	0010	0101	0110

By using this code, it is only necessary to memorize the natural binary numbers from 0 to 9; then very large decimal numbers are easily expressed, for example, $59,564_{10} = 0101\ 1001\ 0101\ 0110\ 0100_{NBCD}$.

Forbidden Numbers. Since each of the four digits can be either a zero or a one, there are sixteen combinations of four-bit numbers. With only ten decimal digits to be represented, there are six combinations that have no meaning. These combinations are called *forbidden numbers*. The forbidden numbers in the 8421 code are 1010, 1011, 1100, 1101, 1110, and 1111. It must be explained that this is not a disadvantage of the NBCD system. Every decimal code has forbidden numbers because at least four binary digits are needed to represent the ten decimal digits.

Addition in NBCD There are two conditions that must be considered when adding in this code: (1) If the sum is less than ten, the correct answer is obtained directly. (2) If the answer is ten or more, the direct sum is incorrect. To get the correct answer, an NBCD $6 = 0110$ is added to the incorrect sum.

Example 3-36: Add 5_{10} to 3_{10} in NBCD.
Solution:

Normal	NBCD
5	0101
3	0011
8	1000

The direct addition of 0101 and 0011 results in the correct answer in Ex. 3-36.

Example 3-37: Add 7_{10} and 4_{10} in NBCD.
Solution:

Normal	NBCD
7	0000 0111
4	0000 0100
11	0000 1011

The NBCD solution is set up using two four-bit groups because the result is a two-digit decimal number. At first, the answer may seem to be correct. However, recall that 1011 is a forbidden number in NBCD. The correct answer is obtained by adding six (0110) to the incorrect sum.

0000 1011	(Incorrect answer.)
0000 0110	(Add 6.)
0001 0001	(Equals eleven in NBCD.)

Example 3-38: Add 9_{10} and 7_{10} in NBCD.
Solution:

Normal	NBCD	
9	0000 1001	
7	0000 0111	16s carry.
16	0001 0000	

The NBCD solution is again set up using two four-bit groups because the result is a two-digit decimal number. Again, the answer may appear to be correct. However, 0001 0000 is the correct answer in binary, but not in NBCD. To obtain the correct answer of any sum greater than nine, add 0110.

$$\begin{array}{ll} 0001\ 0000 & \text{(Incorrect answer.)} \\ \underline{0000\ 0110} & \text{(Add 6.)} \\ 0001\ 0110 & \text{(Equals sixteen in NBCD.)} \end{array}$$

The computer recognizes a sum greater than nine by looking for a forbidden number or, if the sum exceeds 1111, by a carry-out from the fourth (8s) column into the fifth (16s in binary) column. The 16s carry is pointed out in the solution above.

Disadvantage of the NBCD System. The simplest way for a computer to complement a number is to change each 1 to 0 and each 0 to 1 (complement each bit). This cannot be done in the NBCD system. For example, the 9s complement of 2 is 7. Since 2 = 0010 in NBCD, changing each bit results in 1101. This not only does not equal 7, but it is also a forbidden number.

If the computer uses the NBCD system, it must have special circuitry to form either the 9s or 10s complement of the coded decimal number.

The Excess Three XS3 Code. In this code each decimal digit is represented by a four-bit number that has an excess of 3. The code numbers for the ten decimal digits are given in Table 3-6. Decimal numbers containing more than one digit are represented by XS3 numbers containing a four-bit group for each decimal digit. Thus, as also shown in Table 3-6, decimal number 10 is 0100 0011 in XS3.

Table 3-6 Decimal—XS3 conversion.

Decimal	XS3 System		
0			0011
1			0100
2			0101
3			0110
4			0111
5			1000
6			1001
7			1010
8			1011
9			1100
10		0100	0011
99		1100	1100
256	0101	1000	1001

Advantage of the XS3 Code. By changing each 0 to 1 and each 1 to 0 (complementing each bit), the 9s complement is formed. This reduces the amount of hardware required compared to the NBCD system. For example, the 9s complement of 2 is 7. In the XS3 code, 0101 = 2. Complementing each bit produces 1010, which is 7 in XS3. A more involved example follows.

Example 3-39: Express 235 in XS3 and verify that its 9s complement is formed by complementing each bit.
Solution:
$$235 = 0101\ 0110\ 1000$$

Complementing each bit produces 1010 1001 0111 which equals decimal 764 in XS3. The nines complement of 235 = 999 − 235 = 764.

A second advantage of the XS3 code can be seen by referring again to Table 3-6. A square wave is available from the highest-order bit output.

Forbidden Numbers in the XS3 System. As explained in the discussion on NBCD forbidden numbers, every decimal code has forbidden numbers. Since the XS3 code is also a four-bit code, it also has six forbidden numbers. They are: 0000, 0001, 0010, 1101, 1110, and 1111.

The 74210 Code. This code is also known as the *two-out-of-five code*. It is a *weighted* code in which, when necessary, a 1 is placed in the 0 position so that each group always contains two 1s. The decimal digit 0 is a special case; it is represented by the one remaining two-out-of-five combination. As shown in Table 3-7, decimal 0 is represented by placing ones in the 7s and 4s positions. The digits 1, 2, 4, and 7 require the addition of a 1 in the 0s position, while 3, 5, 8, and 9, which already have two 1s, have a 0 in the 0s position.

The advantage of this code is that it aids in detecting errors. If a check shows anything other than two 1s, an error has been introduced. A disadvantage is the extra bit required.

Table 3-7 Decimal—74210 conversion.

Decimal	74210 System		
0			11000
1			00011
2			00101
3			00110
4			01001
5			01010
6			01100
7			10001
8			10010
9			10100
10		00011	11000
99		10100	10100
256	00101	01010	01100

Biquinary Codes. Biquinary codes are weighted codes that provide a square-wave output. Examples of these codes are given in Table 3-8. The 50–43210 code also aids in detecting errors, while the 2'421 code is *self-complementing*.

As shown in the 50–43210 portion of Table 3-8, each decimal digit is represented by using a single one in each section of the seven-bit number. Anything other than a single one means that an error has been introduced. An obvious disadvantage of this code is the seven bits required to represent each decimal digit. A square wave is available from either output of the two-bit group.

Table 3-8 Decimal-Biquinary conversion.

Decimal	50–43210	5421	4'421	2'421
0	01 00001	0000	0000	0000
1	01 00010	0001	0001	0001
2	01 00100	0010	0010	0010
3	01 01000	0011	0011	0011
4	01 10000	0100	0100	0100
5	10 00001	1000	1001	1011
6	10 00010	1001	1010	1100
7	10 00100	1010	1011	1101
8	10 01000	1011	1100	1110
9	10 10000	1100	1101	1111

As shown in the 2'421 portion of the table, the decimal digits 0 through 4 are written using the natural binary weights 4, 2, and 1. Then by using the 2' position to represent the digits 5 through 9, the 9s complement of each coded number is formed by simply changing each 0 to 1 and each 1 to 0. For example, the 9s complement of 4 is 5. Decimal 4 is 0100 in the 2'421 code and changing each bit yields 1011, which is 5 in this code. A square wave is available from the 2' output, as it is from the 5 and 4' outputs of the 5421 and 4'421 codes.

3-6 Boolean Algebra

Boolean algebra is used in conjunction with logical design procedures to develop equations describing all the mathematical operations performed by the computer. It is a mathematics of logic and gets it name from George Boole. In a text published in 1854 Boole devised rules that permit logical conclusions to be made by combining certain *propositions*. Others, such as A. Whitehead, B. Russell, C. Shannon, and A. De Morgan, later added to Boole's principles.

This section covers the principles, theorems, and postulates of Boolean algebra. The less obvious theorems and laws are explained using *switching circuits* and *truth tables*.

Terminology. Following are several terms that must be defined before the discussion is continued:

Propositions are algebraic statements that are said to be either *true* T or *false* F. Propositions are represented by letters. Usually, beginning letters of the alphabet are used for individual propositions and later letters are used for combined propositions. The binary digit "1" represents *true* and "0" represents *false*.

Propositions may be combined by the words *AND* and *OR*. The *AND* combination of two or more propositions is true only if *all* the individual propositions are true. For example, take the following proposition: *Two is an even number, four is an even number, and five is an even number*. This proposition is false even though the first two individual elements are true. The *OR* combination of two or more propositions is true *if at least one* of the individual propositions is true. For example, the following proposition is true even though only one of the individual elements is true: *Three is an even number or four is an even number or five is an even number*.

All of the algebraic symbols used to indicate *multiplication* (parentheses, brackets, braces, ×, raised dot, etc.) are used to represent the *AND* connector. The *AND* function is sometimes referred to as the *logical product* and the *OR* function is sometimes called the *logical sum*.

The addition sign + is used to represent the *OR* connector. The *AND* and *OR* functions are discussed again below.

Propositions may be negated by the addition of the word *NOT*. The negation of a proposition is true only if the original proposition is false. For example, *two is an odd number* is true if *NOT* is added. That is, *two is not an odd number* is a true statement.

The meaning of the terms just defined is better understood with the aid of switching circuits.

Switching Circuit Notation. As previously mentioned, switching circuits help illustrate some of the less obvious theorems, but first it is necessary to learn the meaning of the various switch symbols.

Figure 3-1 shows the standard symbols for any normally open (NO) or normally closed (NC) switch. The letter symbol next to the switch indicates the signal condition necessary to close the switch. In Fig. 3-1(a) signal X *closes* the switch. In Fig. 3-1(b) the switch is normally closed (NC) and opens when signal X is present. Therefore, $\bar{X}$ (the absence of signal X) must be present for this switch to be closed.

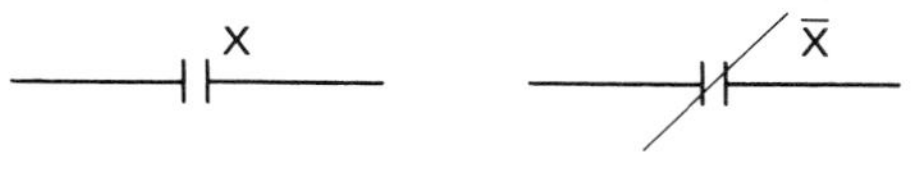

Fig. 3-1 Standard symbol for (a) a normally open switch and (b) a normally closed switch.

A thorough background in the theory and laws of simple series and parallel circuits is essential in electronics. In Boolean algebra it is equally important to understand the theory and laws of series and parallel switching circuits. Therefore, the laws of series and parallel circuits are discussed next.

Series Circuit Laws. Figure 3-2 shows several series switching circuits. The Boolean equation for each circuit expresses the signal conditions required to obtain transmission through the circuit. In Fig. 3-2(a) signal X and signal Y are needed. Thus, the transmission function of Fig. 3-2(a) is $f_a = X \text{ AND } Y = XY$.

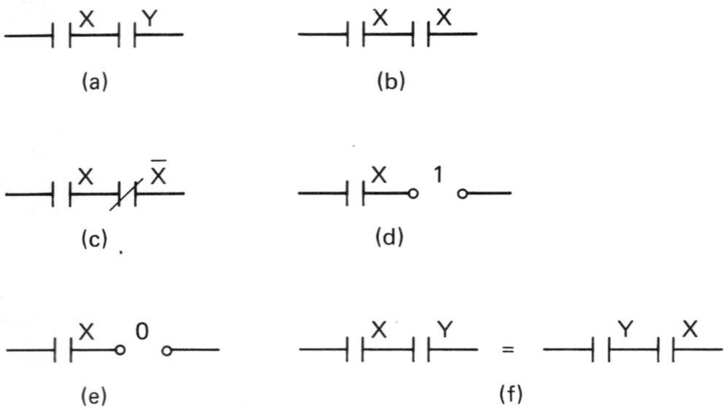

Fig. 3-2 Switches in series.

In Fig. 3-2(b) two switches that operate in the same manner in the presence of signal X are placed in series. The transmission function for this circuit is $f_b = X \text{ AND } X = X \cdot X$. It should be obvious that the same control may be obtained with a single switch that closes with signal X. Thus, $f = X \cdot X$ can be reduced to $f = X$.

In Fig. 3-2(c) two switches that operate *inversely* with signal X are placed in series. The transmission function for this circuit is $f_c = X \cdot \bar{X}$. It should also be obvious that transmission can never occur in this circuit. When signal X is present to close the normally open (NO) switch, it opens the normally closed (NC) switch. Thus, $f = X\bar{X} = 0$ (an *always-open* circuit).

In Fig. 3-2(d) a NO switch is in series with an *always-closed* circuit (represented by a 1).

The transmission function for this circuit is $f_d = X \cdot 1 = X$.

In Fig. 3-2(e) a NO switch is in series with an always-open circuit (represented by a 0).

The transmission function $f_e = X \cdot 0 = 0$ because transmission can never occur.

The *series commutative law* is illustrated in Fig. 3-2(f). This law states that logical multiplication can be performed in any order. Thus, $f_f = XY = YX$.

Fig. 3-3 The logic symbol for an *AND* circuit.

The logic symbol used in this text to represent any *AND* circuit is shown in Fig. 3-3.

Parallel Circuit Laws. Figure 3-4 shows several parallel switching circuits. The transmission function is written for each, and any simplifications are explained.

Figure 3-4(a) shows two switches in parallel. In order for transmission to occur, either signal *X OR* signal *Y* is needed. Thus, $f_a = X + Y$.

In Fig. 3-4(b) two switches that operate in the same manner are connected in parallel. The transmission function is $f_b = X + X$. It is obvious that the same control could be obtained with a single switch that closes with signal *X*. Thus, $f = X + X$ can be simplified to $f = X$.

Two switches that operate inversely with signal *X* are connected in parallel in Fig. 3-4(c). The transmission function is $f_c = X + \bar{X}$. Examination of the circuit shows that with signal *X* the top path is closed. But in the absence of signal *X*, the bottom path is closed. Thus, transmission always occurs and $f = X + \bar{X}$ can be reduced to $f = 1$.

In Fig. 3-4(d) a normally open (NO) switch that closes with signal *X* is placed in parallel with an always-closed circuit. It is obvious that the transmission function for this circuit $f_d = X + 1$ can be reduced to $f = 1$.

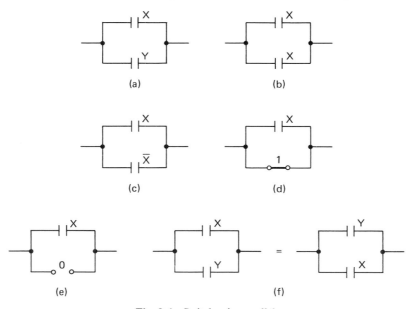

Fig. 3-4 Switches in parallel.

The NO switch is placed in parallel with an always-open circuit in Fig. 3-4(e). Transmission occurs only if signal X is present to close the top path. Thus, $f_e = X + 0$ can be reduced to $f = X$.

The *parallel commutative law* is illustrated in Fig. 3-4(f). This law states that logical addition can be performed in any order. Thus, $f_f = X + Y = Y + X$.

The logic symbol used in this text to represent any OR circuit is given in Fig. 3-5.

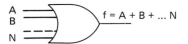

Fig. 3-5 The logic symbol for an OR circuit.

The Distributive Laws. The distributive laws are two of the basic laws of series-parallel circuits: They are explained with the aid of Figs. 3-6 and 3-7. To effect transmission through the top half of Fig. 3-6(a), signals A and B are needed. Transmission could also take place through the bottom half of this circuit if signal A and signal C are present. Hence, the transmission function is $f_a = AB + AC$, which reads *the function of "a" equals the quantity* A *AND* B *OR the quantity* A *AND* C. But closer examination of the circuit shows that as soon as signal A appears, both A switches close to permit transmission through the first half of each path. Transmission then takes place if either signal B or signal C is applied. Hence, the circuit of Fig. 3-6(a) can be reduced to the circuit of Fig. 3-6(b), and the $f_a = AB + AC$ can be simplified to $f_b = A(B + C)$. This is known as the *first distributive law*.

An expansion of this rule is illustrated in Fig. 3-6(c). If signals A and B are simultaneously applied, transmission occurs if either signal C is applied to close the top line or if signals D and E are applied to close the bottom line.

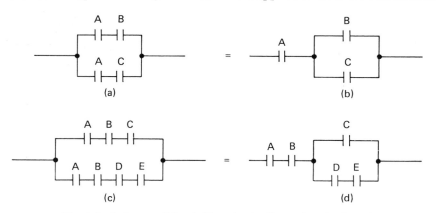

Fig. 3-6 Series-parallel switching circuits illustrating the first distributive law.

Thus, $f_c = ABC + ABDE$ can be simplified to $f_d = AB(C + DE)$. The simplified switching circuit is shown in Fig. 3-6(d).

A second series-parallel circuit rule, known as the *second distributive law*, is illustrated in Fig. 3-7. Transmission occurs through the first half of Fig. 3-7(a) if either signal A or signal B is applied. Then, transmission through the second half of the circuit takes place if either A or C is applied. Therefore, $f_a = (A + B)(A + C)$, which reads *the function of "a" equals the quantity* A *OR* B *AND the quantity* A *OR* C. But closer examination of the circuit shows that as soon as signal A is applied, both A switches close to permit transmission through the entire top half of the circuit. Transmission will also occur, through the bottom half, if signals B and C are applied simultaneously. Hence, the circuit of Fig. 3-7(a) can be reduced to the circuit of Fig. 3-7(b), and the $f_a = (A + B)(A + C)$ can be simplified to $f_b = A + BC$.

An expansion of the second distributive law is shown in Fig. 3-7(c). Transmission takes place through the top line as soon as signals A and B and C are simultaneously applied, and transmission takes place through the bottom half if signals D, E, F, and G are simultaneously applied. Thus, the circuit of Fig. 3-7(c) is reduced to the circuit of Fig. 3-7(d), and $f_c = (ABC + DE)(ABC + FG)$ is reduced to $f_d = ABC + DEFG$.

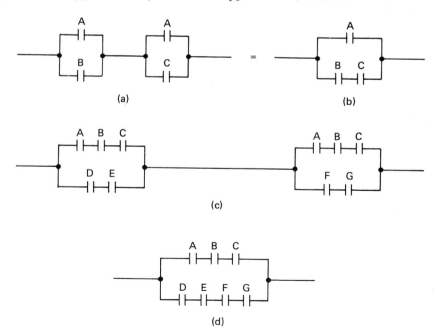

Fig. 3-7 Series-parallel switching circuits illustrating the second distributive law.

The Associative Laws. These laws, like the first distributive law and the two commutative laws, are also true in ordinary algebra. One associative law states that it is permissible to group any two terms of a sum. For example, $A + B + C$ can be written $(A + B) + C$, $A + (B + C)$, or $(A + C) + B$. The second associative law states that it is permissible to group any two factors of a product. Hence, $ABC = (AB)C = A(BC) = (AC)B$.

The Negation Theorem. This theorem states that if the complement of a term is negated or complemented, it will equal the original term. That is,

$$\bar{\bar{A}} = A$$

$$\overline{\overline{AB}} = AB$$

The Absorption Theorem. The absorption theorem states that if any term is *O*Red to another term that is made up of itself *AND*ed to some other term, the first term absorbs the second. For examples,

(1) A + AB = (A) + (A) · (B) = A

                      Any term *O*Red to itself *AND*ed to some other term equals the first term.

(2) ABC + ABCDE = (ABC) + (ABC) · (DE) = ABC

To further explain this theorem, the switching circuits for these equations are given in Fig. 3-8. In the circuit for $A + AB$, shown in Fig. 3-8(a), it can be seen that transmission either occurs through the top line if A is present or through the bottom line if both A and B are present simultaneously. It is obvious that whenever the bottom line is closed, the top line is simultaneously closed to offer a parallel path. Since only one transmission path is necessary, the bottom line may be omitted, that is, A absorbs AB.

The switching circuits of Fig. 3-8(b) show that whenever transmission

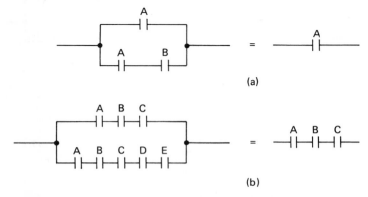

Fig. 3-8 Applications of the absorption theorem.

occurs through the bottom ($ABCDE$) leg, the top (ABC) circuit is also closed to offer a parallel path. Again, since one path is sufficient, the bottom leg may be omitted.

Further proof of the absorption theorem may be obtained by applying some of the laws of Boolean algebra that have already been explained.

$A + AB = A(1 + B)$ The first distributive law says that it is permissible to factor.

$A(1 + B) = A(1)$ A basic law of parallel circuits says that $1 + B = 1$.

$A(1) = A$ A basic law of series circuits.

Hence,
$$A + AB = A$$

The Complementary Absorption Theorem. The complementary absorption theorem states that if any term is ORed to another term consisting of its complement ANDed to some other term, the complement is absorbed. For example,

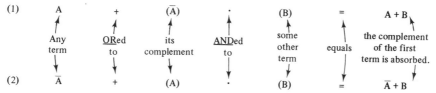

This theorem is further explained by the switching circuits of Fig. 3-9. In Fig. 3-9(a) transmission can occur through the top line if signal A is

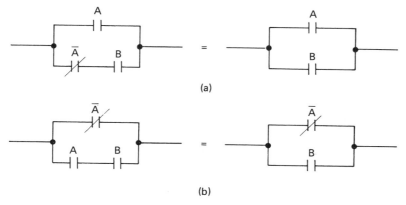

Fig. 3-9 Applications of the complementary absorption theorem.

present or through the bottom line if A is absent and B is present. Note, though, that whenever the A line is open, the $\bar{A}$ switch is closed and only signal B is required to complete the bottom line. Therefore, the two-switch $A + B$ circuit can replace the three-switch $A + \bar{A}B$ circuit. The complement of A is absorbed.

The circuit of Fig. 3-9(b) shows that transmission takes place through the top line if $\bar{A}$ is present or through the bottom line if A and B are both present. This circuit can be simplified because whenever the top line is open, the A switch is closed and only B is required to complete the bottom line. Therefore, $\bar{A} + AB = \bar{A} + B$. The complement of $\bar{A}$ is absorbed.

The Truth Table (Table of Combinations). The above theorems may also be proved by setting up a table of all the combinations for the two expressions making up the equation. If transmission and hindrance occur at the same times for the two expressions, they are equal. A truth table verifying that $A + \bar{A}B = A + B$ is shown in Table 3-9. The first two columns contain all combinations of the two signals A and B. There are 2^n combinations where n is the number of variables. Given two variables A and B, there are

Table 3-9 A truth table verifying that for all signal possibilities $A + \bar{A}B$ equals $A + B$.

A	B	$\bar{A}$	$\bar{A}B$	$A + \bar{A}B$	$A + B$
0	0	1	0	0	0
0	1	1	1	1	1
1	0	0	0	1	1
1	1	0	0	1	1

$2^2 = 4$ combinations. The next two columns are included because they make up part of the $A + \bar{A}B$ expression. The $\bar{A}B$ term is a 1 (closed circuit) only when both $\bar{A}$ and B are 1s. The $A + \bar{A}B$ expression is a 1 if either A or $\bar{A}B$ is a 1. The $A + B$ term is a 1 if either A or B is a 1. Note that the $A + \bar{A}B$ and $A + B$ columns are equal for every combination of the A and B signals.

De Morgan's Theorems. These theorems state

$$\overline{A + B} = \bar{A}\bar{B} \qquad (1)$$

$$\overline{AB} = \bar{A} + \bar{B} \qquad (2)$$

(1) The complement of the sum of two (or more) terms equals the product of the complements of those terms.

(2) The complement of the product of two (or more) terms equals the sum of the complements of those terms.

The truth tables for De Morgan's theorems are shown in Table 3-10. The (a) part of Table 3-10 proves that $\overline{AB} = \bar{A} + \bar{B}$, and the (b) part of the table proves that $\overline{A + B} = \bar{A}\bar{B}$.

Table 3-10 Truth tables for De Morgan's theorems.

A	B	$\bar{A}$	$\bar{B}$	AB	$\overline{AB}$	$\bar{A} + \bar{B}$
0	0	1	1	0	1	1
0	1	1	0	0	1	1
1	0	0	1	0	1	1
1	1	0	0	1	0	0

(a)

A	B	$\bar{A}$	$\bar{B}$	A + B	$\overline{A + B}$	$\bar{A}\bar{B}$
0	0	1	1	0	1	1
0	1	1	0	1	0	0
1	0	0	1	1	0	0
1	1	0	0	1	0	0

(b)

The Redundancy Theorem. Figure 3-10 helps to explain this theorem. In Fig. 3-10(a) it can be seen that transmission can take place through the bottom line if both *B* and *C* are present. However, this would also close the *B* and *C* switches in the top two lines. This action permits transmission

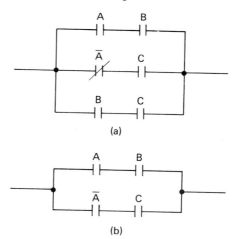

Fig. 3-10 The switching circuits for the redundancy theorem.

through the top line if *A* is present or through the middle line if *A* is not present. Since either *A* or $\bar{A}$ is always present, the bottom line is redundant. The simplified switching circuit is shown in Fig. 3-10(b). The redundant term is recognized in the following way:

1. Look for a *complementary pair* of variables. In the expression below, *A* and $\bar{A}$ are complements.

2. Multiply their coefficients. B is the coefficient of A. C is the coefficient of $\bar{A}$.

3. If one of the other terms in the expression equals the product of the coefficients, that term is redundant. The third term BC in the expression below equals the product of the coefficients of the complementary pair and is therefore redundant.

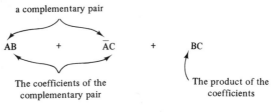

Hence,
$$AB + \bar{A}C + BC = AB + \bar{A}C$$

In some expressions more than one term may be redundant. In other expressions one or the other of two terms may be redundant, but not both. For example, in the expression

$$A\bar{C} + AB + \bar{B}\bar{C} + BC$$

either $A\bar{C}$ or AB is redundant. When either one of these terms is removed, the other term is no longer redundant.

3-7 Basic Logical Design

This section describes a basic procedure for the logical design of digital circuits. The problem is stated and then the following logical design procedures are performed:

1. A truth table is set up. All of the signal combinations that can appear at the inputs to the digital circuit are listed.
2. The correct output (or outputs) for each input combination is determined from the information given in the problem statement.
3. A Boolean equation called the *transmission function* is written for each output line.
4. Each equation is simplified using the Boolean theorems and laws of Sec. 3-6.
5. A logic diagram is drawn for each simplified equation.

Several examples are given to explain these procedures.

Logical Design of the EXCLUSIVE-OR Circuit. The *EXCLUSIVE-OR* circuit is frequently encountered in digital computers. It has two inputs and one output. The output is high (transmission occurs) only if one *or* the other input is high. If both inputs are high, transmission does *not* take place as it does in the *INCLUSIVE-OR* circuit of Fig. 3-4(a).

Example 3-40: A block diagram of a digital circuit having two inputs and one output is shown in Fig. 3-11(a). The output is to be high only if one or the other input is high. Draw the logic diagram for this circuit.

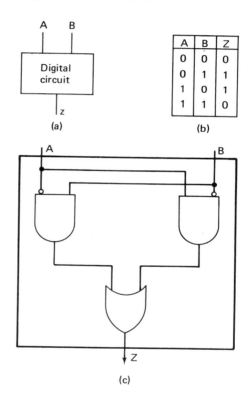

Fig. 3-11 The block diagram, truth table, and logic diagram of the *EXCLUSIVE-OR* circuit.

Solution: With two inputs, there are $2^2 = 4$ combinations. They are shown in the truth table of Fig. 3-11(b). The output z terminal is low (no transmission) when both inputs are low and when both inputs are high. When A is low and B is high, or when A is high and B is low, z is high.

The transmission function for the output line f_z is determined by writing in a Boolean equation each input combination that makes the output high. The output line is high when A and $\bar{B}$ are present or when $\bar{A}$ and B are present. Hence,

$$f_z = A\bar{B} + \bar{A}B$$

118 / Computer Math and Logic

Since this equation cannot be simplified, the logic diagram can now be drawn. It is shown in Fig. 3-11(c). The small circles at the input to each *AND* gate represent inverters or *NOT* gates. They indicate that the input is complemented before going into the gate. Other inverter symbols are shown in Fig. 5-1.

Logical Design of the Equivalence Circuit. Another circuit frequently encountered in digital computers is the *EQUIVALENCE* circuit. This circuit allows transmission only if the two inputs are the same, both 0s or both 1s.

Example 3-41: A block diagram of a digital circuit having two inputs and one output is shown in Fig. 3-12(a). The output is to be high whenever both inputs are the same.

Solution: The two-input table of combinations is shown in Fig. 3-12(b). The

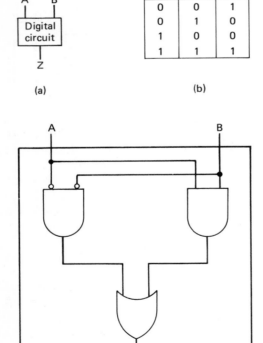

Fig. 3-12 The block diagram, truth table, and logic diagram for the *EQUIVALENCE* circuit.

output terminal z is a 1 when A and B are both 0s or when A and B are both 1s. The transmission function is

$$f_z = \bar{A}\bar{B} + AB$$

which cannot be simplified. The logic diagram for this circuit is shown in Fig. 3-12(c).

It should be noted that the *EQUIVALENCE* function is the complement of the *EXCLUSIVE-OR* function. Although this may not be immediately recognized by looking at the two expressions, it can be seen by examining the two truth tables of Figs. 3-11(b) and 3-12(b). The function number of terminal z in the *EXCLUSIVE-OR* table is 0110, and it is 1001 in the *EQUIVALENCE* circuit table. It can also be proved that they are complements by using theorems. The complement of the *EXCLUSIVE-OR* function is

$$\overline{A\bar{B} + \bar{A}B}$$

Remove the large overbar by applying De Morgan's theorems.

$$\overline{A\bar{B} + \bar{A}B} = (\bar{A} + B)(A + \bar{B})$$

Logical multiplication of the two terms on the right side of the equation yields

$$0 + \bar{A}\bar{B} + AB + 0$$

Hence, the complement of the *EXCLUSIVE-OR* function is $\bar{A}\bar{B} + AB$.

Logical Design of Encoders. There are many times when one set of binary digits must be changed to another set. These circuits, which can have any number of inputs and outputs, are known as *encoders* and *decoders*. An example of an encoder is a circuit that takes decimal inputs and provides NBCD outputs. The circuit that converts the NBCD digits back to decimal is called a *decoder*. Decoders are shown in Chap. 12. A simple encoder is designed in the following example.

Example 3-42: Design an encoder that has three inputs that represent binary numbers N and that provides an output equal to $2N + 3$.

Solution: The block diagram of the circuit is shown in Fig. 3-13(a). With three inputs there are eight input combinations. Since the largest binary number N that can exist at the input is $111 = 7$, the largest output is $2N + 3 = 2(7) + 3 = 17$. Five outputs are required because $17 = 10001$ in binary. The outputs for every input number are given in the truth table of Fig. 3-13(b).

Three of the output transmission functions can be determined by inspection. The z output is a 1 for all input combinations; hence, $f_z = 1$. The y output is the complement of C, regardless of the input combination; hence,

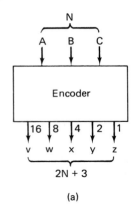

N	A	B	C	V	W	X	Y	Z	2N + 3
0	0	0	0	0	0	0	1	1	3
1	0	0	1	0	0	1	0	1	5
2	0	1	0	0	0	1	1	1	7
3	0	1	1	0	1	0	0	1	9
4	1	0	0	0	1	0	1	1	11
5	1	0	1	0	1	1	0	1	13
6	1	1	0	0	1	1	1	1	15
7	1	1	1	1	0	0	0	1	17

(b)

Fig. 3-13 (a) The block diagram of a three-input encoder with (b) its truth table for an output equal to $2N + 3$.

$f_y = C$. The v output is a 1 only when all inputs are high; hence, $f_v = ABC$. The two remaining functions must be determined by applying the Boolean laws of Sec. 3-6.

The w output is high if the binary numbers 3, 4, 5, and 6 are applied. Thus,

$$f_w = \bar{A}BC + A\bar{B}\bar{C} + A\bar{B}C + AB\bar{C}$$

Applying the first distributive law, $A\bar{B}$, which is common to the second and third terms, and $A\bar{C}$, which is common to the second and fourth terms, may be factored. This yields

$$f_w = \bar{A}BC + A\bar{B}(\bar{C} + C) + A\bar{C}(\bar{B} + B)$$

But $\bar{C} + C = 1$ and $\bar{B} + B = 1$. Hence,

$$f_w = \bar{A}BC + A\bar{B}(1) + A\bar{C}(1)$$

which can be further reduced to

$$f_w = \bar{A}BC + A\bar{B} + A\bar{C}$$

The x output is high if the binary numbers 1, 2, 5, and 6 are applied. Thus,

$$f_x = \bar{A}\bar{B}C + \bar{A}B\bar{C} + A\bar{B}C + AB\bar{C}$$

Applying the first distributive law, $\bar{B}C$, which is common to the first and third terms, and $B\bar{C}$, which is common to the second and fourth terms, may be factored. This yields

$$f_x = \bar{B}C(\bar{A} + A) + B\bar{C}(\bar{A} + A)$$

But $\bar{A} + A = 1$. Hence,

$$f_x = \bar{B}C(1) + B\bar{C}(1)$$

which reduces to

$$f_x = \bar{B}C + B\bar{C}$$

The five output functions are, therefore,

$$f_v = ABC \qquad f_w = \bar{A}BC + A\bar{B} + A\bar{C}$$
$$f_x = \bar{B}C + B\bar{C} \qquad f_y = \bar{C} \qquad f_z = 1$$

The logic diagram for this encoder is shown in Fig. 3-14. All of the A, B, and C inputs to the AND gates are connected to the corresponding A, B, and C inputs of the block diagram of Fig. 3-13(a). The complemented inputs $\bar{A}$, $\bar{B}$, and $\bar{C}$ are produced by the inverters, each symbolized by a triangle-enclosed I. The z output is simply connected to a source equal to the 1-level voltage of the system.

3-8 NAND, NOR, INHIBIT, and IMPLICATION Gates

Up to this point only AND, OR, and NOT gates have been discussed. Two other circuits commonly used in digital computers are the $NAND$ (NOT AND) and the NOR (NOT OR) gates. Not as common, but still useful, are the $INHIBIT$ and $IMPLICATION$ gates. Each of these gates is now discussed.

NAND Gates. The $NAND$ gate is an AND gate followed by an inverter. The logic symbol for a $NAND$ gate is shown in Fig. 3-15(a). The $NAND$ function is called a *primitive function* because it can be used to produce any Boolean equation. It is shown in Fig. 3-15(b) producing $\bar{A}$ from A. All unused inputs should be tied to a 1 level, V_{CC}, or the used input. In Fig. 3-15(c) the $EXCLUSIVE\text{-}OR$ function is produced using only $NAND$ gates.

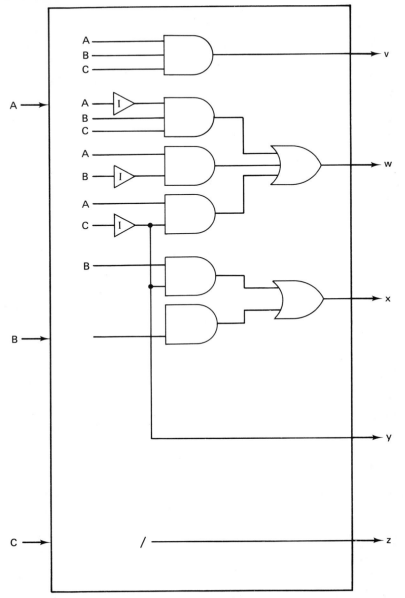

Fig. 3-14 The logic diagram of the three-input encoder of Ex. 3-42.

NOR Gates. The *NOR* gate is an *OR* gate followed by an inverter. The logic symbol for the *NOR* gate is shown in Fig. 3-16(a). The *NOR* function is also a primitive function because it too can be used to produce any Boolean equation. It is shown producing $\bar{A}$ from A in Fig. 3-16(b). All unused inputs

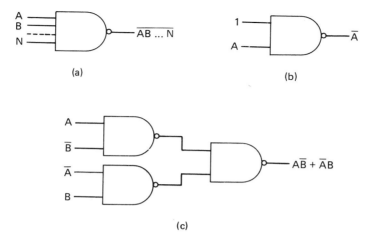

Fig. 3-15 (a) The *NAND* gate logic symbol. (b) A *NAND* gate used as an inverter. (c) The *EXCLUSIVE-OR* function produced with *NAND* gates.

should be tied to a 0 level, ground, or the used input. In Fig. 3-16(c) it is used to produce the *EXCLUSIVE-OR* function.

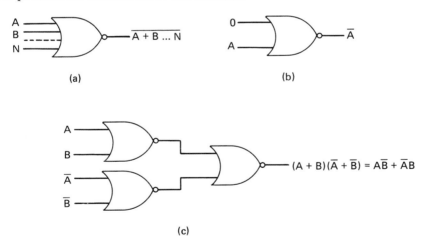

Fig. 3-16 (a) The logic symbol for a *NOR* gate. (b) The *NOR* gate used an an inverter. (c) The *NOR* gate used to produce the *EXCLUSIVE-OR* function.

The INHIBIT and IMPLICATION Functions. An *AND* gate with one input inverted is called an *INHIBIT* gate. A 1 on the inverted input line *inhibits* the gate. That is, the output will be 0 regardless of the other inputs. The logic symbol for an *INHIBIT* gate is shown in Fig. 3-17(a).

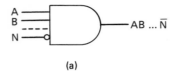

(a)

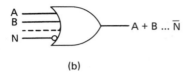

(b)

Fig. 3-17 The logic symbols for (a) an *INHIBIT* gate and (b) an *IMPLICATION* gate.

An *OR* gate with one input inverted is called an *IMPLICATION* gate. A 1 on the inverted input line *implies* that a 1 must be applied to one of the other inputs to produce a 1 at the output. The logic symbol for an *IMPLICATION* gate is shown in Fig. 3-17(b).

3-9 Summary

The two basic requirements for computer design are a system of numbers that can easily be used by the computer and a system of mathematics that can describe every operation to be performed. The binary (base 2) number system is used. Base 8 and base 16 are also used for certain applications because it is easy to convert between these bases and binary. Computers frequently use the decimal system, but the decimal numbers are expressed in some form of binary code.

Negative numbers can be stored in *complement form*. This permits a computer to subtract using the same circuits that it uses for adding. Either the radix (10s in decimal or 2s in binary) complement system or the radix minus one (9s in decimal or 1s in binary) complement system is used.

Boolean algebra is a mathematics of logic. It is used along with logical design techniques to produce equations describing every operation to be performed.

Problems

3-1 (For Sec. 3-1: Number Systems.)
    A. Convert each of the following to decimal:
        (a) 110110.1101_2 (f) 457.2_8
        (b) 1100111.11011_2 (g) $A635.E_{16}$
        (c) 10111011011.011_2 (h) 532.41_6
        (d) 1111011.101_2 (i) 233.22_4
        (e) 1111111.111_2 (j) 284.65_9

B. Convert each of the following numbers to binary:
   (a) 85.625_{10}
   (b) 130.8_{10}
   (c) 187.75_{10}
   (d) 214.2_{10}
   (e) 2000.8125_{10}
   (f) 264.05_8
   (g) 325.12_8
   (h) $A56C.1D_{16}$
   (i) $586E.F_{16}$
   (j) 24.33_5

C. Convert the following binary numbers to base 8 and to base 16:
   (a) 10111011.1011
   (b) 10111100.10111
   (c) 111011101.0101
   (d) 1101101111.11
   (e) 111111101.0111

3-2 (For Sec. 3-3: Binary Arithmetic.)
   A. Add the following numbers:
      (a) 1011011.011_2 and 1101111.11_2
      (b) 11111111.1101_2 and 111111011.1_2
      (c) 1110111.11011_2 and 100110.111_2
      (d) 10111101.1111_2 and 111111.1111_2
      (e) 110111101.101_2 and 10101010.01_2
   B. Do the following subtraction problems:
      (a) 110111000.11_2 less 10110.1111_2
      (b) 11100000.01_2 less 111011.11_2
      (c) 1000000.0001 less 111101.111
      (d) 10111011.10111 less 111111.111_2
      (e) 110111011.101_2 less 100101.11111_2
   C. Multiply
      (a) 101010.11_2 by 101.1_2
      (b) 11011.111_2 by 11.1_2
      (c) 11111011.101_2 by 110.1_2
      (d) 1101101.11_2 by 1.101_2
      (e) 10111.1_2 by 11.011_2
   D. Divide
      (a) 101111.01_2 by 11.1_2
      (b) 1010001001.11 by 10.111_2
      (c) 10011001.10101 by 10.11_2
      (d) 100111011 by 111
      (e) 11101011 by 101

3-3 (For Sec. 3-4: Complements.)
   A. Give the Rs and R − 1 complements of
      (a) 1011_2
      (b) 1111_2
      (c) 1000_2
      (d) 0000_2
      (e) 0001_2
      (f) 76_{10}
      (g) 235_{10}
      (h) 107_{10}
      (i) 24_8
      (j) 16_8

B. Using $1 = -$, $0 = +$, and the $R - 1$ complement system, add
  (a) $+.0001_2$ and $+.1001_2$
  (b) $+.1110_2$ and $-.0101_2$
  (c) $+.1000_2$ and $-.1001_2$
  (d) $-.1001_2$ and $-.0010_2$
  (e) $-.1010_2$ and $-.0011_2$

C. Using $1 = -$, $0 = +$, and the Rs complement system, add
  (a) $+.1000_2$ and $+.0100_2$
  (b) $+.1100_2$ and $-.1000_2$
  (c) $+.1000_2$ and $-.1111_2$
  (d) $-.0111_2$ and $-.1000_2$
  (e) $-.1000_2$ and $-.0001_2$

3-4 (For Sec. 3-5: Decimal Codes.)
A. Express 205_{10} and 468_{10} in the following codes:
  (a) NBCD   (d) 74210
  (b) XS3   (e) 2'421
  (c) 5421

B. Add the following decimal numbers in NBCD:
  (a) 6 and 2   (d) 756 and 235
  (b) 7 and 5   (e) 362 and 904
  (c) 9 and 8

3-5 (For Sec. 3-6: Boolean Algebra.)
A. Using the AND gate and OR gate logic symbols of Figs. 3-3 and 3-5 and the inversion-ball symbol for a NOT gate, draw the logic diagrams for the following Boolean expressions:
  (a) $AB + AC$ (f) $\overline{AB}$
  (b) $A(B + C)$ (g) $\overline{A + B}$
  (c) $(A + B)(A + C)$ (h) $AB + A\bar{C} + \bar{A}C$
  (d) $A + BC$ (i) $AC + \bar{A}B$
  (e) $A + \bar{A}B$ (j) $A\bar{C} + BC + \bar{B}\bar{C}$

B. Simplify each of the following by using the Boolean laws and theorems:
  (a) $f_{(a)} = \bar{A}\bar{B}\bar{C} + \bar{A}\bar{B}C + \bar{A}BC + ABC$
  (b) $f_{(b)} = \bar{A}\bar{B}\bar{C} + \bar{A}\bar{B}C + A\bar{B}C + AB\bar{C} + ABC$
  (c) $f_{(c)} = \bar{A}\bar{B}\bar{C} + \bar{A}\bar{B}C + \bar{A}B\bar{C} + \bar{A}BC + A\bar{B}\bar{C}$
  (d) $f_{(d)} = \bar{A}\bar{B}\bar{C} + \bar{A}B\bar{C} + \bar{A}BC + AB\bar{C} + A\bar{B}C + ABC$
  (e) $f_{(e)} = \bar{A}\bar{B}\bar{C} + \bar{A}\bar{B}C + \bar{A}B\bar{C} + \bar{A}BC + A\bar{B}\bar{C} + A\bar{B}C + AB\bar{C}$

3-6 (For Sec. 3-7: Basic Logical Design.) Do the logic design for each of the following circuits. Show the truth table, original and simplified Boolean equations for each output line, and the logic diagram for:
  (a) a circuit with three inputs and one output. The output is to be high whenever an odd number of inputs are high or if all inputs are low.
  (b) an encoder that has three inputs that represent a binary number N and produces an output equal to $N^2 + 3$.

(c) an encoder that has three inputs that represent a binary number N and produces an output equal to $N^2 - 15$. Let the minus sign be represented by a 1 and the plus sign by a 0. Also let the sign bit be the leftmost digit.

(d) a circuit that has four inputs that represent binary numbers and produces NBCD outputs.

(e) a two-bit by two-bit multiplier.

3-7 (For Sec. 3-8: *NAND, NOR, INHIBIT,* and *IMPLICATION* Gates.)

(a) By using De Morgan's theorem, give a second logic symbol that may be used for each of the above listed gates.

(b) Using the *NAND* gate logic symbol of Fig. 3-15(a), draw the logic circuit for $f_{(b)} = A\bar{B} + AC + \bar{A}B$. (The variables are available in true form only. The complements must be produced.)

(c) Using only *NOR* gates, draw the logic circuit for $f_{(c)} = (A + B)(\bar{A} + C)(\bar{A} + \bar{B})$. (Complements must be produced.)

(d) Using the *INHIBIT* gate symbol of Fig. 3-17(a), draw the logic circuit for $f_{(d)} = (A + B)(\bar{A} + \bar{B})$. (Complements must be produced.)

(e) Using the *IMPLICATION* gate symbol of Fig. 3-17(b), draw the logic circuit for $f_{(e)} = A\bar{C} + \bar{A}C$. (Complements must be produced.)

4

DIODE LOGIC

The most complex digital computer performs only a few basic functions. These functions are performed many times by only a few different circuits. Two of these circuits are known as the *AND* gate and the *OR* gate. These gates may be made by using any kind of bistate device. This chapter deals with the static and transient analysis of diode *AND* gates and *OR* gates individually and in *cascade*.

The other basic circuits found in digital computers are the *NOT* gate and the *FLIP-FLOP*. These and other circuits that combine the *AND* and *NOT* operations (*NAND* gate) and the *OR* and *NOT* operations (*NOR gate*) are discussed in subsequent chapters.

4-1 Static Analysis of Diode AND Gates

The *AND* gate is a circuit with two or more inputs and only one output. The output is in the 1 state only if *all* inputs are in the 1 state. Several logic symbols are used to represent the *AND* circuit, as shown in Fig. 4-1. The electrical circuit for a two-input positive-logic* *AND* gate is given in Fig. 4-2(a), and the output voltage, for every combination of input voltage, is given in Fig. 4-2(b). The operation of this circuit is explained first using ideal diodes.

*When the high level of voltage (V_H) represents a logical high (1) and when the low level of voltage (V_L) represents a logical low (0), it is called positive logic.

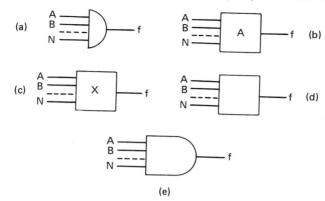

Fig. 4-1 Logic symbols for $f = AB \ldots N$.

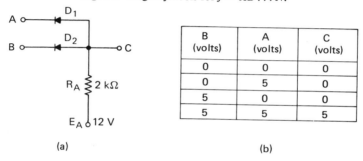

Fig. 4-2 (a) The AND gate circuit. (b) The voltage table for a high input voltage $V_H = 5$ V and a low input voltage $V_L = 0$ V.

Both Inputs Low. The anodes are connected through a 2-kΩ resistor to +12 V. Thus, with 0 V applied to each cathode the diodes see an equal amount of forward bias. Both diodes conduct with current flowing from E_A, through R_A, and then dividing equally through D_1 and D_2. If the ideal diodes are assumed to be short circuits when conducting, the 0 V potential at A and B is connected directly to terminal C. Thus, with *both* inputs *low*, the output is low.

$$I_{R_A} = \frac{E_A - V_C}{R_A} = \frac{12 \text{ V} - 0 \text{ V}}{2 \text{ k}\Omega} = 6 \text{ mA}$$

Each input circuit acts as a *sink* for 3 mA.

A High B Low. When there are different inputs, the diode that sees the greatest difference of potential between its cathode and E_A conducts and cuts off the other diode. At the instant 5 V is applied to input A, D_1 becomes reverse biased. D_2 conducts with current flowing from E_A, through R_A and D_2. The 0 V potential at B is connected directly to C through the short circuit offered by D_2. D_1 sees 5 V at its cathode and 0 V at its anode causing 5 V reverse bias. With different inputs, the output is low.

A Low B High. This condition is similar to the second one except that now D_2 is cut off. D_1 conducts and applies the 0 V potential at A directly to C. D_2 sees 5 V at its cathode and 0 V at its anode for 5 V reverse bias. Thus, with one input low the output is low and

$$I_{RA} = \frac{E_A - V_C}{R_A} = \frac{12 \text{ V} - 0 \text{ V}}{2 \text{ k}\Omega} = 6 \text{ mA}$$

The full 6 mA flows into one driver circuit.

Both Inputs High. Since the two diodes see an equal amount of forward bias, they both conduct. Current flows from E_A, through R_A, and then divides equally through D_1 and D_2. The 5 V potential at A and B is connected directly to C through the short circuits offered by the diodes. In this state

$$I_{RA} = \frac{E_A - V_C}{R_A} = \frac{12 \text{ V} - 5 \text{ V}}{2 \text{ k}\Omega} = 3.5 \text{ mA}$$

Each input source acts as a sink for 1.75 mA. Hence, the *AND* function has been implemented. The output is high when and only when *all* inputs are high.

If negative logic* is used, the circuit shown in Fig. 4-2(a) performs the *OR* function. This may be understood by studying the truth tables in Table 4-1. In the positive-logic truth table a logical 1 is written every time that 5 V

Table 4-1 The truth tables for the voltage chart in Fig. 4-2(b). (a) The positive-logic table. (b) The negative-logic table.

B	A	C		B	A	C
0	0	0		1	1	1
0	1	0		1	0	1
1	0	0		0	1	1
1	1	1		0	0	0
(a)				(b)		

is seen in the voltage chart and a logical 0 is written every time that 0 V is seen in the voltage chart. The output C is 1 only when both A and B are 1; thus, the positive logic $f_c = AB$. In the negative-logic truth table a logical 0 is written every time that 5 V is seen in the voltage chart and a logical 1 is

*When the low level of voltage (V_L) represents a logical high (1) and when the high level of voltage (V_H) represents a logical low (0), it is called negative logic.

written every time that 0 V is seen in the voltage chart. The output C is now 1 when any one or both of the inputs are high; thus, the negative logic $f_c = AB + \bar{A}B + A\bar{B}$ which can be simplified to $f_c = A + B$.

The Minimum Anode Supply. For an AND gate to function properly, at least one diode must conduct at all times. This is accomplished in Fig. 4-2(a) by making $E_A = 12$ V. The diodes still conduct with the worst-case input condition of $V_A = V_B = V_H$ and apply 5 V directly to the output. In this case the AND gate has a low output impedance when the output is a high level. A load connected to point C of Fig. 4-2(a) sees R_f/n in parallel with R_A, where R_f is the forward resistance of the diodes and n is the number of conducting diodes.

The circuit also functions properly with an anode supply as low as V_H. With this condition the diodes are unable to conduct when both inputs are high. However, the correct output is still produced because with no current $V_{RA} = 0$ V and $V_C = E_A = V_H$. Now the AND gate has a *high* output impedance. A load looking back sees R_A in parallel with the cutoff diodes.

When a load is connected, the anodes no longer see E_A. Instead, they see some Thévenin equivalent voltage which must be equal to or greater than V_H. If the load produces a $V_{TH} < V_H$, the diodes are unable to conduct when all inputs are high and the output is less than the required V_H. For example, in Fig. 4-3 the load R_A and E_A have a Thévenin equivalent circuit of $R_{TH} = 1.2$ kΩ and $V_{TH} = 2.4$ V. When A and B are both at V_H, the diodes are reverse biased and $V_C = V_{TH} = 2.4$ V.

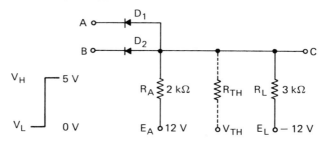

Fig. 4-3 An inoperative AND gate. V_{TH} is 2.4 V which prevents the diodes from conducting when $V_A = V_B = V_H$.

4-2 A Practical Diode AND Gate

To understand the effects of the AND gate on the input signal and the effects of a load on the AND gate, the forward and reverse resistance and the cutin voltage of the diodes must be considered. In this analysis the piecewise linear representation of the diode explained in Sec. 2-1 is used. The exact output voltage is calculated for each input combination. The output voltages

132 / Diode Logic

for the three loads shown in Fig. 4-4(a) are given in Fig. 4-4(b). Calculations are shown only for $R_L = 20 \text{ k}\Omega$ and for $E_L = 5$ V. The same procedure is used to determine the output voltages for the other loads.

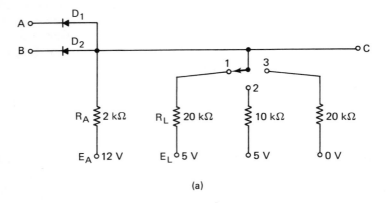

(a)

B (volts)	A (volts)	C_1 (volts)	C_2 (volts)	C_3 (volts)
0	0	0.647	0.639	0.641
0	5	0.79	0.775	0.778
5	0	0.79	0.79	0.778
5	5	5.579	5.571	5.573

(b)

Fig. 4-4 (a) A positive-logic *AND* gate with three possible loads. (b) The voltage chart for 0 V and 5 V input levels. Columns C_1, C_2, and C_3 are the output voltages with the switch in each of the three corresponding positions.

Example 4-1: The diodes in Fig. 4-4(a) have a cutin voltage $V_\gamma = 0.5$ V, a forward resistance $R_f = 50 \, \Omega$, and a back resistance $R_b = 10$ M. Neglecting capacitances and any input source resistance, determine the output voltage with $R_L = 20$ k and $E_L = 5$ V for each input combination shown in Fig. 4-4(b).

Solution: The first step is to determine whether or not the circuit can function properly under the given load conditions. This is done by finding the Thévenin equivalent seen by the anodes. Using Eq. (1-4), it may be seen that

$$V_{TH} = \frac{E_A R_L + E_L R_A}{R_L + R_A} = \frac{(12)(20) + (5)(2)}{20 + 2} = 11.36 \text{ V} \qquad (1\text{-}4)$$

Since $V_{TH} > V_H = 5$ V, the diodes conduct even when *all* inputs are at V_H.

The output will go toward the inputs of 5 V, proving that the circuit can function properly.

$$R_{TH} = \frac{R_A R_L}{R_A + R_L} = \frac{(2)(20)}{20 + 2} \text{ k}\Omega = 1.82 \text{ k}\Omega$$

For the first set of inputs given in Fig. 4-4(b), both diodes conduct; thus, the equivalent circuit is as shown in Fig. 4-5(a). Using Millman's theorem, it may be seen that

$$V_C = \frac{\dfrac{V_{\gamma_1}}{R_{f_1}} + \dfrac{V_{\gamma_2}}{R_{f_2}} + \dfrac{V_{TH}}{R_{TH}}}{\dfrac{1}{R_{f_1}} + \dfrac{1}{R_{f_2}} + \dfrac{1}{R_{TH}}} = \frac{\dfrac{0.5}{50} + \dfrac{0.5}{50} + \dfrac{11.36}{1.82 \text{ k}}}{\dfrac{1}{50} + \dfrac{1}{50} + \dfrac{1}{1.82 \text{ k}}} = 0.647 \text{ V}$$

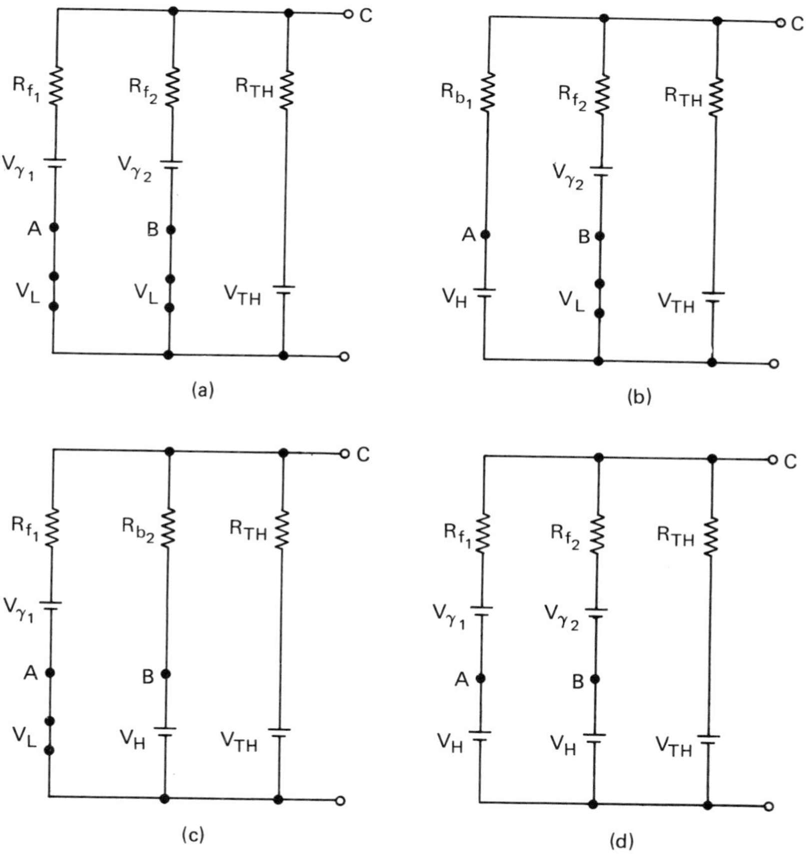

Fig. 4-5 The equivalent circuits of Fig. 4-3 for all of the input combinations.

The two conditions in which one input is low and the other is high provide the same output. The equivalent circuits are shown in Fig. 4-5(b) and (c). In both cases the output is

$$V_C = \frac{\frac{0.5}{50} + \frac{5}{10\,M} + \frac{11.36}{1.82\,k}}{\frac{1}{50} + \frac{1}{10\,M} + \frac{1}{1.82\,k}} = 0.79 \text{ V}$$

When both inputs are high, both diodes conduct and the equivalent circuit is as shown in Fig. 4-5(d). Now the output voltage is

$$V_C = \frac{\frac{5.5}{50} + \frac{5.5}{50} + \frac{11.36}{1.82\,k}}{\frac{1}{10} + \frac{1}{10} + \frac{1}{1.82\,k}} = 5.579 \text{ V}$$

If the inputs of Ex. 4-1 are applied as pulses instead of dc voltages, the signals are as shown in Fig. 4-6.

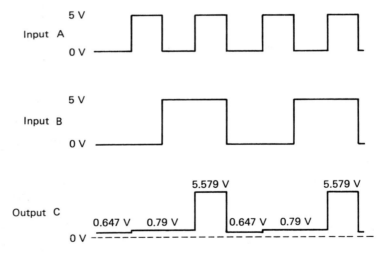

Fig. 4-6 The output signal C for 0 V and 5 V inputs applied as pulses to inputs A and B with the switch in position 1.

The Input Source Resistance. In Ex. 4-1 the input source resistance R_s was neglected. The equivalent circuits show R_s in series with each diode, and the calculations of output voltage have slightly different results. For example, if $R_s = 40\ \Omega$, the equivalent circuit for all inputs equal to V_H is as shown in

Fig. 4-7 and the output is

$$V_C = \frac{\frac{5.5}{90} + \frac{5.5}{90} + \frac{11.36}{1.82\text{ k}}}{\frac{1}{90} + \frac{1}{90} + \frac{1}{1.82\text{ k}}} = 5.64 \text{ V}$$

Note that this value is *not* much different from the output of 5.579 V obtained without R_s in Ex. 4-1.

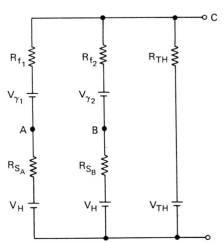

Fig. 4-7 The equivalent circuit of Fig. 4-3(a) with all inputs having a source resistance equal to R_S and a voltage equal to V_H.

Load Effects. When the load is varied, V_{TH} and R_{TH} are changed. But as long as the diodes conduct ($V_{TH} > V_H$), the output voltage still equals the lowest input voltage plus the drop across the forward-biased diode. Since the voltage across a forward-biased junction is relatively constant over a wide range of currents, the output voltage remains relatively constant.

Calculations have been made for Fig. 4-4(a) with the switch in positions 2 and 3 and the results are tabulated in Fig. 4-4(b). Note that the load variations have very little effect on the output voltage.

Summary. Because of the voltage drop on the forward biased diodes, the *AND* gate shifts the signal in a positive direction and reduces its peak-to-peak amplitude. The output of an *AND* gate is independent of the load. As long as the diodes conduct, the output voltage equals the lowest input voltage plus the drop across the forward-biased diode. The diodes must be able to conduct under the *worst-case* condition of all inputs at V_H. This occurs if the anodes see a $V_{TH} > V_H$. If the *AND* gate is driven by an *OR* gate, a saturated transistor or any other *low* resistance source, the output voltage may be calculated by considering the source resistance negligible.

136 / Diode Logic

4-3 Transient Analysis of AND Gates

In the preceding discussion of *AND* gates it was assumed that the output instantaneously rises and falls from one logic level to the other. In practice, it takes time for these changes in output levels to occur. The reason for these delays is shown in Fig. 4-8. There is capacitance between the output lead and ground that requires time to charge and discharge. In the analysis that follows it will be shown that the fall time of an *AND* gate is negligible but the rise time is much longer and must be considered when the gate is designed.*

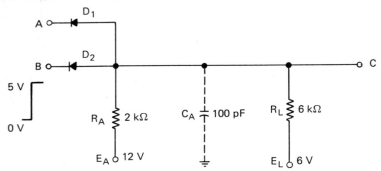

Fig. 4-8 A positive-logic *AND* gate with an output capacitance C_A.

Fall Time. Figure 4-9 shows the circuit seen by the *AND* gate capacitance C_A when all inputs have been high for a long time. The diodes see a Thévenin equivalent

$$V_{TH} = \frac{E_A R_L + E_L R_A}{R_A + R_L} \tag{1-4}$$

$$= \frac{(12)(6) + (6)(2)}{2 + 6}$$

$$V_{TH} = 10.5 \text{ V}$$

and

$$R_{TH} = \frac{R_A R_L}{R_A + R_L}$$

$$= \frac{(2)(6)}{2 + 6}$$

$$R_{TH} = 1.5 \text{ k}\Omega$$

---

*As explained in Sec. 2-5 the actual rise and fall times are the times between the 10 percent and 90 percent values of the leading and trailing edges of the pulse. The equation for the rise time of an *AND* gate is developed in Appendix 4-A. The rise time of an *AND* gate is calculated in Ex. 4-2.

Transient Analysis of AND Gates / 137

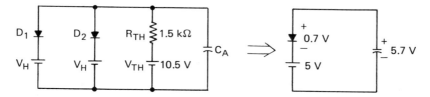

Fig. 4-9 The circuit seen by C_A when all inputs have been high for a long time.

Since the anodes see a voltage V_{TH} that is more positive than the inputs applied to their cathodes, they conduct. Diode voltages are relatively constant over a wide range of currents and it may be assumed that the inputs are supplied from constant voltage sources. Therefore, the capacitor sees a constant voltage equal to $V_D + V_H = 5.7$ V, to which it will have charged.

Now when one or more inputs go low, the voltage across C_A remains the same for an instant. The diode with the low input conducts harder and the circuit seen by the capacitor changes to that shown in Fig. 4-10. The worst

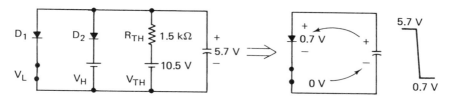

Fig. 4-10 The circuit seen by C_A when one of the inputs goes low.

case is when only one input goes low. The capacitor then sees a low resistance path through which it can discharge to $V_D + V_L = 0.7$ V. If two inputs go low, there will be two diodes drawing current to discharge the capacitor even more quickly.

Rise Time. Figure 4-11 shows the circuit seen by C_A at the instant both inputs again go high. The voltage across C_A remains 0.7 V momentarily and causes both diodes to cut off. The capacitor sees the Thévenin equivalent circuit and begins to charge slowly through R_{TH} toward V_{TH}. When V_{C_A}

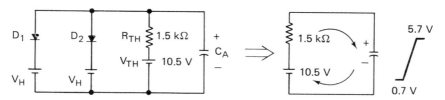

Fig. 4-11 The circuit seen by C_A at the instant both inputs go high.

becomes positive enough to allow the diodes to come on $(V_H + V_\gamma)$, the circuit returns to the one shown in Fig. 4-9. The output is clamped at $V_D + V_H = 5.7$ V. The rise time of an AND gate is calculated in Ex. 4-3.

Methods of Improving Rise Time. The complete output waveform of the AND gate is shown in Fig. 4-12. Figure 4-12(a) shows the rise and fall time for the circuit of Fig. 4-8. Rise time may be reduced by increasing E_A (to increase V_{TH}) and/or reducing R_A (to reduce R_{TH} and increase V_{TH}). The effect of an increase in V_{TH} is shown in Fig. 4-12(b). The capacitance charges from 0.7 V toward the higher V_{TH}, but it still gets clamped at $V_H + V_D = 5.7$ V. The actual change in V_{C_A} is a smaller percentage of the change from 0.7 V to V_{TH} with the higher Thévenin voltage.

The improvement in rise time is limited because increasing E_A or decreasing R_A causes the diode current to increase. Diodes with heavier

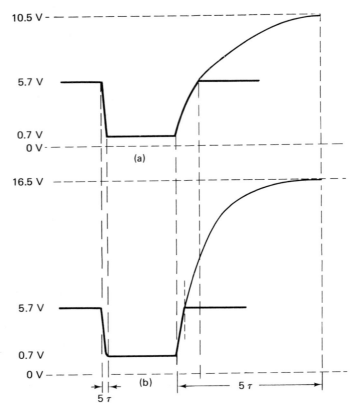

Fig. 4-12 (a) The waveform showing the charge and discharge time for C_A in Fig. 4-8. (b) The waveform showing improved rise time due to a larger V_{TH}.

current-handling capability must be used. Also, the driver will have to sink more current from the AND gate and the power consumption of the AND gate is increased.

4-4 Static Analysis of Diode OR Gates

The OR gate is a circuit with two or more inputs and only one output. The output is in the 1 state if one or more inputs are in the 1 state. Some of the logic symbols used to represent OR gates are shown in Fig. 4-13. The electrical circuit for a two-input positive-logic OR gate is given in Fig. 4-14(a) and the output voltage, for every combination of input voltage, is given in Fig. 4-14(b). The operation of this circuit is explained by following the same

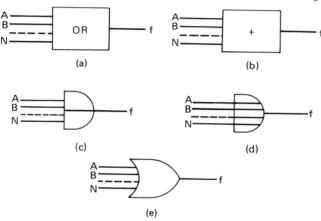

Fig. 4-13 Logic symbols commonly used for $f = A + B + \ldots + N$.

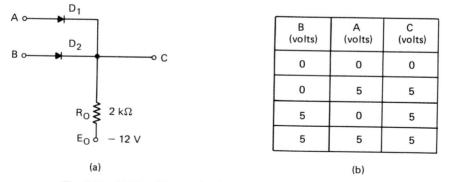

(a) (b)

Fig. 4-14 (a) The OR gate circuit. (b) The voltage table for a high input voltage $V_H = 5$ V and a low input voltage $V_L = 0$ V.

procedure used for the AND gate. First, perfect diodes are assumed and then practical diodes are used.

Both Inputs Low. The cathodes are connected through a 2-k resistor to -12 V; thus, with 0 V applied to each anode the diodes see the same amount of forward bias. They conduct equally with currents leaving the input sources, passing through the diodes, and then joining to return through R_O and E_O. With perfect diodes the 0-V potential at A and B is connected directly to terminal C. Thus, if all inputs are low, the output is low and

$$I_{R_O} = \frac{V_C - E_O}{R_O} = \frac{0\text{ V} - -12\text{ V}}{2\text{ k}} = 6\text{ mA}$$

Each input circuit supplies 3 mA.

A High B Low. With different inputs, the diode that sees the greatest difference of potential conducts causing the other diode to be held OFF. D_1 sees a 17-V difference of potential while D_2 sees only 12-V difference; thus, D_1 conducts with current flowing from A and returning through R_O and E_O. The 5-V potential at A is connected directly to C through the short circuit offered by D_1. D_2 sees 0 V on its anode and 5 V on its cathode for 5 V reverse bias.

A Low B High. This condition is similar to A high B low except that D_1 is OFF. Since D_2 now sees the larger difference of potential, it conducts and applies the 5 V potential at B directly to C. D_1 sees 0 V on its anode and 5 V on its cathode for 5 V reverse bias. Thus, with one input high the output is high and

$$I_{R_O} = \frac{V_C - E_O}{R_O} = \frac{5\text{ V} - -12\text{ V}}{2\text{ k}} = 8.5\text{ mA}$$

This current is supplied by only one of the driver circuits.

Both Inputs High. Since both diodes see a 17-V difference of potential, they conduct equally. Current flows from A and B, passes through the diodes, and returns through R_O and E_O. The 5-V potential at A and B is connected directly to C through the short circuits offered by the diodes.

$$I_{R_O} = \frac{V_C - E_O}{R_O} = \frac{5\text{ V} - -12\text{ V}}{2\text{ k}} = 8.5\text{ mA}$$

Each input circuit acts as a source of 4.25 mA. Since the output is high as long as at least one input is high, the OR function has been implemented.

If negative logic is used, the circuit shown in Fig. 4-14(a) performs the AND function. In Fig. 4-10(a) a 1 is written every time a 5 V level is seen in Fig. 4-14(b) and a 0 is written every time 0 V is seen. The output is a 1 if either A or B or both are 1. The positive logic $f_c = A\bar{B} + \bar{A}B + AB$ which can be simplified to $f_c = A + B$. However, in the (b) part of Table 4-2 it may

Table 4-2 The truth tables for the voltage chart in Fig. 4-15(b). (a) The positive-logic table. (b) The negative-logic table.

B	A	C
0	0	0
0	1	1
1	0	1
1	1	1

(a)

B	A	C
1	1	1
1	0	0
0	1	0
0	0	0

(b)

be seen that reversing the logic results in the output being a 1 only if A and B are 1, so the negative logic $f_c = AB$.

The Minimum Cathode Supply. For an OR gate to function properly, at least one diode must conduct at all times. The worst case occurs when both inputs are at the low voltage V_L. The cathodes must see a potential more negative than V_L. This is accomplished in Fig. 4-14(a) by making the supply voltage $E_o = -12$ V. Even when both inputs are 0 V, the diodes can conduct and make the output 0 V. In this state the OR gate has a low output impedance. A load looking back sees R_f/n in parallel with R_o, where R_f is the forward resistance of the diodes and n is the number of conducting diodes.

The circuit also functions properly if the supply voltage E_o is exactly equal to the low-level input V_L. With $E_o = V_L$ and both inputs at 0 V the diodes cannot conduct. However, the correct output is still obtained because with no current $V_{R_o} = 0$ V and $V_C = E_o = V_L$. The OR gate now appears as a high impedance source. The load sees R_o in parallel with the reverse resistance of the cutoff diodes.

When a load is connected, the cathodes must see a Thévenin equivalent voltage equal to or more negative than V_L. If the load produces a V_{TH} that is less negative than V_L, the diodes are unable to conduct when all inputs are low. This results in an output not equal to V_L. For example, in Fig. 4-15 the

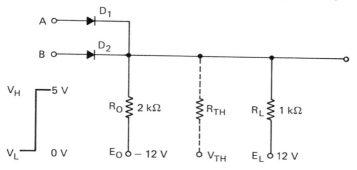

Fig. 4-15 An inoperative OR gate. V_{TH} is $+2.4$ V which prevents the diodes from conducting when $V_A = V_B = V_L$.

load R_o and E_o have a Thévenin equivalent circuit of $R_{TH} = 0.67$ k and $V_{TH} = 2.4$ V. When A and B are both at V_L, the diodes are reverse biased and $V_C = V_{TH} = 2.4$ V.

4-5 A Practical Diode OR Gate

The effects of the OR gate on the input signal and the effects of a load on the OR gate are explained by analyzing the circuit shown in Fig. 4-16(a). The piecewise linear representation of a practical diode is used. The output voltages for the three loads shown in Fig. 4-16(a) are given in Fig. 4-16(b). They are obtained by applying Millman's theorem to each of the four

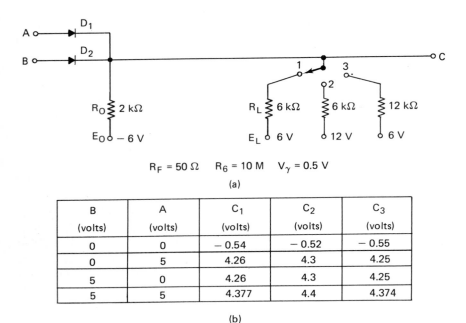

B (volts)	A (volts)	C_1 (volts)	C_2 (volts)	C_3 (volts)
0	0	−0.54	−0.52	−0.55
0	5	4.26	4.3	4.25
5	0	4.26	4.3	4.25
5	5	4.377	4.4	4.374

(b)

Fig. 4-16 (a) A positive-logic OR gate with three possible loads. (b) The voltage chart for 0 V and 5 V input levels. Columns C_1, C_2, and C_3 are the output voltages with the switch in each of the three corresponding positions.

equivalent circuits shown in Fig. 4-17. The procedure is similar to that given for the AND gate in Ex. 4-1.

If the inputs are applied as pulses instead of dc voltages, the signals appear as shown in Fig. 4-18. Note the slight difference in the high outputs of the OR gate.

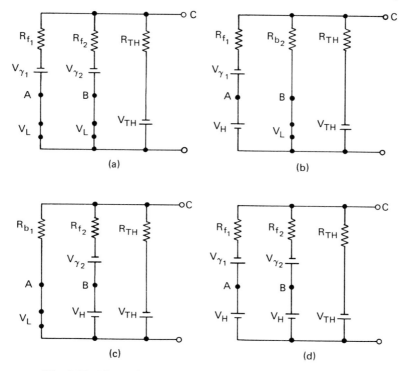

Fig. 4-17 The equivalent circuits of Fig. 4-17 for all of the input combinations.

The Input Source Resistance. The source resistance R_S was not considered when the output voltages of Fig. 4-16 were calculated. If it is considered, the equivalent circuit shows R_S in series with the diodes and the output voltages change slightly. For example, if $R_S = 40\ \Omega$, the equivalent circuit for all inputs equal to V_H appears as in Fig. 4-19, and the output with the load switch in position 1 is

$$V_C = \frac{\frac{4.5}{90} + \frac{4.5}{90} + \frac{-3}{1.5\mathrm{k}}}{\frac{1}{90} + \frac{1}{90} + \frac{1}{1.5\mathrm{k}}} = 4.28\ \mathrm{V}$$

This value is not much different from the 4.377 V calculated without R_S and shown in Fig. 4-16(b).

Load Effects. A change in the load results in different values for R_{TH} and V_{TH}. However, as long as V_{TH} permits the diodes to conduct, the output voltage must equal the *highest* input voltage minus the drop across the forward-biased diode. Since the diode junction voltage remains relatively

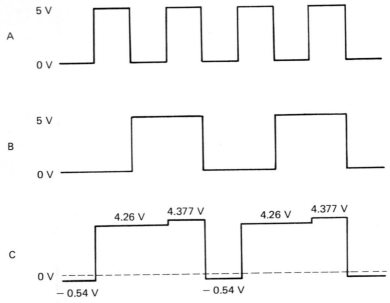

Fig. 4-18 The output signal C for 0 V to 5 V inputs applied as pulses to A and B with the switch in position 1.

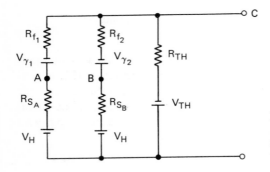

Fig. 4-19 The equivalent circuit of Fig. 4-17(a) with all inputs having a source resistance equal to R_S and a voltage equal to V_H.

constant over a wide range of currents, the output voltage is almost constant. The output voltages tabulated in Fig. 4-16(b) for the three different loads of Fig. 4-16(a) verify this.

Summary. Because the diode is reversed in the OR gate it shifts the signal in a negative direction. As was the case in the AND gate, the peak-to-peak amplitude of the signal is reduced and the output is independent of the load. As long as the diodes conduct, the output voltage equals the *highest* input voltage minus (because of the way the diodes are connected) the drop across the forward-biased diode. The diodes must be able to conduct with the worst-input condition of all inputs at V_L. This happens when the cathodes see a V_{TH}

more negative than V_L. If the OR gate is driven by an AND gate, a saturated transistor or any other low-resistance source, the output voltage will be affected very little by source resistance.

4-6 Transient Analysis of OR Gates

As is the case with AND gates, it takes time for the output of an OR gate to rise and fall. The reason for this is the capacitance between the output terminal and ground. This capacitance C_o is shown in Fig. 4-20. In the analysis that follows it will be shown that rise time is negligible but fall time is long and must be considered when the OR gate is designed.*

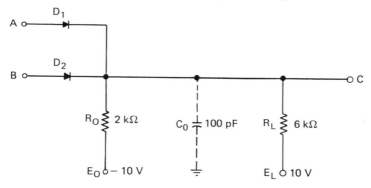

Fig. 4-20 A positive-logic OR gate with an output capacitance C_0.

Rise Time. Figure 4-21 shows the circuit of Fig. 4-20 after both inputs have been low for a long time. The diodes see a Thévenin equivalent

$$V_{TH} = \frac{E_o R_L + E_L R_o}{R_o + R_L} \tag{1-4}$$

$$= \frac{(-10)(6) + (10)(2)}{2 + 6}$$

$$V_{TH} = -5 \text{ V}$$

and

$$R_{TH} = \frac{R_o R_L}{R_o + R_L}$$

$$= \frac{(2)(6)}{1 + 6}$$

$$R_{TH} = 1.5 \text{ k}$$

*As explained in Sec. 2-5 the actual rise and fall times are the times between the 10 percent and 90 percent values of the leading and trailing edges of the pulse. The equation for the fall time of an OR gate is developed in Appendix 4-B. The fall time of an OR gate is calculated in Ex. 4-2.

146 / Diode Logic

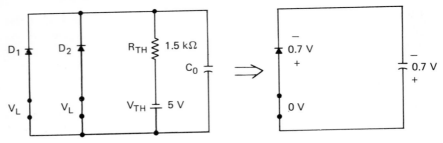

Fig. 4-21 The circuit seen by C_o after all inputs have been low for a long time.

Since the cathodes see a voltage that is negative compared to the inputs applied to their anodes, they conduct. The capacitor then sees a constant voltage equal to $-V_D + V_L$, regardless of the values of R_{TH} and V_{TH}. It will be charged to -0.7 V.

Now when one or more inputs go high, the capacitor voltage remains -0.7 V for an instant. The diode with the high input conducts harder and the capacitor sees the circuit shown in Fig. 4-22. It sees a constant voltage equal to $V_H - V_D = 4.3$ V. It quickly charges to this voltage through the low resistance of the diode with the high input. The worst case is when only one input goes high. Then only one diode is on to supply charging current for C_o.

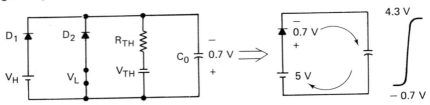

Fig. 4-22 The circuit seen by C_o when one input goes high.

Fall Time. Figure 4-23 shows the circuit at the instant all inputs again go low. The voltage across C_o remains 4.3 V momentarily. The diodes therefore cut off and C_o sees the Thévenin equivalent circuit. It discharges the

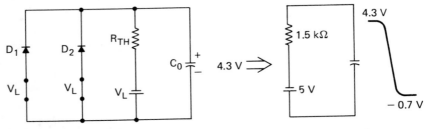

Fig. 4-23 The circuit seen by C_o at the instant both inputs again go low.

4.3 V and begins to charge slowly through R_{TH} toward $V_{TH} = -5$ V when the diodes conduct. The circuit then returns to the one shown in Fig. 4-21 and the output is clamped at $-V_D + V_L = -0.7$ V. Fall time of an OR gate is calculated in Ex. 4-2.

Methods of Improving Fall Time. The output waveform for the circuit of Fig. 4-20 is shown in Fig. 4-24(a). Fall time may be improved by reducing R_o (to decrease R_{TH} and increase V_{TH}) or by increasing E_o (to increase V_{TH}). As shown in Fig. 4-24(b), the capacitor then aims for a higher voltage and reaches -0.7 V more quickly.

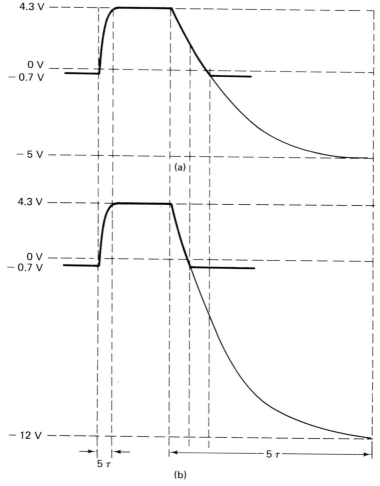

Fig. 4-24 (a) The waveform showing the charge and discharge time for C_o in Fig. 4-21. (b) The waveform showing improved fall time due to a larger V_{TH}.

148 / Diode Logic

The improvement in fall time is limited, however, because decreasing R_O or increasing E_O increases the diode current. The driver will have to supply this higher current and the diodes will have to be able to handle it. In addition, the power consumption of the gate is increased.

4-7 Static Analysis of Cascaded Diode Gates

The function $AB + CD$ may be implemented with two diode AND gates driving a diode OR gate. The function $(A + B)(C + D)$ may be implemented with two diode OR gates driving a diode AND gate. The logic diagrams of these cascaded gates are shown in Fig. 4-25. Cascaded diode gates are explained by applying the same techniques as for individual gates. In this section an AND-to-OR cascaded system is analyzed. The same procedure is followed for an OR-to-AND cascaded system.

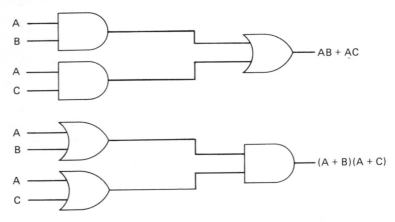

Fig. 4-25 Two logic diagrams of cascaded diode gates.

Figure 4-26(a) shows the electrical circuit for an AND gate driving an OR gate. Figure 4-26(b) gives the output voltage of each gate and the conduction state of each diode (assuming that $V_D \approx 0.7$ V) for every input combination. As long as the OR gate permits the AND gate to see a $V_{TH} > V_H$, the output of the AND gate (Y) will equal its lowest input plus V_D. The four combinations for A and B can be present with C low or with C high, so there are eight combinations for the three-input system. With C and Y the inputs to the OR gate, the state of D_3 and D_4 can be determined by following the rules given in Sec. 4-4. The diode with the highest input conducts and cuts off the other diode. The positive logic $f_z = AB\bar{C} + \bar{A}\bar{B}C + A\bar{B}C + \bar{A}BC + ABC$ which can be simplified to $C + AB$. If negative logic is used, f_z

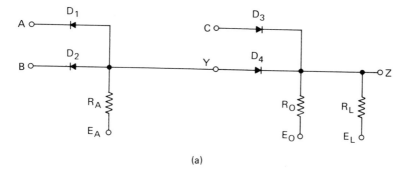

C (volts)	B (volts)	A (volts)	D_1	D_2	Y (volts, approx.)	D_3	D_4	Z (volts, approx.)	Positive Logic output
0	0	0	ON	ON	0.7	OFF	ON	0	0
0	0	5	OFF	ON	0.7	OFF	ON	0	0
0	5	0	ON	OFF	0.7	OFF	ON	0	0
0	5	5	ON	ON	5.7	OFF	ON	5	1
5	0	0	ON	ON	0.7	ON	OFF	4.3	1
5	0	5	OFF	ON	0.7	ON	OFF	4.3	1
5	5	0	ON	OFF	0.7	ON	OFF	4.3	1
5	5	5	ON	ON	5.7	OFF	ON	5	1

(b)

Fig. 4-26 (a) The electrical circuit of a positive-logic AND gate driving a positive-logic OR gate. (b) The output voltage for each gate and the state of each diode (assuming $V_D = 0.7$ V) for every input combination.

$= ABC + \bar{A}BC + A\bar{B}C$ which can be simplified to $C(A + B)$. It would then be an OR-to-AND system.

4-8 Level Shifting

The level shift introduced by an AND gate (OR gate) may be partially counteracted by applying the output to a one-input OR gate (AND gate). Fig. 4-27 shows a three-input OR gate with its output shifted in a negative direction. It is followed by a one-input AND gate that shifts the signal in a positive direction. Complete cancellation does not occur because the level shifts are never exactly the same. The additional gate attenuates the peak-to-

150 / *Diode Logic*

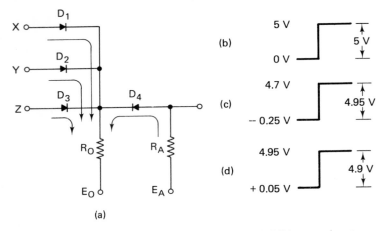

Fig. 4-27 (a) An *OR* gate followed by a level-shifting one-input *AND* gate. (b) The input to the *OR* gate. (c) The output of the *OR* gate. (d) The output of the *AND* gate. Germanium diodes are used.

peak voltage only slightly, but it does cause a decrease in the load current that this circuit can sink.

4-9 Transient Analysis of Cascaded Diode Gates

When diode *AND* gates and *OR* gates are cascaded, both rise and fall times are affected. In Fig. 4-28 it can be seen that the *OR* gate sees a capacitance C_O that affects fall time. If the worst case is considered, it can also be seen that an ideal diode D_4 places C_O in parallel with C_A. Consequently, the *AND* gate sees a capacitance $C_O + C_A$ which affects rise time. The diode gates are usually close together on one plug-in unit, however the output gate may be driving a circuit on another plug-in unit. The method used to connect the

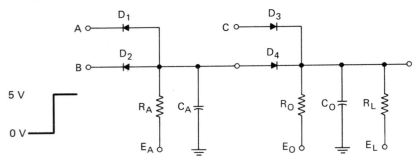

Fig. 4-28 An *AND-to-OR* system taking into account any output capacitances.

output gate to its load can easily cause its capacitance to be much larger than the capacitance between the two diode gates.

Although the input gate sees $C_o + C_A$, its transient response is not necessarily worse than that of the output gate. The resistance loading on the input gate is also increased, reducing the Thévenin resistance through which its capacitance must charge (or discharge if it is an OR-to-AND system) and the V_{RS}/V_{RF} quantities are different. A sample calculation of the rise and fall time for an AND-to-OR system is now given.

Example 4-2: Neglecting source resistance and assuming ideal diodes, determine the rise and fall times for the circuit of Fig. 4-28 if $V_H = 5$ V, $V_L = 0$ V, $E_A = 12$ V, $R_A = 1$ k, $E_o = -12$ V, $R_o = 3$ k, $C_A = 50$ pF, $C_o = 100$ pF, $E_L = 12$ V, and $R_L = 6$ k.

Solution: First determine the fall time of the system. If the diodes and R_S are neglected, the AND gate has no fall time. Simply calculate t_f for the OR gate only.

In Appendix 4-B at the end of this chapter it can be seen that the fall time of an OR gate is

$$t_f = 2.3 R_{THo} C_o \log \frac{V_{THo} - 0.1 V_L - 0.9 V_H}{V_{THo} - 0.9 V_L - 0.1 V_H} \quad (4\text{-}B\text{-}7)$$

where

$$R_{THo} = \frac{R_o R_L}{R_o + R_L}$$

$$= \frac{(3)(6)}{3 + 6}$$

$$R_{THo} = 2 \text{ k}$$

and

$$V_{THo} = \frac{E_o R_L + E_L R_o}{R_L + R_o} \quad (1\text{-}4)$$

$$= \frac{(-12)(6) + (12)(3)}{6 + 3}$$

$$V_{THo} = -4 \text{ V}$$

By substituting these values and the given data in Eq. (4-B-7), the fall time is found to be

$$t_f = 2.3 \times 2 \times 10^3 \times 10^{-10} \log \frac{-4 - 0 - 4.5}{-4 - 0 - 0.5} \quad (4\text{-}B\text{-}7)$$

$$= 4.6 \times 10^{-7} \log \frac{-8.5}{-4.5}$$

$$t_f = 127 \text{ ns}$$

152 / Diode Logic

The *OR* gate has no rise time if the diodes are assumed to have no forward resistance. Hence, simply calculate t_r for the *AND* gate only.

In Appendix 4-A it can be seen that the rise time of an *AND* gate is

$$t_r = 2.3 R_{TH_A} C_A \log \frac{V_{TH_A} - 0.9 V_L - 0.1 V_H}{V_{TH_A} - 0.1 V_L - 0.9 V_H} \qquad (4\text{-}A\text{-}7)$$

where

$$R_{TH_A} = \frac{(R_{TH_o})(R_A)}{R_{TH_o} + R_A}$$

$$= \frac{(2)(1)}{2 + 1}$$

$$R_{TH_A} = 0.67 \text{ k}$$

and

$$V_{TH_A} = \frac{E_A R_{TH_o} + V_{TH_o} R_A}{R_{TH_o} + R_A} \qquad (1\text{-}4)$$

$$= \frac{(12)(2) + (-4)(1)}{2 + 1}$$

$$V_{TH_A} = 6.67 \text{ V}$$

The *AND* gate, in this case, sees a total capacitance equal to $C_A + C_o = 150$ pF. By substituting these values in Eq. (4-A-7), it can be seen that the rise time is

$$t_r = 2.3 \times 6.7 \times 10^2 \times 1.5 \times 10^{-10} \log \frac{6.67 - 0 - 0.5}{6.67 - 0 - 4.5} \qquad (4\text{-}A\text{-}7)$$

$$= 2.31 \times 10^{-7} \log \frac{6.17}{2.17}$$

$$t_r = 105 \text{ ns}$$

4-10 Summary

Diode gates may be connected in cascade with either *AND* gates driving an *OR* gate or *OR* gates driving an *AND* gate. However, gates designed to operate one way cannot be reversed. To calculate the output voltage, first determine the state of each diode. Then draw the equivalent circuit and apply Millman's theorem (see Sec. 1-4).

The output gate, which has its own load, is the load on the input gate. When gates are cascaded, both rise and fall times are affected. The output gate is sometimes nothing more than a one-input circuit used to counteract the level shift of the input gate.

APPENDIX 4-A: AND GATE RISE TIME

The output circuit of a diode AND gate is shown in Fig. 4-A-1(a) and (b). The capacitor charges, as shown in Fig. 4-A-1(c), from V_L toward V_{TH}, but it stops charging at V_H when the diodes turn on. In Fig. 4-A-1(b) it can be seen that the voltage on the Thévenin resistance R_{TH} at the start of rise time is

$$V_{RS} = V_{TH} - V_{CS} \tag{4-A-1}$$

where

$$V_{CS} = V_L + 0.1(V_H - V_L)$$
$$= V_L + 0.1 V_H - 0.1 V_L$$
$$V_{CS} = 0.9 V_L + 0.1 V_H \tag{4-A-2}$$

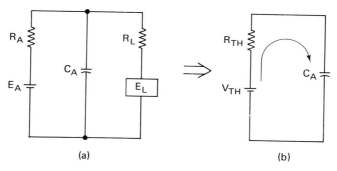

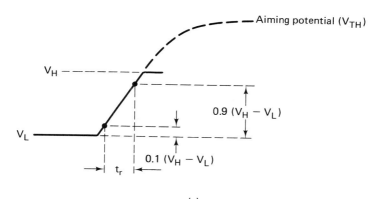

Fig. 4-A-1 (a) The output circuit of a diode AND gate with (b) its equivalent circuit. (c) The charging waveform of the capacitor C_A showing rise time t_r.

154 / Diode Logic

Substituting Eq. (4-A-2) in Eq. (4-A-1) yields

$$V_{RS} = V_{TH} - 0.9V_L - 0.1V_H \qquad (4\text{-A-}3)$$

The voltage on R_{TH} at the finish of rise time is

$$V_{RF} = V_{TH} - V_{CF} \qquad (4\text{-A-}4)$$

where

$$\begin{aligned} V_{CF} &= V_L + 0.9(V_H - V_L) \\ &= V_L + 0.9V_H - 0.9V_L \\ V_{CF} &= 0.1V_L + 0.9V_H \end{aligned} \qquad (4\text{-A-}5)$$

Substituting Eq. (4-A-5) in Eq. (4-A-4) yields

$$V_{RF} = V_{TH} - 0.1V_L - 0.9V_H \qquad (4\text{-A-}6)$$

Finally, substituting R_{TH}, C_A, Eqs. (4-A-3), and (4-A-6) in Eq. (1-7) produces the equation for calculating the rise time of an *AND* gate

$$t_r = 2.3 R_{TH} C_A \log \frac{V_{TH} - 0.9V_L - 0.1V_H}{V_{TH} - 0.1V_L - 0.9V_H} \qquad (4\text{-A-}7)$$

APPENDIX 4-B: OR GATE FALL TIME

The output circuit of a diode *OR* gate is shown in Fig. 4-B-1(a) and (b). The capacitor discharges, as shown in Fig. 4-B-1(c), from V_H toward V_{TH}, but it stops at V_L when the diodes turn on. In Fig. 4-B-1(b) it can be seen that the voltage on the Thévenin resistance at the start of fall time is

$$V_{RS} = V_{TH} - V_{CS} \qquad (4\text{-B-}1)$$

where

$$\begin{aligned} V_{CS} &= V_L + 0.9(V_H - V_L) \\ &= V_L + 0.9V_H - 0.9V_L \\ V_{CS} &= 0.1V_L + 0.9V_H \end{aligned} \qquad (4\text{-B-}2)$$

Substituting Eq. (4-B-2) in Eq. (4-B-1) yields

$$V_{RS} = V_{TH} - 0.1V_L - 0.9V_H \qquad (4\text{-B-}3)$$

The voltage on R_{TH} at the finish of fall time is

$$V_{RF} = V_{TH} - V_{CF} \qquad (4\text{-B-}4)$$

Appendix 4-B: OR Gate Fall Time | 155

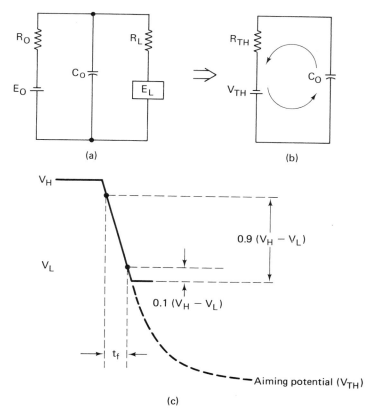

Fig. 4-B-1 (a) The output circuit of a diode *OR* gate with (b) its equivalent circuit. (c) The discharge waveform of the capacitor C_O showing fall time t_f.

where

$$V_{CF} = V_L + 0.1(V_H - V_L)$$
$$= V_L + 0.1V_H - 0.1V_L$$
$$V_{CF} = 0.9V_L + 0.1V_H \quad (4\text{-B-}5)$$

Substituting Eq. (4-B-5) in Eq. (4-B-4) yields

$$V_{RF} = V_{TH} - 0.9V_L - 0.1V_H \quad (4\text{-B-}6)$$

Finally, substituting R_{TH}, C_O, Eq. (4-B-3), and Eq. (4-B-6) in Eq. (1-7) produces the equation for calculating the fall time of an *OR* gate.

$$t_f = 2.3 R_{TH} C_O \log \frac{V_{TH} - 0.1V_L - 0.9V_H}{V_{TH} - 0.9V_L - 0.1V_H} \quad (4\text{-B-}7)$$

Problems

4-1 The three-input AND gate shown in Fig. 4-29 has the following inputs: $V_A = -2$ V, $V_B = -4$ V, and $V_C = -6$ V.
(a) What is the output voltage if $E_A = 12$ V, $R_A = 1$ k, $E_L = -12$ V, $R_L = 2$ k, and if ideal diodes are used?
(b) What is the output voltage of the same circuit if $V_A = 5$ V, $V_B = 6$ V, and $V_C = 7$ V?

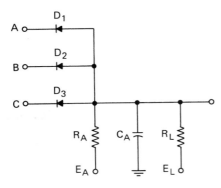

Fig. 4-29

4-2 What minimum value of R_L may be used if the circuit of Fig. 4-3 is to function properly?

4-3 Verify the output voltages given in columns C_2 and C_3 in Fig. 4-4(b).

4-4 Determine the steady-state values of I_{D_1} and I_{D_2} for each output voltage given in column C_1 in Fig. 4-4(b). Refer to data given in Ex. 4-1.

4-5 Determine the rise time of the circuit in Fig. 4-29 if $E_A = 13$ V, $R_A = 2.2$ k, $E_L = 6.5$ V, $R_L = 4.3$ k, $C_A = 80$ pF, $V_H = 6.5$ V, and $V_L = 0$ V. Assume ideal diodes.

4-6 The three-input OR gate shown in Fig. 4-30 has the following inputs: $V_A = 2$ V, $V_B = 4$ V, and $V_C = 6$ V.

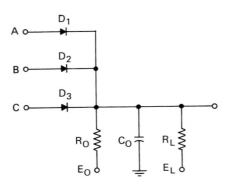

Fig. 4-30

(a) What is the output voltage if $E_o = -12$ V, $R_o = 1$ k, $E_L = 12$ V, $R_L = 2$ k, and if ideal diodes are used?
(b) What is the output of the same circuit if $V_A = -5$ V, $V_B = -6$ V, and $V_C = -7$ V?

4-7 What is the minimum value of R_L that may be used if the circuit of Fig. 4-15 is to function properly?

4-8 Verify the output voltages given in Fig. 4-16(b). Refer to data given in Fig. 4-16(a).

4-9 Determine the steady-state value of I_{D_1} and I_{D_2} for each output voltage given in column C_1 of Fig. 4-16(b). Refer to data given in Fig. 4-16(a).

4-10 Determine the fall time of the circuit in Fig. 4-30 if $E_o = -10$ V, $R_o = 2$ k, $E_L = -6$ V, $R_L = 6$ k, $C_o = 120$ pF, $V_H = 6$ V, and $V_L = 0$ V. Assume ideal diodes.

4-11 The input voltages to the cascaded gates of Fig. 4-27(a) are $V_x = V_y = 0$ V and $V_z = 5$ V. $R_o = 1$ k, $E_o = -9$ V, $R_A = 2$ k, $E_A = 9$ V, $R_b = \infty$, $R_f = 40$ Ω, and $V_y = 0.2$ V. Calculate
(a) The output of the OR gate if the level-shifting gate (D_4, R_A, and E_A) is not used.
(b) The output of the OR gate when D_4, R_A, and E_A are used.
(c) The output of the system when D_4, R_A, and E_A are used.
(d) The voltage drop across each of the four diodes.

4-12 Referring to Fig. 4-31 and assuming ideal diodes, determine all diode currents when
(a) All inputs are low.
(b) All inputs are high.

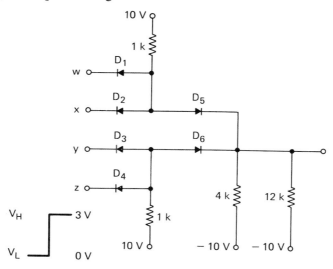

Fig. 4-31

(c) $V_w = V_y = V_L$ and $V_x = V_z = V_H$.
(d) $V_w = V_L$ and $V_x = V_y = V_z = V_H$.
(e) $V_w = V_H$ and $V_x = V_y = V_z = V_L$.

4-13 Assuming ideal diodes, determine the output voltage of the circuit of Fig. 4-32 with
(a) One or more inputs low.
(b) Both inputs high.

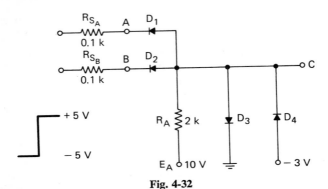

Fig. 4-32

4-14 If D_3 is removed and E_A is made 0 V, determine the output voltage of Fig. 4-32 with
(a) One or more inputs low.
(b) Both inputs high.

4-15 Verify that the cascaded gates designed to work in the *AND*-to-*OR* system of Ex. 4-2 cannot be interchanged.

5

THE
INVERTER
(NOT GATE)

The *inverter* is an essential element of digital computers. In its basic form, it is a single-input transistor circuit whose output is the *complement* or *inverse* of the input. Six logic symbols commonly used to represent the inverter circuit are shown in Fig. 5-1. The IEEE standard employs a circle placed at the point where inversion occurs; it takes place at the input in Fig. 5-1(e) and at the output in Fig. 5-1(f). Because of the importance of the *NOT* gate, consideration is given to several of its variations.

5-1 The Basic Inverter

The basic inverter, shown in Fig. 5-2(a), operates in the saturation mode. When V_H (+10 V) is applied to the input, Q_1 goes into saturation and its collector-to-emitter voltage drops to 0.2 V. By making $V_{EE} = V_L$, the output voltage

$$v_o = (V_{CE_{sat}} + V_{EE}) = (0.2 \text{ V} + 0 \text{ V}) \approx V_L$$

When V_L (0 V) is applied to the input, Q_1 cuts off and the collector-to-ground voltage V_C rises to V_{CC}. By making $V_{CC} = V_H$, the output voltage $v_o = V_H$. The input and output voltage waveforms, shown in Fig. 5-2(b), illustrate the signal inversion that occurs. The turn-on and turn-off delays are explained in Sec. 2-5.

160 / *The Inverter (NOT Gate)*

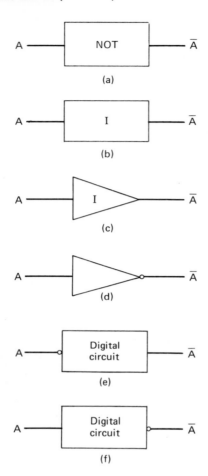

Fig. 5-1 Six logic symbols commonly used to represent the inverter (NOT) circuit.

The values of R_1, R_2, and V_{BB} are chosen to assure cutoff with $v_i = V_L$, and saturation with $v_i = V_H$. When *on*, these components must permit an $I_B \geq I_{B_{min}}$. When *off*, these components must maintain a $V_{BE} \leq V_{BEco}$ (the edge of cutoff) at the highest expected temperature.

Operating the transistor in the cutoff and saturation regions simplifies design procedures. It eliminates the dependence on critical transistor parameters and reduces power dissipation. In the cutoff state, V_{CE} is high, but I_C is low. I_C varies from $I_{CEO} = (1 + \beta)I_{CO}$ with the $I_B = 0$, to I_{CO} with a reverse base current $-I_B = I_{CO}$. In the saturation region, I_C is high, but V_{CE} is low. $V_{CE_{sat}}$ ranges from approximately 0.1 V for germanium to approximately 0.4 V for silicon. The cutoff and saturation regions are discussed in detail in Sec. 2-4.

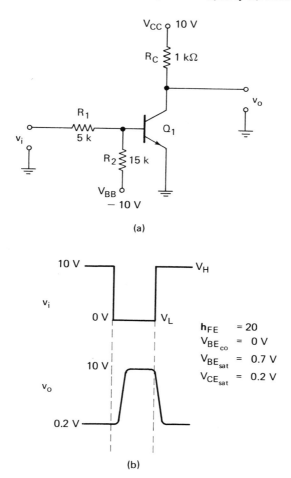

Fig. 5-2 (a) A saturation mode inverter with (b) its input and output voltage waveforms.

5-2 Steady-state Analysis of the Basic Inverter

The inverter circuit of Fig. 5-2(a) is analyzed in two steps: First, it is verified that Q_1 is cut off with $v_i = 0$ V and $I_{CO_{max}} = 200$ μA. Second, it is verified that Q_1 is in saturation when $v_i = 10$ V. All quiescent currents and voltages are calculated for both conditions. Leakage current is neglected in the saturation state because Q_1 must saturate at any temperature.

Example 5-1: First verify that Q_1 is cut off with $v_i = 0$ V and $I_{CO_{max}} = 200$ μA; then solve for all steady-state currents and voltages.

Solution: Referring to Fig. 5-3(a) and using Eq. (1-4), solve for the open-circuited base voltage $V_{B_{TH}}$ and $R_{B_{TH}}$.

$$V_{B_{TH}} = \frac{v_i R_2 + V_{BB} R_1}{R_1 + R_2} \text{ V} \tag{1-4}$$

$$= \frac{(0)(15) + (-10)(5)}{5 + 15} \text{ V}$$

$$= \frac{-50}{20} \text{ V}$$

$$V_{B_{TH}} = -2.5 \text{ V}$$

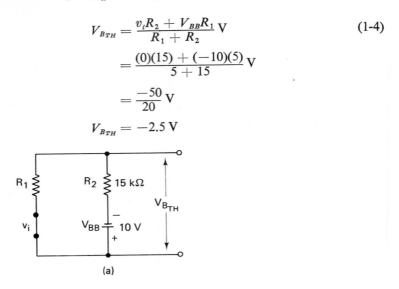

(a)

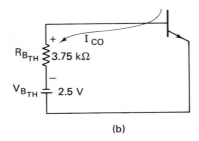

(b)

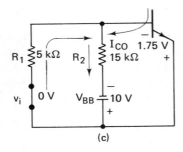

(c)

Fig. 5-3 (a) The input circuit with the base open-circuited. (b) The Thévenin equivalent of the input connected to the base-emitter circuit with I_{CO} flowing as shown. (c) The input circuit current paths with Q_1 cut off and $-I_B = I_{CO}$.

and

$$R_{B_{TH}} = \frac{R_1 R_2}{R_1 + R_2}$$
$$= \frac{(5)(15)}{20} \text{ k}\Omega$$
$$R_{B_{TH}} = 3.75 \text{ k}\Omega$$

The input circuit is shown with these values in Fig. 5-3(b). With $I_{CO_{max}} = 200 \ \mu\text{A}$

$$V_{R_{B(TH)}} = (I_{CO})(R_{B_{TH}})$$
$$= (0.2 \text{ mA})(3.75 \text{ k}\Omega)$$
$$V_{R_{B(TH)}} = 0.75 \text{ V}$$

The actual base voltage is therefore

$$V_B = V_{R_{B(TH)}} + V_{B_{TH}}$$
$$= (0.75) + (-2.5)$$
$$V_B = -1.75 \text{ V}$$

Since $V_{EE} = 0$, $V_{BE} = -1.75$ V and Q_1 is definitely cut off. Fig. 5-3(b) shows that as I_{CO} increases, V_{R_B} increases, and V_{BE} becomes less negative. At a sufficiently high temperature, V_{BE} goes positive and Q_1 comes on.

The input currents shown in Fig. 5-3(c) are

$$I_{R_1(CO)} = \frac{v_i - V_B}{R_1}$$
$$= \frac{0 - (-1.75)}{5 \text{ k}}$$
$$I_{R_1(CO)} = +0.35 \text{ mA}$$

and

$$I_{R_2(CO)} = I_{R_1CO} + I_{CO}$$
$$= 0.35 + 0.2$$
$$I_{R_2(CO)} = 0.55 \text{ mA}$$

Since Q_1 is cut off, $I_C = I_{CO}$ and $V_C \approx V_{CC} = 10$ V. Hence, $v_o \approx V_H$ for $v_i = V_L$.

164 / The Inverter (NOT Gate)

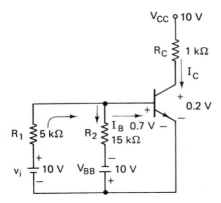

Fig. 5-4 Current paths with Q_1 in saturation.

Now verify that the transistor is saturated with $v_i = 10$ V. The current paths, with $V_B = V_{BE_{sat}}$ and $V_C = V_{CE_{sat}}$, are shown in Fig. 5-4. If $I_B \geq I_{B_{min}}$, Q_1 is saturated.

$$I_{R_1(sat)} = \frac{v_i - V_B}{R_1}$$

$$= \frac{10 - 0.7}{5 \text{ k}\Omega} \text{ V}$$

$$I_{R_1(sat)} = 1.86 \text{ mA}$$

and

$$I_{R_2(sat)} = \frac{V_B - V_{BB}}{R_2}$$

$$= \frac{0.7 - (-10)}{15 \text{ k}\Omega} \text{ V}$$

$$I_{R_2(sat)} = 0.713 \text{ mA}$$

Thus,

$$I_B = I_{R_1} - I_{R_2}$$
$$= 1.86 - 0.713$$
$$I_B = 1.147 \text{ mA}$$

If $V_{CE_{sat}} = 0.2$ V, $V_C = 0.2$ V and

$$I_{C_{sat}} = \frac{V_{CC} - V_{CE_{sat}}}{R_C}$$

$$= \frac{10 - 0.2}{1 \text{ k}\Omega} \text{ V}$$

$$I_{C_{sat}} = 9.8 \text{ mA}$$

Then

$$I_{B_{min}} = \frac{I_{C_{sat}}}{h_{FE}}$$

$$= \frac{9.8 \text{ mA}}{20}$$

$$I_{B_{min}} = 0.49 \text{ mA}$$

Since $I_B > I_{B_{min}}$, Q_1 is definitely in saturation. In fact, this circuit is operating with substantial overdrive. The *overdrive factor* od $= I_B/I_{B_{min}} = 2.34$. As explained in Sec. 2-6, this overdrive improves the turn-on-time of a circuit. Since $v_o \approx V_L$ for $v_i = V_H$, and $v_o = V_H$ for $v_i = V_L$, the *NOT* function is implemented.

This circuit will still function properly as long as the minimum h_{FE} is

$$h_{FE_{min}} = \frac{I_{C_{sat}}}{I_B}$$

$$= \frac{9.8 \text{ mA}}{1.147 \text{ mA}}$$

$$h_{FE_{min}} = 8.55$$

However, turn-on time will be longer without overdrive.

5-3 A Saturation Mode Inverter with Collector Clamping

The addition of the clamping diode in the collector circuit of Fig. 5-5(a) serves two purposes: It improves switching time (see also Sec. 8-9), and it provides a measure of voltage regulation under varying load conditions. It must be emphasized that this diode does *not* prevent saturation, as does the diode of Fig. 2-21(a). Notice that although both transistors are of the *npn* type, the diode of Fig. 5-5(a) is reversed.

When $v_i = 6$ V, Q_1 saturates and, with $V_{CE_{sat}} = 0.3$ V, $V_{D_1} = -5.7$ V to keep D_1 cut off. Thus, $v_o = 0.3$ V $\approx V_L$ with $v_i = V_H$. When $v_i = 0$ V, Q_1 cuts off and V_C rises exponentially toward V_{CC}. However, D_1 (forward biased above 6 V) conducts and clamps V_C at $V_{D_{fwd}} + V_{CL}$. By making $V_{CL} = V_H$, the output voltage $v_o = 0.3$ V $+ 6$ V $= 6.3$ V $\approx V_H$. Because the collector voltage exponentially seeks a value of $V_{CC} > V_H$, it reaches V_H in less time than without a diode clamp.

This improvement in transient response is shown in Fig. 5-5(b). Without D_1, t_2 is the time that it would take for V_C to go from 0.3 V when Q_1 is saturated, to 6.0 V when Q_1 is cut off and with $V_{CC} = V_H$ as in the basic

166 / The Inverter (NOT Gate)

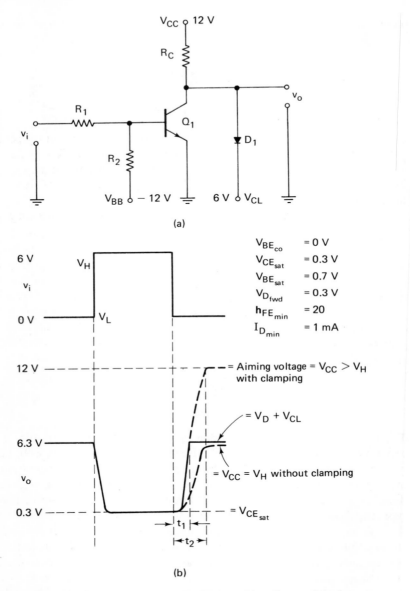

Fig. 5-5 (a) A saturation mode inverter with collector clamping and (b) its input and output waveforms.

inverter. With D_1, t_1 is the time that it takes for V_c to rise from 0.3 V when Q_1 is saturated, to 6.3 V when Q_1 is cut off and with $V_{CC} > V_H$ as in the collector clamped inverter shown in Fig. 5-5(a).

The additional advantage of voltage regulation provided by both the diode and the transistor is explained in the following section.

5-4 The Load (FAN-IN, FAN-OUT, and the UNIT-LOAD Concept)

As explained in Sec. 3-7, it is usually necessary to cascade several logic circuits in order to perform a logic function. The maximum load current capability of a circuit must be considered when cascading gates. Loading effect is frequently measured in terms of a *unit load*. A unit load may be defined as that amount of current required by the basic circuit of a system. The maximum number of unit loads that a circuit can drive under worst-case conditions is called the *fan-out* or *pyramiding factor*. Circuits have both a dc and an ac fan-out. The prime effect of a dc load is a decrease in the 0- and 1-level noise immunity, whereas ac loads increase switching times. Inputs are classified either as *sink loads* or *source loads*. The inputs of positive-logic diode OR gates are called source loads because they must be driven by a source of current. The inputs of positive-logic diode AND gates are called sink loads because they must be driven by a current sink. A circuit with a fan-out of ten unit loads can be used to drive ten different circuits with unit load inputs, five different circuits with two-unit-load inputs, twenty different circuits with half-unit-load inputs, etc. Mathematically the fan-out

$$FO = \frac{I_{L_{max}}}{I_u} \tag{5-1}$$

Fan-in is the maximum number of inputs that can be connected to a circuit.

5-5 dc Noise Margins

In order to ensure reliability, digital circuits have a built-in noise immunity. Worst-case conditions are considered when the circuits are designed. Variations in V_{CC} and the level shifts introduced by the diode gates discussed in Chap. 4 are two examples of dc noise voltages. If these and other factors are not considered when the circuits are designed, the circuits will fail.

Acceptance Levels. If an inverter is designed to saturate with $v_i = V_H$ and it is driven by an OR gate, it will actually operate in the active region because the OR gate *high* output is equal to $V_H - V_D$ instead of V_H. If the inverter is designed to cut off with $v_i = V_L$ and it is driven by an AND gate, it also will operate in the active region because the AND gate *low* output is equal to $V_L + V_D$ instead of V_L. The circuit of Fig. 5-6 has an input signal with $V_L = 0$ V and $V_H = 6$ V, but it has been designed to cut off with

168 / The Inverter (NOT Gate)

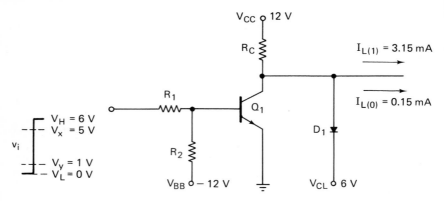

Fig. 5-6 A saturation mode inverter with a clamped collector and source-type loads.

$v_i = 1$ V and to saturate with $v_i = 5$ V. In this circuit, 5 V is the *upper acceptance level* V_x and 1 V is the *lower acceptance level* V_y. For this circuit, the 1-level noise margin is

$$N_{(1)} = V_H - V_x \qquad (5\text{-}2)$$
$$= 6 - 5$$
$$N_{(1)} = 1 \text{ V}$$

and the 0-level noise margin is

$$N_{(0)} = V_y - V_L \qquad (5\text{-}3)$$
$$= 1 - 0$$
$$N_{(0)} = 1 \text{ V}$$

Problems

5-1 Verify that the transistor of Fig. 5-7 is saturated with $v_i = 6$ V.

5-2 Verify that the transistor of Fig. 5-7 is cut off with $v_i = 0$ V.

5-3 Determine the upper acceptance level V_x for the circuit of Fig. 5-2(a). Note: The *upper* acceptance level in *npn* circuits is the *least positive* input voltage that will keep the transistor saturated.

5-4 Determine the lower acceptance level V_y if the circuit of Fig. 5-2(a) is to function properly at 100°C. The transistor is of the silicon type and has a leakage current $I_{co} = 2.7$ mA at 25°C. Note: The *lower* acceptance level in *npn* circuits is the *most positive* input voltage that will keep the transistor cut off.

5-5 The circuit of Fig. 5-2(a) sees a sink load when Q_1 is saturated. What is the greatest load current it can sink when its input voltage $v_i = 10$ V?

5-6 The circuit of Fig. 5-2(a) sees a source-type load when Q_1 is cut off. What is the greatest load current it can supply when its input voltage $v_i = 0$ V, if the upper acceptance level V_x of the load is 7 V?

5-7 The circuit of Fig. 5-7 sees a sink-type load when Q_1 is saturated. What is the greatest load current it can handle when its input voltage $v_i = 6$ V?

5-8 The circuit of Fig. 5-7 sees a source load when Q_1 is cut off. What is the greatest load current it can supply when its input voltage $v_i = 0$ V, if the output voltage must not fall below 6 V?

5-9 Using $V_x = 5$ V and $V_y = 1$ V, what minimum h_{FE} is required if a sink-type load of 10 mA is connected to the circuit of Fig. 5-7?

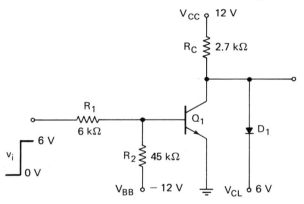

$h_{FE_{min}} = 10$ $V_{BE_{co}} = 0$ V
$V_{BE_{sat}} = 0.7$ V $I_{D_{min}} = 1$ mA
$V_{CE_{sat}} = 0.3$ V $V_D = 0.3$ V

Fig. 5-7

6

RESISTOR-TRANSISTOR LOGIC RTL
AND RESISTOR-CAPACITOR
TRANSISTOR LOGIC RCTL

There are two forms of RTL. The configuration shown in Fig. 6-1 was widely used before monolithic circuits were developed. At that time resistors were less expensive and more reliable than semiconductor diodes and transistors. The number of inputs is limited to three or less with this single-transistor version. The RTL circuit of Fig. 6-2(a) was used in the early days of monolithic circuitry because the transistor was the smallest and probably the least expensive component. This type operates faster and with greater reliability than the circuit of Fig. 6-1. RTL and RCTL (the latter is simply RTL with speed-up capacitors) circuits can be made to perform either the *NOR* or *NAND* operation.

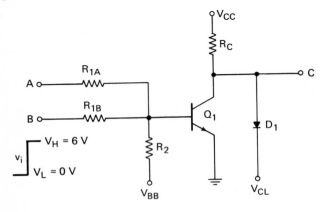

Fig. 6-1 One type of RTL Circuit.

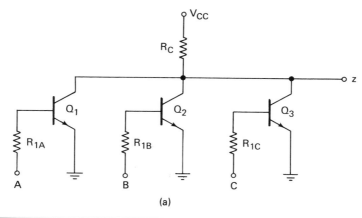

C	B	A	Q_1	Q_2	Q_3	V_Z (approx.)
V_L	V_L	V_L	CO	CO	CO	V_H
V_L	V_L	V_H	Sat	CO	CO	V_L
V_L	V_H	V_L	CO	Sat	CO	V_L
V_L	V_H	V_H	Sat	Sat	CO	V_L
V_H	V_L	V_L	CO	CO	Sat	V_L
V_H	V_L	V_H	Sat	CO	Sat	V_L
V_H	V_H	V_L	CO	Sat	Sat	V_L
V_H	V_H	V_H	Sat	Sat	Sat	V_L

(b)

Fig. 6-2 (a) A three-input positive-logic *NOR* gate and (b) its state chart.

The TTL circuits of Chap. 8 are by far the most common logic circuits used today, and the DTL and ECL circuits of Chap. 7 are also frequently used. The multitransistor version of RTL will be briefly discussed because these circuits may be encountered in older equipment. Also, digital circuits that use field-effect transistors FET use similar configurations. The circuit of Fig. 6-1, which performs the *NOR* function, will not be explained here because it is unlikely that this version of RTL will be encountered by the reader.

6-1 RTL NOR Gates

The state chart for the multitransistor RTL circuit of Fig. 6-2(a) is given in Fig. 6-2(b). A voltage equal to V_H at any input turns that transistor on and makes the output voltage $V_z = V_{CE_{sat}} \approx V_L$. When all inputs go low, all transistors cut off. The output V_z then rises to the 1 level for which the

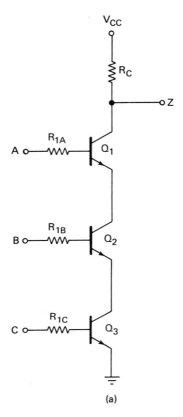

C	B	A	Q_1	Q_2	Q_3	V_Z (approx.)
V_L	V_L	V_L	CO	CO	CO	V_H
V_L	V_L	V_H	Sat	CO	CO	V_H
V_L	V_H	V_L	CO	Sat	CO	V_H
V_L	V_H	V_H	Sat	Sat	CO	V_H
V_H	V_L	V_L	CO	CO	Sat	V_H
V_H	V_L	V_H	Sat	CO	Sat	V_H
V_H	V_H	V_L	CO	Sat	Sat	V_H
V_H	V_H	V_H	Sat	Sat	Sat	V_L

(b)

Fig. 6-3 (a) A three-input positive-logic *NAND* gate and (b) its state chart.

circuit is designed. Since the output voltage is high only when all inputs are low, the positive-logic $f_z = \overline{A}\overline{B}\overline{C} = \overline{A + B + C}$. This circuit is, therefore, a *NOR* gate.

A second three-input RTL circuit is given in Fig. 6-3(a). Again, a voltage equal to V_H at any input turns that transistor on, but, unless all transistors are on, $I_{C_{sat}}$ is limited to the small leakage current of the other transistors. Since the voltage drop on R_C is approximately zero, $V_z \approx V_{CC} = V_H$. When all inputs are high, all transistors conduct and V_z drops to

$$(V_{CE_{sat_1}} + V_{CE_{sat_2}} + V_{CE_{sat_3}}) \approx V_L$$

The state chart is shown in Fig. 6-3(b). Since the output voltage is low only when all inputs are high, $f_z = \overline{ABC}$. Therefore, this arrangement performs the positive-logic *NAND* function.

The parallel version is preferred because the voltage drop across the series transistors causes an offset of the input turn-on voltage. The circuits of Figs. 6-2 and 6-3 are examples of *direct-coupled transistor logic* DCTL.

6-2 Analysis of the Multitransistor RTL NOR Gate

The circuits of Fig. 6-4 show just the Q_1 section of the *NOR* gate of Fig. 6-2(a) with its loads. It is assumed that the driver and load circuits are identical to the Q_1 circuit.

In Fig. 6-4(a) the A driver is cut off and its output is the minimum 1 level of the system. This voltage $V_{(1)min}$ exceeds the upper acceptance level V_x by the desired 1-level noise margin $N_{(1)}$; consequently, all of the loads saturate with base currents $I_B > I_{B_{min}}$. Therefore, when they are cut off, the circuits must act as current sources supplying $I_{L(1)} = (FO)(I_B)$. The fan-out is reduced as $N_{(1)}$ is increased.

In Fig. 6-4(b) the A driver is saturated and its output goes to the maximum 0 level of the system. This voltage $V_{(0)max}$, which is less than the lower acceptance level V_y by the desired 0-level noise margin $N_{(0)}$, cuts off all of the loads. Therefore, when saturated, the circuits act as current sinks for only $(FO)(I_{CO})$, where I_{CO} is the leakage current of each load.

R_C is designed to maintain an output voltage equal to $V_{(1)min}$ when the transistor is cut off. R_1 is designed to allow a base current equal to $I_{B_{min}}$ with an input equal to V_x.

The cutoff output voltage can be determined from Fig. 6-4(a). The leakage current of Q_1 will be neglected so that the transistor may be considered an open circuit. Since all loads are the same, they may be replaced by a Thévenin equivalent load: $R_{L(TH)} = R_1/FO$ and $E_{L(TH)} = V_{BE_{sat}}$. Applying Eq. (1-4) to this circuit yields

$$V_{(1)min} = \frac{V_{CC}\dfrac{R_1}{FO} + V_{BE_{sat}}(R_C)}{R_C + \dfrac{R_1}{FO}} \qquad (6\text{-}1)$$

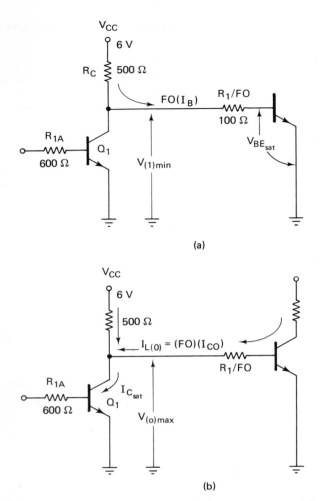

Fig. 6-4 The output circuits of (a) a cutoff RTL *NOR* gate driving *FO* gates exactly like itself and (b) a saturated RTL *NOR* gate driving *FO* gates exactly like itself.

The upper acceptance level is

$$V_x = I_{B_{min}} R_1 + V_{BE_{sat}} \tag{6-2}$$

The difference between Eq. (6-1) and (6-2) is the 1-level noise margin $N_{(1)}$.

$$N_{(1)} = V_{(1)_{min}} - V_x \tag{6-3}$$

The values of V_x, $V_{(1)_{min}}$, and $N_{(1)}$ are calculated in the following examples.

Example 6-1: Determine the upper acceptance level V_x of the circuit of Fig. 6-2 if that circuit drives and is driven by other circuits exactly like itself. Use

Analysis of the Multitransistor RTL NOR Gate / 175

the values shown on the simplified circuit of Fig. 6-4, and $h_{FE_{min}} = 25$, $V_{CE_{sat}} = 0.3$ V, $V_{BE_{sat}} = 0.7$ V, a fan-out $FO = 6$, and $I_{CO} = 40$ μA at the highest operating temperature.

Solution: First $I_{B_{min}}$ must be calculated. By referring to Fig. 6-4(b), it can be seen that

$$I_{C_{sat}} = I_{R_C} + I_{L(0)}$$
$$= \frac{V_{CC} - V_{CE_{sat}}}{R_C} + FO(I_{CO})$$
$$= \frac{6 - 0.3}{0.5\text{ k}} + 6(0.04 \text{ mA})$$
$$I_{C_{sat}} = 11.64 \text{ mA}$$

and with $h_{FE_{min}} = 25$,

$$I_{B_{min}} = \frac{I_{C_{sat}}}{h_{FE_{min}}}$$
$$= \frac{11.64}{25}$$
$$I_{B_{min}} = 0.466 \text{ mA}$$

Thus, the upper acceptance level is

$$V_x = I_{B_{min}} R_1 + V_{BE_{sat}} \quad (6\text{-}2)$$
$$= (0.466)(0.6) + 0.7$$
$$V_x = 0.98 \text{ V}$$

Example 6-2: Determine the minimum 1-level output voltage of the circuit of Ex. 6-1.

Solution: Applying Eq. (6-1) to Fig. 6-4(a) yields

$$V_{(1)min} = \frac{V_{CC}\frac{R_1}{FO} + V_{BE_{sat}}(R_C)}{R_C + \frac{R_1}{FO}} \quad (6\text{-}1)$$

$$= \frac{6\frac{0.6}{6} + 0.7(0.5)}{0.5 + \frac{0.6}{6}}$$

$$= \frac{0.6 + 0.35}{0.5 + 0.1}$$

$$V_{(1)min} = 1.58 \text{ V}$$

Example 6-3: Calculate the 1-level noise margin of the circuit of Exs. 6-1 and 6-2.

Solution: Substitute the answers of Exs. 6-1 and 6-2 in Eq. (6-3).

$$N_{(1)} = V_{(1)_{min}} - V_X \qquad (6\text{-}3)$$
$$= 1.58 \text{ V} - 0.98 \text{ V}$$
$$N_{(1)} = 0.6 \text{ V}$$

With this noise margin, I_B will be greater than $I_{B_{min}}$, sending the loads deep into saturation which results in long storage delays.

The circuit of Fig. 6-5(a) shows the output circuit when Q_1 is on. If the required I_C is greater than $(h_{FE})(I_{B_{min}})$, Q_1 is pulled out of saturation and V_{CE} rises above $V_{CE_{sat}}$. If $V_{CE} + I_{CO}R_1$ rises above the cutin V_{BE} of the loads, they are turned on. Therefore, the 0-level noise immunity is affected by I_{CO}; it may

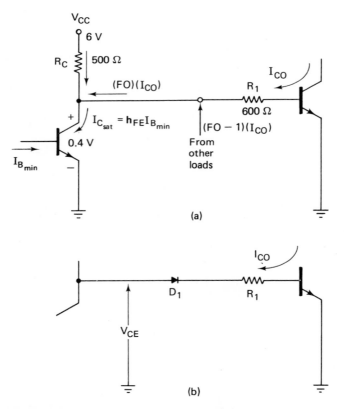

Fig. 6-5 (a) The output circuit of Q_1 when it is saturated. (b) Use of a series diode to improve the 0-level noise immunity.

be improved by inserting a diode in series with R_1. This is shown in Fig. 6-5(b). $V_{CE} + I_{CO}R_1$ must now equal two junction voltages before turn-on occurs. As proved in Prob. 6-5, this improvement in $N_{(0)}$ reduces the fan-out, if the same $N_{(1)}$ is desired.

6-3 Resistor-Capacitor-Transistor Logic RCTL

The circuits of Figs. 6-1 and 6-2 are often modified by the addition of speed-up capacitors across each R_1. The speed-up action is explained in Sec. 2-6. Each capacitor is one-half the C_1 of an inverter (1/3 for three inputs, etc.). The use of speed-up capacitors reduces propagation delays from as high as 50 ns in RTL to as low as 10 ns in RCTL.

RCTL has several disadvantages; as mentioned in Sec. 2-6, the speed-up capacitor slows down the driver because it increases the load capacitance of that circuit, noise pulses are coupled to the base without attenuation, and cost is increased.

The latter is especially true in integrated circuits because the surface area required for each capacitor limits the total number of components on a given size chip. For example, monolithic capacitors are simply reverse biased pn junctions. Capacity varies with the reverse voltage and the physical size of the junction. For a typical EB junction, a capacitance of approximately 0.4 pF per square mil can be expected. Thus, a 30-pF capacitor requires approximately 75 mil² of surface area.

6-4 Summary

RTL may be produced with two or three inputs driving a single transistor, or with several transistors each with its own input; in either case, *NOR* gates are recommended when *npn* transistors are used. The multitransistor version is preferred because it is faster and has a greater fan-in capability than the single-transistor circuit. By adding speed-up capacitors, RCTL circuits are produced. This reduces propagation delays to less than half that of RTL.

Problems

6-1 Determine the minimum h_{FE} required for the circuit of Fig. 6-1 to function as a *NOR* gate if $R_{1A} = R_{1B} = 4.6$ k, $R_2 = 40$ k, $R_C = 5$ k, $I_{L(0)} = 10$ mA (sink-type load), $I_{L(1)} = 0.4$ mA (source-type load), $V_{CC} = 10$ V, $V_{BB} = -10$ V, $V_{CL} = 6$ V, $V_{BE_{sat}} = 0.7$ V, $V_D = 0.7$ V, and $V_{CE_{sat}} = 0.3$ V.

178 / Resistor-Transistor Logic RTL and Resistor-Capacitor Transistor Logic RCTL

6-2 Given $R_C = 900\ \Omega$, $R_1 = 500\ \Omega$, $V_{CC} = 6\ V$, $V_{BE_{sat}} = 0.7\ V$, $V_{CE_{sat}} = 0.3\ V$, $FO = 5$, and $I_{CO} = 40\mu A$ per input and $h_{FE_{min}} = 20$, determine the upper acceptance level V_x for the circuit of Fig. 6-2. Assume that the circuit drives and is driven by other circuits exactly like itself.

6-3 What is the lowest 1-level output voltage of the circuit of Prob. 6-2?

6-4 Using the answers to Probs. 6-2 and 6-3, determine the 1-level noise immunity $N_{(1)}$.

6-5 When the series diode of Fig. 6 5(b) is added to the circuit of Ex. 6-1, the fan-out is reduced if the same 1-level noise margin (found in Ex. 6-3) is maintained. Determine the fan-out with the diode. Note: Since it is obvious from Ex. 6-1 that $I_{L(o)}$ has a negligible effect on $I_{C_{sat}}$, the fan-out need not be known in order to solve for $I_{B_{min}}$. Use the same $I_{C_{sat}}$ found in Ex. 6-1.

6-6 Repeat Prob. 6-5 for the circuit of Probs. 6-2, 6-3, and 6-4.

6-7 If input levels equal to $-6\ V$ and $0\ V$ are applied to the X and Y inputs, what is the positive-logic function f_z for the circuit of Fig. 6-6?

6-8 If the transistors of Fig. 6-6 are changed to *npn* and if V_{CC} is made $+6\ V$, what is the positive-logic function f_z for input levels equal to $+6\ V$ and $0\ V$?

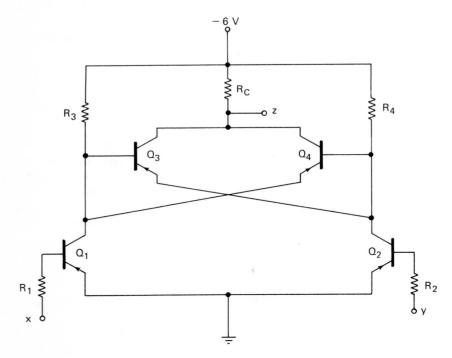

Fig. 6-6

6-9 Determine the positive-logic function f_c of the circuit of Fig. 6-7(a) for input levels of 0 V and 6 V.

6-10 Determine the positive-logic function f_c of the circuit of Fig. 6-7(b) for input levels of 0 V and -6 V.

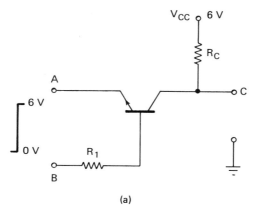

(a)

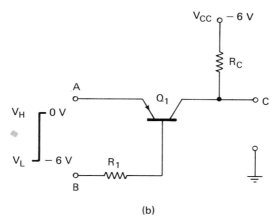

(b)

Fig. 6-7

7

DIODE-TRANSISTOR LOGIC DTL
EMITTER-COUPLED LOGIC ECL
WIRED LOGIC

When diode gates are combined with transistor inverters, it is called *diode-transistor logic* DTL. A diode gate may drive an inverter (*NAND* and *NOR* functions) or an inverter may drive one input to a diode gate (*INHIBIT* and *IMPLICATION* functions). Diode transistor logic has several advantages compared to RTL. It has better noise immunity, greater speed, a higher fan-in and a higher fan-out. However, it does have higher (but not excessive) power dissipation. Diode logic is sometimes called *hybrid logic*. In its early form it consisted of diode *AND* or *OR* gates such as those shown in Figs. 4-2(a) and 4-15(a) and an inverter such as the one shown in Fig. 5-2(a). The modern integrated circuit version which is a modification of the early DTL circuits is discussed in this chapter. *Emitter-coupled logic*, which uses current switching and is used when very high speed is needed, and *wired logic*, which is produced by connecting the outputs of two or more gates in parallel, are also discussed in this chapter.

7-1 The Modern Integrated Circuit Version of Diode-Transistor Logic

The circuit of Fig. 7-1 is the conventional integrated circuit DTL *NAND* gate. The input diodes are usually formed by shorting the collector to the base by metalization and by using the EB junctions (see Fig. 8-2, Sec. 8-1). This method reduces storage time and, thus, permits faster diode turn-off times. In the analysis procedures that follow, it is assumed that the drivers and loads are identical to this circuit.

The Modern Integrated Circuit Version of Diode-Transistor Logic / 181

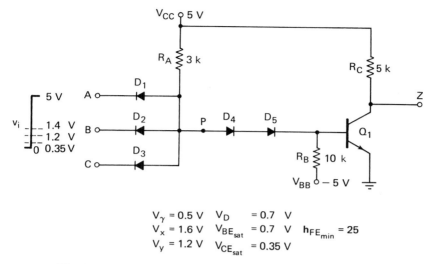

$V_\gamma = 0.5$ V $V_D = 0.7$ V
$V_x = 1.6$ V $V_{BE_{sat}} = 0.7$ V $h_{FE_{min}} = 25$
$V_y = 1.2$ V $V_{CE_{sat}} = 0.35$ V

Fig. 7-1 A conventional integrated circuit version of DTL.

Circuit Analysis. The three input diodes, R_A, and V_{CC} form the basic AND gate, where V_p tries to go to the lowest input voltage plus the forward voltage drop on the diode. The transistor is cut off unless V_p exceeds three junction voltage drops (D_4, D_5, and the EB junction of Q_1). When one or more inputs equal V_L, $V_P = V_L$ plus the AND gate diode voltage V_D. Q_1 then is cut off. The unloaded output goes to V_{CC} which is well above the logical 1 acceptance level V_x. Current flows from V_{CC} through R_A and then splits. Part of the current flows through the *offset diodes* D_4 and D_5 and through R_B. The remaining current flows through the conducting AND gate diodes.

When all inputs equal V_H, the AND gate output rises and permits Q_1 to come on. Current then flows from V_{CC} through R_A and the diodes D_4 and D_5. It then splits, with part of the current flowing through R_B and the remaining current flowing into the base. The transistor goes into saturation and the output drops to $V_{CE_{sat}}$, which is well below the logical 0 acceptance level V_y. The AND gate output V_P rises to $V_{D_4} + V_{D_5} + V_{BE_{sat}} = 2.1$ V, which cuts off the AND gate diodes. The driver supplies only the small reverse current I_{CO} to the diodes when its output is high.

The transfer characteristic for this circuit is shown in Fig. 7-2. At point 1, Q_1 is just off. Assuming that V_{BE} equals the cutin voltage $V_\gamma = 0.5$ V and that the diode forward drops equal 0.7 V, the lower acceptance level for this circuit is

$$V_y = -V_{D(AND)} + V_{D_4} + V_{D_5} + V_{BE} \tag{7-1}$$
$$= -0.7 + 0.7 + 0.7 + 0.5$$
$$V_y = 1.2 \text{ V}$$

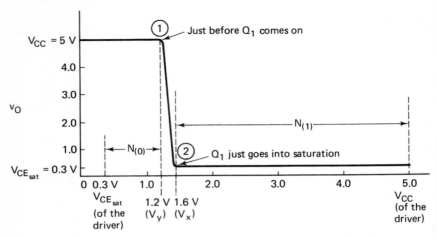

Fig. 7-2 The transfer characteristic for the circuit of Fig. 7-1 showing noise immunity.

At point 2, the AND gate diodes are just off with $V_D = V_y = 0.5$ V and Q_1 just goes into saturation. Since the offset diode forward voltages and $V_{BE_{sat}}$ are all equal to 0.7 V, the upper acceptance level is

$$V_x = -V_{D(AND)} + V_{D_4} + V_{D_5} + V_{BE_{sat}} \qquad (7\text{-}2)$$
$$= -0.5 + 0.7 + 0.7 + 0.7$$
$$V_x = 1.6 \text{ V}$$

The 0-level noise immunity $N_{(0)}$ is the difference between Eq. (7-1) and the actual 0-level output $V_{CE_{sat}}$.

$$N_{(0)} = V_y - V_{o(0)} \qquad (7\text{-}3)$$
$$N_{(0)} = (-V_{D(AND)} + V_{D_4} + V_{D_5} + V_{BE}) - V_{CE_{sat}}$$

where V_{BE} equals the cutin voltage $V_y = 0.5$ V. Hence,

$$N_{(0)} = 1.2 - 0.3$$
$$N_{(0)} = 0.9 \text{ V}$$

The 1-level noise immunity $N_{(1)}$ is the difference between the minimum 1-level output voltage of Eq. (7-2). From Fig. 7-3

$$V_{o(1)min} = V_{CC} - (FO)(I_{CO})(R_C)$$
$$= 5 - (10)(0.04)(5) \qquad (7\text{-}4)$$
$$= 5 - 2$$
$$V_{o(1)min} = 3 \text{ V}$$

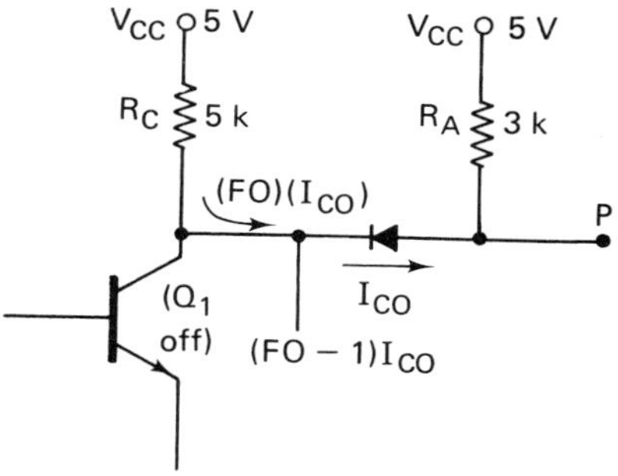

Fig. 7-3 The circuit for determining the 1-level output voltage $V_{0(1)}$.

Hence,

$$N_{(1)} = V_{o(1)} - V_x$$
$$= V_{o(1)} - (V_{D(AND)} + V_{D_4} + V_{D_5} + V_{BE_{sat}})$$
$$= 3 - 1.6$$
$$N_{(1)} = 1.4 \text{ V}$$

Improving Noise Immunity. The 0-output noise immunity may be increased by adding a third offset diode. This increases the value of Eq. (7-1) to 1.9 V and raises $N_{(0)}$ to 1.6 V. It also increases the value of Eq. (7-2) to 2.3 V, which reduces $N_{(1)}$ to 0.7 V. Noise immunity is normally one of the factors that must be considered when designing any digital circuit.

An improved circuit is shown in Fig. 7-4. This circuit has both a high noise immunity and a high fan-out. It consists of a diode *AND* gate, an emitter follower Q_1, and an inverter Q_2. The emitter-base junction of Q_1 provides the offset voltage of D_4 in Fig. 7-1; hence, V_P is still 2.1 V when the output transistor Q_2 is saturated. The current amplification provided by the emitter follower increases the base current of Q_2. Thus, for the same h_{FE}, the fan-out capability is increased. Negative voltage feedback is provided by R_2 to stabilize the operating point of Q_1.

An expansion node (point X in Fig. 7-4) is often provided in DTL gates to increase the flexibility of the system. If a greater fan-in is required, the number of inputs can be increased by connecting a *gate expander* as shown by the dotted lines in Fig. 7-4.

Transient Response of DTL. The basic DTL integrated circuit is repeated, for convenience, in Fig. 7-5(a). The transistor is in saturation when all inputs are high. When any input goes low, the *AND* gate output goes low and pulls Q_1 out of saturation. However, because of the charge stored in the shunt capacitance and in the base it takes time for Q_1 to come out of saturation (see Sec. 8-2). This delay is shown at the leading edge of the v_o waveform in Fig. 7-5. In this circuit, R_B is returned to a negative supply to help remove the charge stored in the base. In other circuits, R_B is grounded and the offset

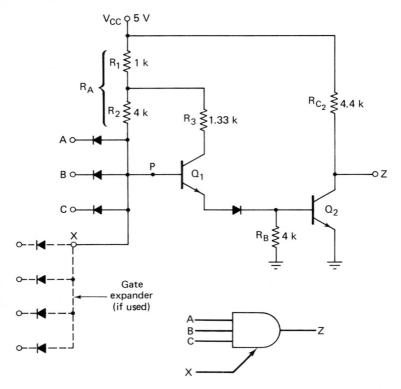

Fig. 7-4 An improved DTL *NAND* gate. (Courtesy of Signetics Corporation.)

diodes are cut off when the *AND* gate diodes conduct. The offset diodes used in the circuits that have R_B returned to ground have a slow recovery time. The recombination that takes place in the diodes uses the charge stored in the base and thus causes the transistor to turn off more quickly. When all inputs again go high, the input diodes cut off and Q_1 comes on. The normal turn-on delays of Q_1 (see Sec. 2-5) cause the delay shown at the trailing edge of the v_o waveform. In the circuits where R_B is grounded, the reverse-biased offset diodes will act as a speed-up capacitor to provide overdrive to the base of Q_1 and reduce turn-on time. Modern integrated circuit DTL gates have propagation delays as low as 10 ns.

A DTL Power Gate and Buffer. When a high fan-out is required, the DTL *power gate* and *buffer* of Fig. 7-6 are useful. In the power gate, Q_1 and Q_2 are emitter followers driving Q_3 a high-current grounded-emitter output. The bare collector of Q_3 increases system flexibility. It permits the designer to choose the collector pull-up resistor R_C for any combination of paralleled collectors (see Sec. 7-3).

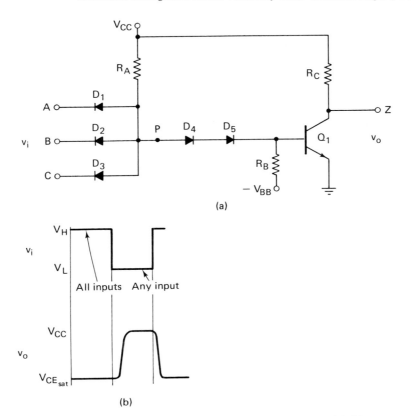

Fig. 7-5 (a) The basic integrated circuit DTL *NAND* gate with (b) its waveforms showing turn-on and turn-off delays.

The buffer is especially useful when high-capacitance loads are encountered. The *totem-pole* output provides the speed of the emitter follower Q_4 for source loads and the speed of the saturated inverter for sink loads. It produces very short RC time constants for the 1 output as well as the 0 output condition.

The phase splitter Q_2 serves two functions: (1) When all inputs are high, Q_1 conducts and turns Q_2 on. This pulls current away from the base of Q_3, thus cutting it off. At the same time, the emitter current of Q_2 develops a positive voltage across the 0.63-k resistor that turns Q_4 on and causes the output to go to $V_{CE_{sat}}$. (2) When one or more inputs go low, Q_1 and Q_2 cut off. This removes the base drive for Q_4, thus cutting it off. At the same time, the collector voltage of Q_2 rises, permitting Q_3 to conduct and send the output to the logical 1 level of the system. Both circuits are provided with input expansion nodes (X).

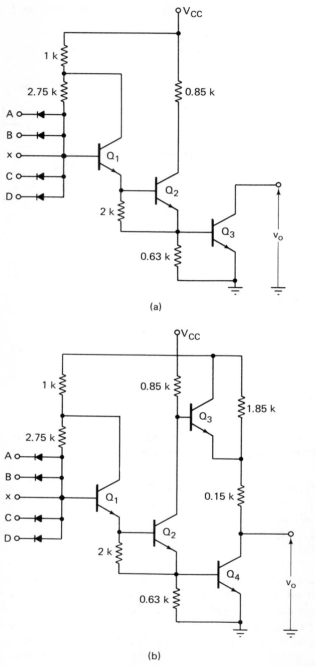

Fig. 7-6 (a) A DTL *Power Gate*. (b) A DTL *Buffer*. (Courtesy Fairchild Semiconductor.)

The combination of the phase splitter with the totem-pole output is frequently encountered in digital circuits. Further discussion is given these circuits in Chap. 8, including examples showing current and power calculations. A variation of the totem-pole output is discussed in Sec. 8-9.

7-2 Emitter-coupled Logic ECL

Emitter-coupled logic, which uses current switching, is used when very high speeds are required. In ECL (also known as *current-mode logic* CML and current-controlled logic CCL), the transistors operate in the active region which permits faster switching. Nonsaturation by circuit design is one of the four nonsaturating techniques explained in Sec. 2-7. Switching times of less than 5 ns are typical with emitter-coupled logic. Some basic gates have speeds in the 1.5 ns to 2.5 ns range.

An ECL circuit is shown in Fig. 7-7. The inputs are applied to the bases of Q_1, Q_2, and Q_3. A relatively constant current is switched between these transistors and Q_4. A fixed reference voltage equal to -1.05 V is applied to the base of Q_4. Two outputs are obtained through the emitter-follower stages

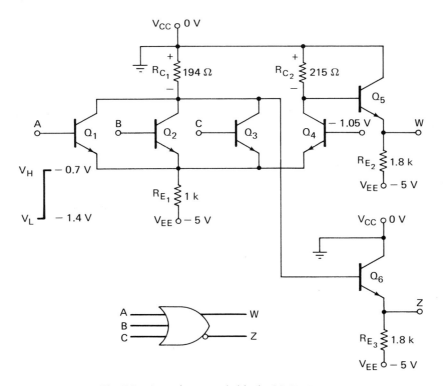

Fig. 7-7 An emitter-coupled logic *OR-NOR* gate.

Q_5 and Q_6: an OR output from W and a NOR output from Z. The state chart and positive-logic truth table for this circuit are given in Table 7-1.

Table 7-1 (a) The state chart for Fig. 7-7. (b) The positive-logic truth table showing $f_W = A + B + C$ and $f_z = \overline{A + B + C}$.

V_C	V_B	V_A	Q_3	Q_2	Q_1	Q_4	V_W	V_Z
−1.4	−1.4	−1.4	Off	Off	Off	On	−1.4 V	−0.7
−1.4	−1.4	−0.7	Off	Off	On	Off	−0.7	−1.4
−1.4	−0.7	−1.4	Off	On	Off	Off	−0.7	−1.4
−1.4	−0.7	−0.7	Off	On	On	Off	−0.7	−1.4
−0.7	−1.4	−1.4	On	Off	Off	Off	−0.7	−1.4
−0.7	−1.4	−0.7	On	Off	On	Off	−0.7	−1.4
−0.7	−0.7	−1.4	On	On	Off	Off	−0.7	−1.4
−0.7	−0.7	−0.7	On	On	On	Off	−0.7	−1.4

(a)

C	B	A	W	Z
0	0	0	0	1
0	0	1	1	0
0	1	0	1	0
0	1	1	1	0
1	0	0	1	0
1	0	1	1	0
1	1	0	1	0
1	1	1	1	0

(b)

The simplified circuit of Fig. 7-8(a) is used to explain the switching action. In this discussion it is assumed that the transistors cut in with $V_{BE} = 0.5$ V and that $V_{BE} = 0.7$ V when the transistors are fully on. The analysis is begun by removing Q_1 in Fig. 7-8(a) and finding the open-circuited emitter voltage. In Fig. 7-8(b) it can be seen that Q_4 conducts when Q_1 is removed; hence,

$$V_{E_{oc}} = V_{EB_4} + V_{B_4} = (-0.7) + (-1.05) = -1.75 \text{ V}$$

Figure 7-8(c) shows the base-emitter circuit of Q_1 when it is replaced with V_L applied to its input. $V_{BE_1} = 0.35$ V which is less than the cutin voltage. Therefore, the open-circuit conditions still exist. From Fig. 7-8(b), with Q_1 open, the current through the emitter of Q_4 is

$$I_{E_4} = \frac{V_E - V_{EE}}{R_{E_1}} \tag{7-5}$$

$$= \frac{(-1.75) - (-5)}{1 \text{ k}}$$

$$I_{E_4} = 3.25 \text{ mA}$$

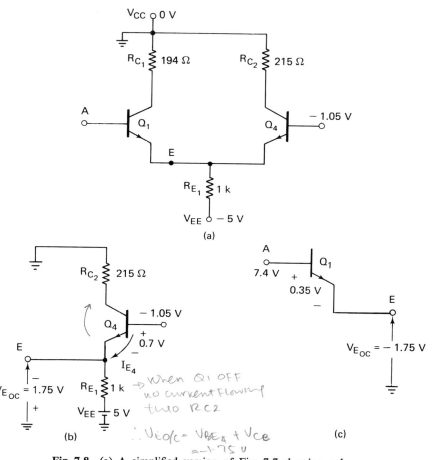

Fig. 7-8 (a) A simplified version of Fig. 7-7 showing only one input and Q_4. (b) The emitter circuit with Q_1 removed to find V_{Eoc}. (c) Q_1 replaced with V_1 applied to its input.

The output at W in Fig. 7-7 is $(V_{EB_5}) + (-V_{Rc_2})$. While $I_{E_4} = I_{C_4} + I_{B_4}$, $I_{Rc_2} = I_{C_4} + I_{B_5}$. Thus, I_{Rc_2} is essentially equal to I_{E_4} and the output at W is

$$v_{ow} = V_{EB_5} - (I_{Rc_2})(R_{C_2}) \tag{7-6}$$
$$\approx -0.7 - (3.25 \text{ mA})(0.215 \text{ k})$$
$$v_{ow} \approx -1.4 \text{ V} = V_L$$

The output at Z in Fig. 7-7 is $(V_{EB_6}) + (-V_{Rc_1})$. Since I_{Rc_1} is only the small base current of Q_6, $V_{Rc_1} \approx 0$, and the output at Z is

$$v_{oz} \approx V_{EB_6} \tag{7-7}$$
$$v_{oz} \approx -0.7 \text{ V} = V_H$$

Figure 7-9(a) shows the simplified circuit at the instant Q_1 is replaced with V_H at its input. Since $V_{BE_1} = +1.05$ V, Q_1 comes on and changes V_E. Figure 7-9(b) illustrates the conditions that exist after Q_1 comes on. V_E rises to $V_{EB_1} + V_A = (-0.7) + (-0.7) = -1.4$ V, making $V_{BE_4} = 0.35$ V. Since this is less than the cutin voltage, Q_4 cuts off and all of the current flows through Q_1. The current through the emitter of Q_1 is

$$I_{E_1} = \frac{V_E - V_{EE}}{R_{E_1}} \tag{7-8}$$

$$= \frac{(-1.4) - (-5)}{1\,\text{k}}$$

$$I_{E_1} = 3.6\,\text{mA}$$

Again by referring to Fig. 7-7, it can be seen that the output at W is still $(V_{EB_5}) + (-V_{R_{C_2}})$. But now, with Q_4 cut off, $I_{R_{C_2}}$ is only the small base current of Q_5. Hence, $V_{R_{C_2}} \approx 0$ and

$$v_{ow} \approx V_{EB_5} \tag{7-9}$$

$$v_{ow} \approx -0.7\,\text{V} = V_H$$

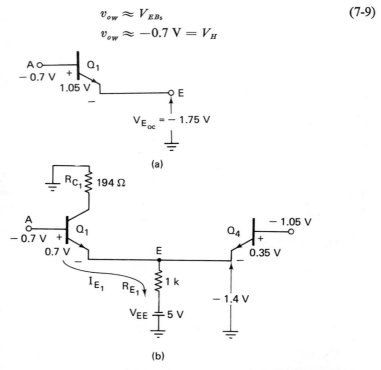

Fig. 7-9 (a) Q_1 at the instant it is replaced with V_H at its input. (b) The circuit after Q_1 comes on. V_E rises to -1.4 V and cuts off Q_4.

The output at Z is still $(V_{EB_6}) + (-V_{RC_1})$. But now, with Q_1 on, $I_{RC_1} = I_{C_1} + I_{B_6}$. Since $I_{E_1} = I_{C_1} + I_{B_1}$, $I_{RC_1} \approx I_{E_1}$. Therefore, the output voltage at Z is

$$v_{oz} = V_{EB_6} - (I_{RC_1})(R_{C_1}) \quad (7\text{-}10)$$
$$\approx -0.7 - (3.6 \text{ mA})(0.194 \text{ k})$$
$$v_{oz} \approx -1.4 \text{ V} = V_L$$

As shown in Table 7-1(a), if one or more of the inputs equal V_H, the transistors with the V_H inputs conduct and cut off Q_4. The W output then equals V_H and the Z output equals V_L. Only when all inputs equal V_L will all of the input transistors be cut off. Then, Q_4 conducts and sends the W output to -1.4 V. With only the small base current of Q_6 flowing through R_{C_1}, the Z output rises to -0.7 V.

The collector resistors are designed to ensure an output equal to V_L, as required.

Example 7-1: Calculate the value of R_{C_1} in Fig. 7-7.
Solution: The output at Z equals $(V_{EB_6}) + (-V_{RC_1})$. However, with the input transistors cut off, V_{RC_1} is approximately zero and $v_{oz} \approx V_{EB_6} = V_H$. R_{C_1} is therefore calculated to produce an output equal to V_L with the input transistors on. From Eqs. (7-8) and (7-10), R_{C_1} must drop 0.7 V with 3.6 mA flowing through it. Thus,

$$R_{C_1} = \frac{0.7 \text{ V}}{3.6 \text{ mA}}$$
$$R_{C_1} = 194 \, \Omega$$

Example 7-2: Calculate the value of R_{C_2} in Fig. 7-7.
Solution: The W output equals $(V_{EB_5}) + (-V_{RC_2})$. With Q_4 cut off, $V_{RC_2} \approx 0$ and $v_{ow} \approx V_{EB_5} = V_H$. Hence, R_{C_2} is calculated to produce an output equal to V_L with Q_4 on. From Eqs. (7-5) and (7-6), R_{C_2} must drop 0.7 V with 3.25 mA flowing through it. Thus,

$$R_{C_2} = \frac{0.7 \text{ V}}{3.25 \text{ mA}}$$
$$R_{C_2} = 215 \, \Omega$$

Noise Immunity. The transfer curves for this circuit are given in Fig. 7-10(a). The input waveform, with the upper and lower acceptance levels, is shown in Fig. 7-10(b). The 1- and 0-level noise immunities $N_{(1)}$ and $N_{(0)}$ both equal 150 mV. From Eq. (7-8), $v_{oz} \approx V_{EB_6}$ as long as Q_1 is cut off. And from Eq. (7-10), $v_{ow} \approx V_{EB_5}$ as long as Q_4 is cut off. Hence, the 1 output at both W and Z are constant at -0.7 V. The 0 output at W is also constant. From Eq.

192 / Diode-Transistor Logic DTL Emitter-Coupled Logic ECL Wired Logic

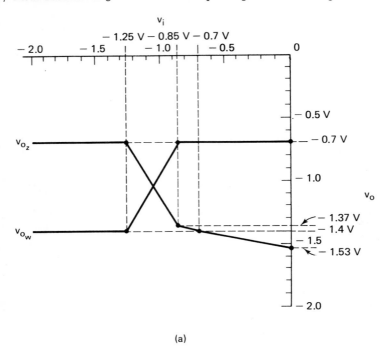

(a)

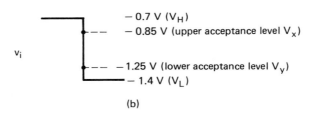

(b)

Fig. 7-10 (a) The transfer curve for the ECL circuit of Fig. 7-7. (b) The input signal with important voltage levels shown.

(7-6), $v_{ow} = -1.4$ V as long as Q_4 is on. However, the 0 output at Z shifts slightly if the 1 input at A varies. The Z output voltages shown on the transfer curve will be calculated after the upper and lower acceptance levels V_x and V_y are determined in Exs. 7-3 and 7-4.

Example 7-3: Determine the input voltage that will just hold Q_1 off when Q_4 is on.

Solution: By referring to Fig. 7-11(a), it can be seen that if Q_4 is on, V_{BE_4}

Emitter-coupled Logic ECL / 193

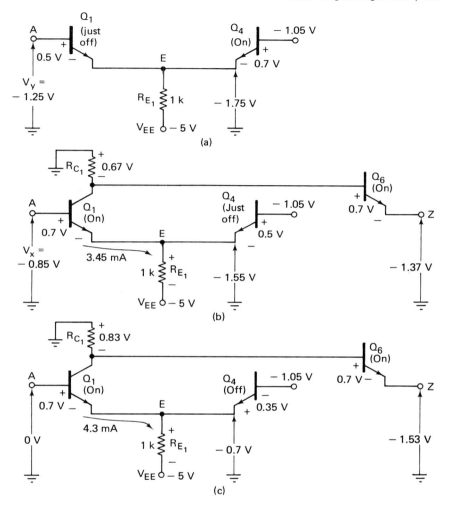

Fig. 7-11 The simplified circuit with (a) Q_4 on and Q_1 just cut off (b), Q_4 just cut off and Q_1 on, and (c) the circuit showing how the 0 output at z only drops to -1.53 V if the 1 input at A rises to 0 V.

$= 0.7$ V and the voltage at the common emitter is

$$V_E = V_{EB_4} + V_{B_4}$$
$$= (-0.7) + (-1.05)$$
$$V_E = 1.75 \text{ V}$$

If Q_1 is just off, the cutin voltage appears across its EB junction. Thus, the input voltage that will just hold Q_1 off is

$$V_A = V_{BE_1} + V_E$$
$$= (0.5) + (-1.75)$$
$$V_A = -1.25 \text{ V}$$

This is the lower acceptance level V_y.

Example 7-4: Determine the input voltage that will keep Q_1 on and just hold Q_4 off.

Solution: By referring to Fig. 7-11(b), it may be seen that if Q_4 is just off, the cutin voltage appears across its EB junction. The voltage at the common emitter is then

$$V_E = V_{EB_4} + V_{B_4}$$
$$= (-0.5) + (-1.05)$$
$$V_E = -1.55 \text{ V}$$

Since Q_1 is on, $V_{EB_1} = 0.7$ V. Therefore, the A input must be

$$V_A = V_{EB_1} + V_E$$
$$= (0.7) + (-1.55)$$
$$V_A = -0.85 \text{ V}$$

This is the upper acceptance level V_x.

As previously mentioned, the 0 output at Z will change if the A input causes Q_1 to conduct harder. However, the following example shows that this variation is small.

Example 7-5: Calculate the output at Z for $V_A = V_x = -0.85$ V (Q_4 is just off), and $V_A = 0$ V (Q_4 is well beyond cutoff).

Solution: Figure 7-11(b) shows the circuit with an input equal to -0.85 V. Q_1 is on and just holds Q_4 off.

$$I_{E_1} = \frac{V_E - V_{EE}}{R_{E_1}} \tag{7-8}$$

$$= \frac{(-1.55) - (-5)}{1 \text{ k}}$$

$$I_{E_1} = 3.45 \text{ mA}$$

$$v_{oz} = V_{EB_6} - (I_{R_{C_1}})(R_{C_1}) \tag{7-10}$$

where, as explained in the development of this equation, $I_{R_{C_1}} \approx I_{E_1}$. Hence,

$$v_{oz} = -0.7 - (3.45 \text{ mA})(194 \text{ }\Omega)$$
$$= -0.7 - 0.67$$
$$v_{oz} = -1.37 \text{ V}$$

Figure 7-11(c) shows the circuit with an input of 0 V. Since Q_1 will surely be on, $V_E = V_{EB} + V_A = -0.7 + 0 = -0.7$ V. This increases I_{E_1} to

$$I_{E_1} = \frac{V_E - V_{EE}}{R_{E_1}}$$
$$= \frac{(-0.7) - (-5)}{1 \text{ k}}$$
$$I_{E_1} = 4.3 \text{ mA}$$

and causes v_{oz} to shift to

$$v_{oz} = -0.7 - (4.3 \text{ mA})(194 \text{ }\Omega)$$
$$= -0.7 - 0.83$$
$$v_{oz} = -1.53 \text{ V}$$

Although this large shift in input voltage has little effect on v_{oz} and has no effect on v_{ow}, it does send Q_1 into saturation (with $V_A = 0$ V and $V_{R_{C_1}} = 0.83$ V, $V_{CB} = -0.83$ V) and it reverse biases the EB junction of Q_4 (with $V_E = -0.7$ V and $V_{B_4} = -1.05$ V, $V_{BE_4} = -0.35$ V). The result will be an increase in switching time.

7-3 Wired Logic

The outputs of two or more gates, with passive pull-up or pull-down elements, may be connected in parallel to produce a *wired-logic* function. Some manufacturers refer to these circuits as *wired-OR* circuits. In reality, they are *wired-AND* circuits. For example, if the outputs of two inverters, one with an input equal to A and the other with an input equal to B, are wired in parallel, they produce the function $f = \bar{A}\bar{B}$. However, by De Morgans theorem $\bar{A}\bar{B} = \overline{A + B}$. Also, if a false output is desired from true inputs, the wired output will be false if either A OR B is true. Hence, the name *wired OR*.

Wired logic is very useful because it frequently results in a reduction in the amount of hardware required to implement a function. Figure 7-12(a) shows the function $A\bar{B} + \bar{A}B$ produced with three *NAND* gates and two levels of propagation delay. *NAND* gate 1 has inputs A and $\bar{B}$, and it has an output equal to $\overline{A\bar{B}}$. *NAND* gate 2 has inputs $\bar{A}$ and B, and it has an output equal to $\overline{\bar{A}B}$. Applying these outputs to *NAND* gate 3 yields an output equal to

$$\overline{\overline{A\bar{B}} \cdot \overline{\bar{A}B}} = A\bar{B} + \bar{A}B$$

196 / Diode-Transistor Logic DTL Emitter-Coupled Logic ECL Wired Logic

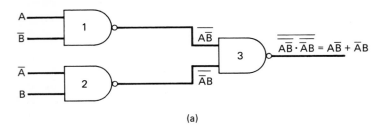

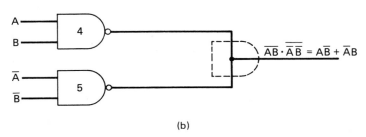

Fig. 7-12 The function $f = A\bar{B} + \bar{A}B$ produced by (a) three NAND gates and (b) two NAND gates with paralleled outputs.

Figure 7-12(b) shows the same function $f = A\bar{B} + \bar{A}B$ implemented with only two NAND gates and one level of propagation delay. The inputs A and B are applied to NAND gate 4; NAND gate 5 has inputs equal to $\bar{A}$ and $\bar{B}$. If their outputs are connected in parallel, the wired output becomes

$$\overline{AB} \cdot \overline{\bar{A}\bar{B}} = \overline{AB + \bar{A}\bar{B}} = A\bar{B} + \bar{A}B$$

Wired logic cannot be used with circuits that have active pull-up and pull-down elements. For example, the buffer circuit of Fig. 7-6(b) does not lend itself to wired logic because its output combines the low output impedance of the emitter follower with the low output impedance of the saturated inverter. Wiring two buffer outputs in parallel can produce an ambiguous condition when one output tries to go high while the other output tries to go low. The actual output will fall somewhere between $V_{(1)}$ and $V_{(0)}$. Even if the inverter can saturate to produce the correct output, the current requirements are too high to be practical.

The outputs of two or more circuits with *bare collectors* can be connected in parallel. The power gate of Fig. 7-6(a) is an example of this type of circuit. The designer chooses the value of R_C.

Range of Values for R_C. Figure 7-13 shows n wired DTL NAND gates driving FO inputs to other gates. The loads are assumed to be identical to the driver gates. The maximum value for the collector resistor $R_{C_{max}}$ is limited

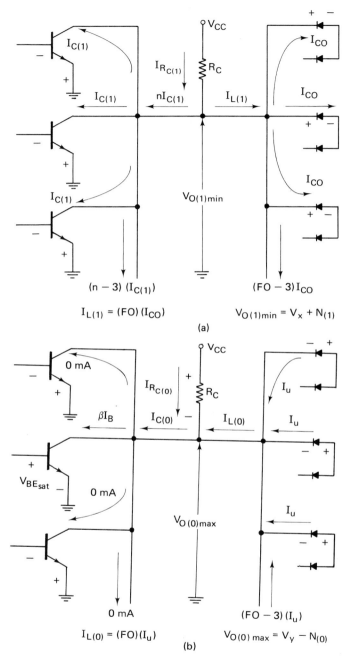

Fig. 7-13 The circuits for determining (a) $R_{C\max}$ and (b) $R_{C\min}$ for DTL *NAND* gates with bare collectors wired in parallel.

by the upper acceptance level V_x and the desired 1-output noise immunity $N_{(1)}$. It must ensure a minimum 1-output voltage $V_{o(1)\min} = V_x + N_{(1)}$ at the common collector node, under the worst-case conditions. This condition is shown in Fig. 7-13(a).

The minimum value of collector resistance $R_{C_{\min}}$ is limited by the lower acceptance level V_y and the desired 0-output noise immunity $N_{(0)}$. It must ensure a maximum 0-output voltage $V_{o(0)\max} = V_y - N_{(0)}$ at the common

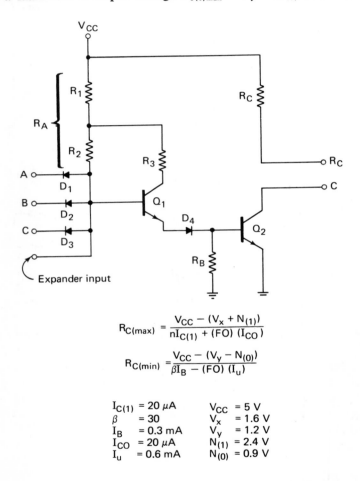

$$R_{C(\max)} = \frac{V_{CC} - (V_x + N_{(1)})}{nI_{C(1)} + (FO)(I_{CO})}$$

$$R_{C(\min)} = \frac{V_{CC} - (V_y - N_{(0)})}{\beta I_B - (FO)(I_u)}$$

$I_{C(1)}$ = 20 μA V_{CC} = 5 V
β = 30 V_x = 1.6 V
I_B = 0.3 mA V_y = 1.2 V
I_{CO} = 20 μA $N_{(1)}$ = 2.4 V
I_u = 0.6 mA $N_{(0)}$ = 0.9 V

n is the number of wired collectors or number of driving gates.
FO is the fan-out or number of driven gates.

(a)

Fig. 7-14(a) A DTL *NAND* gate.

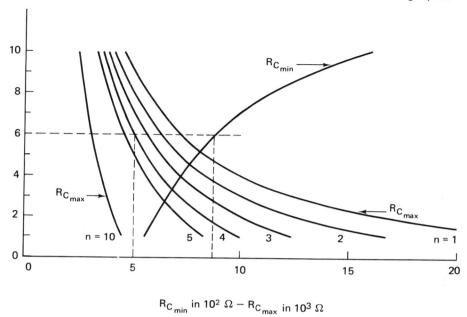

Fig. 7-14(b) The range of values for the DTL gate shown in (a).

collector node, under the worst-case conditions. This condition is shown in Fig. 7-13(b).

It is important that the permissible range of values of R_C be known in wired-logic systems. Figure 7-14 shows a DTL *NAND* gate with an optional pull-up resistor. Equations (7-11) and (7-12) are used to calculate $R_{C_{max}}$ and $R_{C_{min}}$. Current, voltage, and noise immunity values for this circuit are also shown. Only the number of collectors to be wired together (n) and the fan-out (*FO*) must be known in order to solve the equations. Manufacturers frequently provide curves for bare-collector circuits that permit the designer to choose R_C for various combinations of wired collectors (n) and fan-out (*FO*). A typical set of curves is given in Fig. 7-14(b). The broken lines are used to show how the range of values of R_C is determined. For four wired collectors ($n = 4$) to drive six unit loads ($FO = 6$), $R_{C_{min}} = 870 \, \Omega$ and $R_{C_{max}} = 5 \, k$. The desired R_C may be obtained by paralleling optional pull-up resistors of two or more gates or by selecting a discrete external resistor. The pins for the output and the optional collector resistor are placed next to each other in order to simplify connections. Although upper-limit values of R_C reduce power consumption, they also reduce speed and ac noise immunity. Lower-limit values provide maximum speed and ac noise immunity at the expense of increased power. The guaranteed dc noise margins are not affected as long as R_C is within its limits.

7-4 Summary

Diode-transistor logic DTL combines the basic diode gates with the transistor inverter.

In the modern integrated circuit versions of DTL, the diodes are produced by shorting the collector-base junction of transistors. This permits faster switching. The basic gates perform the $NAND$ or NOR functions. In some circuits emitter followers are added to increase fan-out capabilities.

Emitter-coupled logic ECL uses current switching. The transistors operate in the active region to permit faster switching.

ECL has several disadvantages: It requires a greater number of components, more power, and an additional supply for the reference voltage.

The outputs of two or more gates may be connected in parallel to produce a wired-logic function. The wired output will equal the outputs of the individual gates ANDed together. This technique, which is limited to circuits that have passive pull-up or pull-down elements, often reduces the number of gates required to produce a function while increasing speed.

Problems

7-1 Calculate I_B in Fig. 7-15(a). The voltage drop on each forward-biased diode and $V_{BE_{sat}}$ equal 0.7 V. Neglect I_{CO}.

7-2 The DTL circuit of Fig. 7-15(a) is modified by splitting R_A and replacing one of the offset diodes with an emitter follower as shown in Fig. 7-15(b). If the base current in the output transistor of Fig. 7-15(a) is doubled by this modification, what are I_B and I_C for the emitter follower? Use 0.7 V for V_D and V_{BE}.

7-3 Verify that all transistors in Fig. 7-7 remain out of saturation for all input combinations.

7-4 Figure 7-16 shows two circuits with their outputs (emitters) wired in parallel. The emitter resistor is common to both circuits. What is the output function for the inputs shown on the diagram?

7-5 Determine the average power consumption of the circuit shown in Fig. 7-1 for a 50-percent duty cycle. Use $v_{i(0)} = V_{CE_{sat}}$ and $v_{i(1)} = V_{CC}$.

7-6 Calculate the fan-out of the DTL gate of Fig. 7-17 if $\mathbf{h}_{FE} = 20$. Assume that the gate is driven by and drives other gates exactly like itself.

7-7 Calculate the actual 1-level output voltage if the leakage current of each load of the DTL gate of Fig. 7-17 is 30 μA maximum and $FO = 10$.

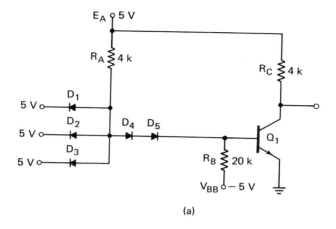

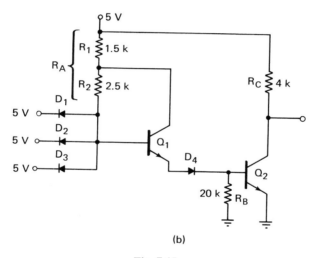

Fig. 7-15

7-8 If $I_{L(0)} = 10$ mA (sink-type load), $v_{i(0)} = 0$ V, and $v_{i(1)} = 5$ V, what is $h_{FE_{min}}$ for the circuit of Fig. 7-17?

7-9 Calculate the average power consumption of the circuit shown in Fig. 7-17 for a 50-percent duty cycle. Use $v_{i(0)} = V_{CE_{sat}} = 0.3$ V and $v_{i(1)} = V_{CC}$.

7-10 Calculate the current through each input diode of Fig. 7-17 if all inputs equal 0 V.

7-11 A DTL voltage-switching package is shown in Fig. 7-18. Neglecting the voltage drop on forward biased junctions and using $V_H = 10$ V

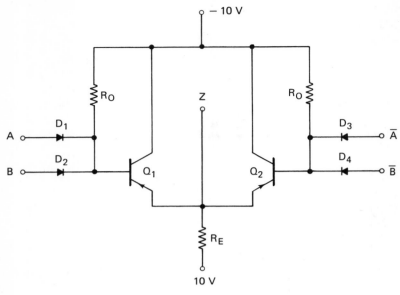

Fig. 7-16

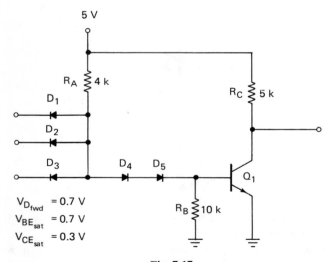

Fig. 7-17

and $V_L = 0$ V, determine the base current and the voltages at points P_1 and P_2 when
(a) All inputs equal V_H.
(b) V_W and $V_X = V_H$ and the other inputs equal V_L.
(c) All inputs equal V_L.

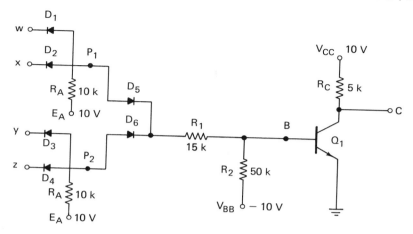

Fig. 7-18

7-12 Using the information from Prob. 7-11, determine the highest current that will ever flow through any *AND* gate diode.

7-13 Referring to Fig. 7-18 and using $V_{BE_{sat}} = 0.7$ V, $V_{CE_{sat}} = 0.3$ V, $V_D = 0.7$ V, $v_{i(0)} = 0.3$ V, $v_{i(1)} = 8$ V, and $h_{FE} = 40$, calculate *FO*. Assume that this circuit is driven by and drives others exactly like itself.

7-14 If $h_{FE_1} = 19$, $h_{FE_2} = 10$, and $I_{CO} = 30$ μA per input, calculate the fan-out *FO* and the minimum 1-level output voltage $V_{o(1)}$ of Fig. 7-15(b). Assume the circuit drives and is driven by others exactly like itself.

8

TRANSISTOR-TRANSISTOR LOGIC TTL OR T²L WITH OUTPUT CIRCUIT VARIATIONS

The switching speed of digital circuits is greatly improved with the use of *multi-emitter transistors*. Logic circuits that use these transistors fall into the category known as *transistor-transistor logic TTL*. The switching speeds of the logic types already discussed in Chaps. 6 and 7 are 50 ns in RTL, 10 ns in RCTL and DTL, and 2 ns (with low noise immunity) in ECL. In TTL speeds usually range between 5 ns and 50 ns (with good noise immunity), but speeds as high as 2 ns or 3 ns are possible. The higher speeds are at the expense of power consumption. For example, a circuit with 50 ns speed consumes approximately 1 mW. But as speed is increased to say 25 ns, 10 ns, and 5 ns, the power consumption increases to approximately 10 mW, 20 mW, and 30 mW, respectively.

TTL is the most common type of logic used in modern digital computers. The various TTL building blocks are discussed in this chapter. In adition to the TTL *OR, AND, NOR, NAND, INVERTER*, and the *AND-OR-INVERT* circuits, the various output configurations, common to both DTL and TTL, are explained.

8-1 Construction of the Multi-emitter Transistor

The construction and schematic symbol of a multi-emitter transistor are shown in Fig. 8-1. The collector is formed by diffusing a large n section in a p substrate(s). A p section is diffused into the collector to form the base, and three emitters are produced by diffusing three separate n sections into the p

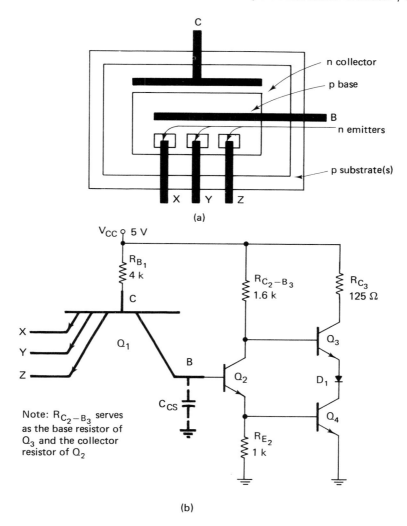

Fig. 8-1 (a) Construction of a multi-emitter transistor. (b) Its schematic symbol is shown in bold lines in a three-input TTL *NAND* gate. (Courtesy Sprague Electric Co.)

base. Metal contacts B, C, X, Y, and Z are brought out to permit connections to the other circuit elements.

The capacitance C_{CS} in Fig. 8-1(b) is due to the reverse-biased junction formed by the collector edges and bottom with the substrate.

$$C_{CS} = \frac{KA_1}{V^n} + \frac{KA_2}{V^{1/2}} \tag{8-1}$$

where

KA_1/V^n = the capacitance between collector edges and the substrate;
$KA_2/V^{1/2}$ = the capacitance between the collector bottom and the substrate;
$n = 1/3$ if the concentration of majority carriers around the edges varies with depth and $n = 1/2$ if there is an abrupt change from acceptor to donor ions;

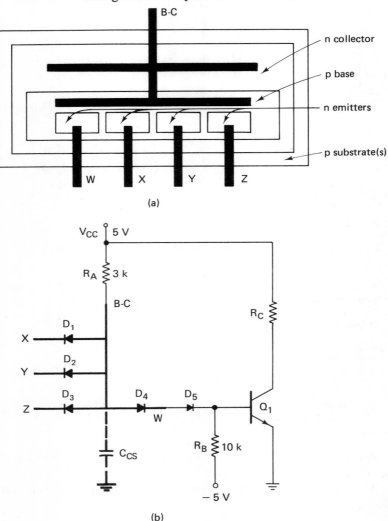

Fig. 8-2 (a) Construction of the four-diode network (shown in bold lines) in the DTL *NAND* gate shown in (b).

A_1 and A_2 = the perimeter and bottom areas of the isolation region;
K = a constant depending on the type of material and doping;
V = the reverse voltage between the collector and the substrate.

For comparison, the construction of the input diode network of a three-input DTL *NAND* gate is shown in Fig. 8-2(a). This network is shown in bold lines in the schematic of Fig. 8-2(b). The circuit is the basic DTL *NAND* gate that is explained in Sec. 7-1. Note that the anodes of the *AND* gate diodes and the offset diode D_4 have a common connection. The basic structure is the same as that shown in Fig. 8-1(a) except that the collector is shorted to the base by metalization to produce the common anode connection *B–C*. The fourth junction is produced by diffusing an additional emitter into the base.

The steady-state operation of the TTL *NAND* gate is discussed in the following section and its faster switching action is explained in Sec. 8-3.

8-2 Steady-state Analysis of a TTL NAND Gate

The TTL *NAND* gate of Fig. 8-1(b) is redrawn with resistance values and voltages in Fig. 8-3. This circuit employs a totem-pole output. Q_3 is an emitter follower that provides a low output impedance (less than 100 Ω) when the

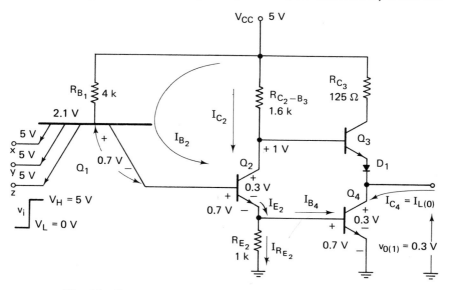

Fig. 8-3 Current paths when all inputs to Q_1 are high and a 0-level output is produced.

circuit is in the 1 state ($v_o \geq V_{(1)\min}$). Q_4 is a saturated common emitter circuit that provides a low output impedance (less than 20 Ω) when the circuit is in the 0 state ($v_o \leq V_{(0)\max}$). Thus, very short RC time constants exist for both the 1 and 0 levels, even with high capacitance loads. Several output configurations, including the totem-pole circuit, are discussed in Sec. 8-9.

The steady-state operation of the basic TTL *NAND* gate is explained in this section. Examples showing the procedures for calculating the circuit currents for both the 1 and 0 states, fan-out, noise margins, and power consumption are included.

0-Level Output Analysis. The circuit is shown in Fig. 8-3 with all of its inputs at $V_H = 5$ V. Since $V_{CC} = 5$ V, all of the emitters of Q_1 are cut off and only its collector-base junction can conduct. A current I_{B_2} flows as shown and sends the phase splitter Q_2 into saturation.

The collector of Q_2 draws current away from the base of Q_3 to turn that transistor off. Part of the emitter current I_{E_2} flows through R_{E_2} and develops voltage $V_{E_2} = (I_{R_{E(2)}})(R_{E_2}) = 0.7$ V to turn Q_4 on. The base current $I_{B_4} = I_{E_2} - I_{R_{E(2)}}$ and causes Q_4 to saturate. Hence, when all inputs are high, the output voltage equals $V_{CE_{4sat}} = 0.2$ V to 0.4 V $\approx V_L$.

The diode in the emitter of Q_3 is used to guarantee that Q_3 is cut off when Q_4 is on. The voltage at the base of Q_3 is

$$V_{B_3} = V_{CE_{2sat}} + V_{BE_{4sat}}$$
$$= 0.3 + 0.7$$
$$V_{B_3} = 1 \text{ V}$$

Without the emitter diode D_1,

$$V_{BE_3} = V_{B_3} - V_{CE_{4sat}}$$
$$= 1 - 0.3$$
$$V_{BE_3} = 0.7 \text{ V}$$

This, of course, is enough to keep Q_3 on. But with D_1, the same 0.7 V is across two junctions and cannot cause conduction.

The high impedance of D_1 and Q_3 in the collector of Q_4, instead of a passive pull-up resistance, keeps power consumption at a minimum.

The *NAND* gate now acts as a sink for current from other loads. The currents I_{B_2}, I_{C_2}, I_{E_2}, I_{R_E}, and I_{B_4}, shown in Fig. 8-3, are calculated in the following example. I_{C_4} depends on the load.

Example 8-1: Calculate the steady-state currents I_{B_2}, I_{C_2}, I_{E_2}, $I_{R_{E(2)}}$, and I_{B_4} shown in Fig. 8-3. Use $V_{BE_{sat}} = 0.7$ V, $V_{CE_{sat}} = 0.3$ V, $V_{BC_1} = 0.7$ V, and $V_i = V_H = 5$ V at all inputs. Neglect leakage currents.

Solution: The base current of Q_2 is

$$I_{B_2} = \frac{V_{R_{B(1)}}}{R_{B_1}}$$
$$= \frac{V_{CC} - (V_{BC_1} + V_{BE_2} + V_{BE_4})}{R_{B_1}}$$
$$= \frac{5 - (0.7 + 0.7 + 0.7)}{4 \text{ k}}$$
$$I_{B_2} = 0.725 \text{ mA}$$

and the collector current of Q_2 is

$$I_{C_2} = \frac{V_{R_{C(2)}}}{R_{C_2}}$$
$$= \frac{V_{CC} - (V_{CE_{sat}} + V_{BE_{sat}})}{R_{C_2}}$$
$$= \frac{5 - (0.3 + 0.7)}{1.6 \text{ k}}$$
$$I_{C_2} = 2.5 \text{ mA}$$

The emitter current of Q_2 is

$$I_{E_2} = I_{C_2} + I_{B_2}$$
$$= 2.5 + 0.725$$
$$I_{E_2} = 3.225 \text{ mA}$$

This current splits into $I_{R_{E(2)}}$ and I_{B_4}

$$I_{R_{E(2)}} = \frac{V_{R_{E(2)}}}{R_{E_2}}$$
$$= \frac{0.7 \text{ V}}{1 \text{ k}}$$
$$I_{R_{E(2)}} = 0.7 \text{ mA}$$

and

$$I_{B_4} = I_{E_2} - I_{R_{E(2)}}$$
$$= 3.225 - 0.7$$
$$I_{B_4} = 2.525 \text{ mA}$$

Because of this high base current the circuit has a high fan-out. It is capable of sinking not only the steady-state current from other gate inputs, but also the high current required to discharge the load capacitance and effect a fast 1-to-0 transition.

1-Level Output Analysis. The circuit is shown in Fig. 8-4 with one of its inputs equal to $V_L = 0$ and the other inputs at $V_H = 5$ V. With one or more of the inputs at V_L, Q_1 goes into saturation and removes the base drive to the phase splitter Q_2. As explained in Sec. 8-3, Q_1 provides a low impedance return to ground to quickly clear the base of its stored charge and cut off Q_2. Consequently, the negative base return voltage in the DTL circuit of Fig. 8-2(b) is not needed.

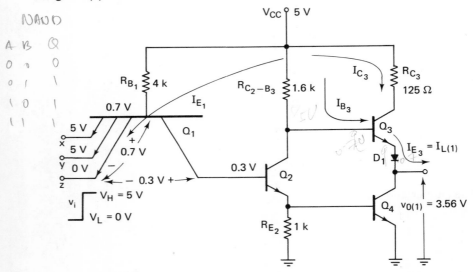

Fig. 8-4 Current paths when one (or more) inputs are low and a 1-level output is produced.

When Q_2 is off, the base current to Q_4 is removed, causing it to cut off. The charge stored in the base of Q_4 is removed by R_{E_2}. At the same time, the voltage at the collector of Q_2 rises and permits Q_3 to conduct.

Q_3 does not saturate; it provides a 1-level output to source-type loads by its emitter-follower action. The 1-level output equals $V_{CC} - (I_{B_3} R_{B_3} + V_{BE_3} + V_{D_1}) \approx 5 - (0.04 + 0.7 + 0.7) = 3.56$ V. The resistance of R_{C_3} and the emitter diode D_1 limit the amount of surge current during transitions.

The currents I_{E_1} and I_{B_3} shown in Fig. 8-4 are calculated in the following example. I_{C_3} and I_{E_3} depend on the load.

Example 8-2: Calculate the steady-state currents I_{E_1} and I_{B_3} shown in Fig. 8-4. Use $V_{BE_{sat}} = 0.7$ V, $V_{CE_{sat}} = 0.3$ V, a 1-level output equal to 3.56 V, $V_i = 0$ V at one input, and $V_i = 5$ V at the other inputs.

Solution: The emitter (z) current of Q_1 equals

$$I_{E_1} = \frac{V_{R_{B(1)}}}{R_{B_1}}$$

$$= \frac{V_{CC} - (V_{BE_{sat}} + V_{E(Z)})}{R_{B_1}}$$

$$= \frac{5 - (0.7 + 0)}{4\text{ k}}$$

$$I_{E_1} = 1.075 \text{ mA}$$

The base current of Q_3 is

$$I_{B_3} = \frac{V_{R_{B(3)}}}{R_{B_3}}$$

$$= \frac{V_{CC} - (V_{BE_3} + V_{D_1} + V_{o(1)})}{R_{B_3}}$$

$$= \frac{5 - (0.7 + 0.7 + 3.56)}{1.6\text{ k}}$$

$$I_{B_3} = 25 \text{ }\mu\text{A}$$

The output voltage equal to 3.56 V is more than enough to cut off the following gate inputs. Therefore, in the steady state, Q_3 only has to supply a small leakage current (approximately 40 μA) to each of its loads. It can, however, also supply the high current required to charge the load capacitance and bring about a fast 0-to-1 transition.

Fan-out. The steady-state fan-out of the TTL *NAND* gate is determined by the 0-level output conditions. In the 0 state it must act as a sink for the input currents of the loads. In the 1 state it supplies the leakage currents of these inputs.

$$FO = \frac{I_{L(0)\text{max}}}{I_u} \tag{8-2}$$

where $I_{L(0)\text{max}}$ is the maximum load current that the saturated inverter Q_4 can sink and I_u is a unit load or the current that Q_4 will have to sink from each of its loads.

Example 8-3: Calculate the fan-out of the TTL *NAND* gate if Q_4 has an $h_{FE_{\text{min}}} = 4$. Use $V_{CC} = 5$ V, $V_{CE_{sat}} = 0.3$ V, $V_{BE_{sat}} = 0.7$ V. Neglect leakage currents.

Solution: From Ex. 8-1, Q_4 has a base current $I_{B_4} = 2.525$ mA. Therefore, the maximum load current that it can sink is

$$I_{L(0)\max} = h_{FE_{\min}} I_{B_4}$$
$$= 4 \times 2.525$$
$$I_{L(0)\max} = 10.1 \text{ mA}$$

As shown in Fig. 8-5, when Q_4 is saturated, it applies $V_{CE_{sat}} = 0.3$ V (not $V_L = 0$ V) to each of its loads. Under worst-case conditions, the other emitters of the multi-emitter transistor will be cut off and Q_4 will have to sink all of the current that flows through R_{B_1}. Hence, if leakage current is neglected, a unit load for this circuit is

$$I_u = \frac{V_{CC} - (V_{BE_{sat}} + V_{i(0)})}{R_{B_1}}$$
$$= \frac{5 - (0.7 + 0.3)}{4\text{ k}} \qquad (8\text{-}3)$$
$$I_u = 1 \text{ mA}$$

This value differs from I_{E_1} calculated in Ex. 8-2 because $V_i = V_{CE_{sat}}$ and not V_L.

Now by applying Eq. (8-2), it can be seen that

$$FO = \frac{I_{L(0)\max}}{I_u}$$
$$= \frac{10.1 \text{ mA}}{1 \text{ mA}}$$
$$FO = 10.1 \text{ (actual } FO = 10)$$

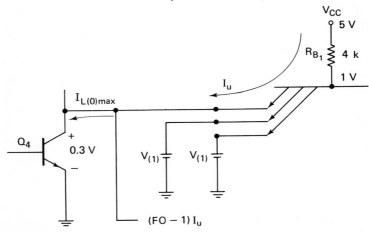

Fig. 8-5 The circuit for the fan-out calculation. The $V_{CE_{sat}}$ of Q_4 is the input $v_{i(0)}$ to the multi-emitter transistor.

As previously explained, this does not include leakage currents or the current required by the load capacitance. A large part of the current handling capability of Q_4 is used to quickly discharge this capacitance and speed up the 1-to-0 transition. A higher h_{FE} increases the current that Q_4 can sink and permits the load capacitance to discharge more quickly. The leakage currents plus the worst-case value of V_{CC}, which may be as high as 5.5 V, reduce the fan-out. If $FO = 10$ is to be maintained, the h_{FE} of Q_4 must be higher than the value of 4 given in this example.

Leakage Currents. When one of the inputs is equal to or greater than $V_{(1)min}$ and another is equal to or greater than $V_{(0)max}$, an *emitter-to-emitter transistor* is formed. This *parasitic transistor* is shown in broken lines in Fig. 8-6(a). The 1-level source must supply a leakage current equal to

$$I_{EE} = (h_{FE_{EE}})(I_{B_1}) \tag{8-4}$$

$h_{FE_{EE}}$ is the h_{FE} of the parasitic emitter-to-emitter transistor and I_{B_1} is the current through R_{B_1} when Q_1 is on. By employing special construction techniques increased spacing between the emitters, and so on, $h_{FE_{EE}}$ can be made as low as 0.01. Hence,

$$I_{EE} = h_{FE_{EE}} \frac{V_{CC} - (V_{BE_{sat}} + V_{i(0)})}{R_{B_1}}$$

$$\approx (0.01)(1 \text{ mA})$$

$$I_{EE} \approx 10 \ \mu A$$

As shown in Fig. 8-6(a), I_{EE} adds to I_{B_1} to increase the amount of current that flows out of the emitter with the 0-level input. Hence, the unit load current is increased by the parasitic transistor. If additional emitters are connected to a 1-level voltage, new parasitic transistors are formed between these emitters and the emitter with the 0-level voltage. A leakage current will flow from each 1-level source through the parasitic transistors to the 0-level source. Thus, although each I_{EE} is small, the total leakage through the 0-input line must be considered, especially in transistors with many emitters.

The test circuit for measuring the 0-input current is shown in Fig. 8-6(b). V_{CC} actually exceeds the 1-level inputs to this circuit, which are the outputs of other gates. By referring to Fig. 8-4a, it can be seen that

$$V_{o(1)} = V_{CC} - (V_{RB_3} + V_{BE_3} + V_{D_1}) \approx 3.6 \text{ V}$$

The 0-input current including leakage is

$$I_{i(0)} = (1 + nh_{FE_{EE}})I_{B_1}$$

where n is the number of emitters connected to the 1-level voltage. The maximum 0-input current for the TTL *NAND* gate is 1.6 mA.

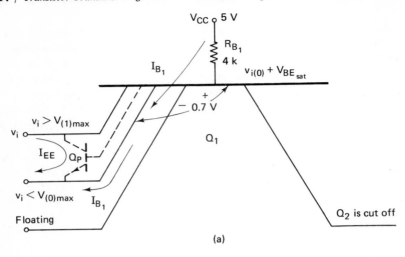

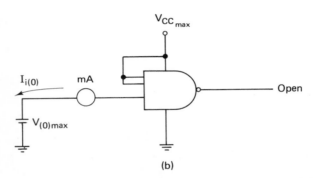

Fig. 8-6 (a) The parasitic transistor that exists between an emitter with a 0-level voltage and another emitter with a 1-level voltage. (b) The test circuit for measuring the 0-input current.

There is also a leakage current when all of the inputs are at the 1 level. This current is related to the h_{FE} of the *inverse transistor* $h_{FE(inv)}$. Figure 8-7(a) shows the current paths through the inverse transistor. The h_{FE} of this transistor is in the order of tenths or less. The current $(h_{FE(inv)})(I_{B_1})$ shown in Fig. 8-7(a) divides among the three emitters, which act as the collector of the inverse transistor.

The one-input current that each source must supply is greatest when one or more inputs are grounded and emitter-to-emitter leakage also occurs. The test circuit for measuring the one-input current is shown in Fig. 8-7(b). The maximum one-input current for the TTL *NAND* gate is 40 μA.

Power Requirements. Power consumption is usually specified in terms of steady-state power. It is also a function of frequency and duty cycle. The

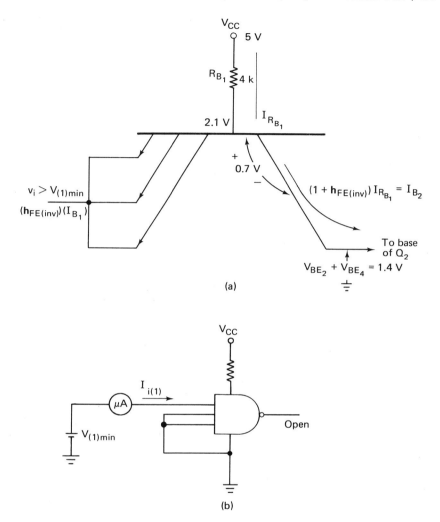

Fig. 8-7 (a) The inverse transistor. The emitters become the collector and the collector becomes the emitter. (b) The test circuit for measuring the 1-input current.

dynamic power, which is often expressed in milliwatts for a 50 percent duty cycle, is greater because of the current spikes that occur in the output circuit during transitions. Worst-case power consumption takes place when resistance values are at their lower limit. It is measured at 25° C with V_{cc} at its normal value. The curve of Fig. 8-8(a) shows how power consumption increases as frequency increases.

The test circuit for measuring the 1- and 0-level power consumptions

216 / *Transistor-Transistor Logic TTL OR T²L with Output Circuit Variations*

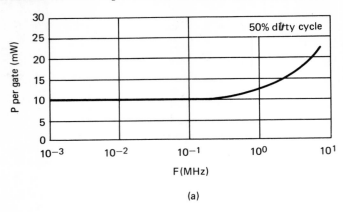

(a)

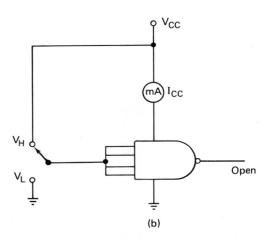

(b)

Fig. 8-8 (a) Gate power consumption as a function of frequency. (b) The test circuit for measuring the 1-level and 0-level power consumptions.

of the gate is shown in Fig. 8-8(b). The power supply current is measured with all inputs first at V_L and then at V_H, with the output open. The 1- and 0-state power consumptions calculated in the following example are for the worst-case load conditions.

Example 8-4: Calculate the power requirements for the circuit of Fig. 8-9 under worst-case load conditions. All inputs equal $V_H = 5$ V when Q_1 is cut off and one or more inputs equal $V_L = 0$ V when Q_1 is on. Use $V_{BE_{sat}} = 0.7$ V, $V_{CE_{sat}} = 0.3$ V, $V_{BC_1} = 0.7$ V, $V_{o(1)} = 3.56$ V, $V_{o(0)} = 0.3$ V, $FO = 10$, and $I_{i(1)} = 40$ μA (this is the maximum input leakage current drawn by each of the loads when this circuit applies a 1-level voltage).

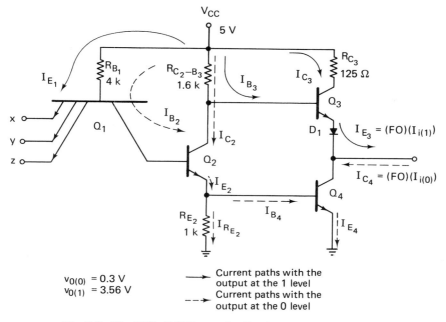

$V_{0(0)} = 0.3$ V
$V_{0(1)} = 3.56$ V

→ Current paths with the output at the 1 level
--→ Current paths with the output at the 0 level

Fig. 8-9 The TTL *NAND* gate for Ex. 8-4. All inputs are high when I_{B_2} flows. One or more inputs are low when I_{E_1} flows.

Solution: The currents shown in solid lines in Fig. 8-9 flow when the output is at the logic 1 level. The V_{CC} source supplies I_{E_1}, I_{B_3}, and I_{C_3}. I_{E_1} and I_{B_3} were calculated in Ex. 8-2.

$$I_{E_1} = \frac{V_{CC} - (V_{BE_1} + V_{i(0)})}{R_{B_1}} = 1.075 \text{ mA}$$

$$I_{B_3} = \frac{V_{CC} - (V_{BE_3} + V_{D_1} + V_{o(1)})}{R_{B_3}} = 25 \text{ μA}$$

With a $FO = 10$ and each load drawing a leakage current equal to $I_{i(1)} = 40$ μA, the current through the emitter of Q_3 is

$$I_{E_3} = I_{L(1)max} = (FO)(I_{i(1)})$$
$$= (10)(40 \text{ μA})$$
$$I_{E_3} = I_{L(1)max} = 400 \text{ μA}$$

The collector current of Q_3 is

$$I_{C_3} = I_{E_3} - I_{B_3}$$
$$= 400 \text{ μA} - 25 \text{ μA}$$
$$I_{C_3} = 375 \text{ μA}$$

Hence the 1-state power supply current is

$$I_{CC(1)} = I_{E_1} + I_{B_3} + I_{C_3}$$
$$= 1.075 + 0.025 + 0.375$$
$$I_{CC(1)} = 1.475 \text{ mA}$$

and the 1-state power dissipation is

$$P_{CC(1)} = (I_{CC(1)})(V_{CC})$$
$$= (1.475 \text{ mA})(5)$$
$$P_{CC(1)} = 7.375 \text{ mW}$$

The currents shown in broken lines in Fig. 8-9 flow when the output is at the logic 0 level. The V_{CC} source supplies only I_{B_2} and I_{C_2} which were calculated in Ex. 8-1.

$$I_{B_2} = \frac{V_{CC} - (V_{BC_1} + V_{BE_2} + V_{BE_4})}{R_{B_1}} = 0.725 \text{ mA}$$

$$I_{C_2} = \frac{V_{CC} - (V_{CE} + V_{BE})}{R_{C_2}} = 2.5 \text{ mA}$$

Hence, the 0-state power supply current is

$$I_{CC(0)} = I_{B_2} + I_{C_2}$$
$$= 0.725 + 2.5$$
$$I_{CC(0)} = 3.225 \text{ mA}$$

and the 0-state power dissipation is

$$P_{CC(0)} = (I_{CC(0)})(V_{CC})$$
$$= (3.225 \text{ mA})(5 \text{ V})$$
$$P_{CC(0)} = 16.125 \text{ mA}$$

If a 50-percent duty cycle is assumed, the average power consumption is

$$P_{CC(av)} = \frac{P_{CC(1)} + P_{CC(0)}}{2}$$
$$= \frac{7.375 + 16.125}{2}$$
$$P_{CC(av)} = 11.75 \text{ mW}$$

Output Short-circuit Current $I_{o(sc)}$. The output short-circuit current of the TTL *NAND* gate ranges from 18 mA to 55 mA. The test circuit of Fig. 8-10(a) shows how $I_{o(sc)}$ is measured. The maximum value is important because it guarantees that the gate will not be damaged if the output terminal is accidently short-circuited (when the emitter follower Q_3 is on) or if heavy capacitance loads are connected. The minimum value is the least amount of current that the output can supply to charge the load capacitance during the 0-to-1 transition.

The 1-level output current depends on the output impedance and voltage. R_{C_3} is added to limit the maximum current to an amount that will not damage Q_3. The impedance seen looking into the output terminal when Q_3 is on is approximately 160 Ω. And the no-load output voltage is approximately 3.6 V. An equivalent circuit is shown in Fig. 8-10(b). A typical value of output short-circuit current can be calculated from this circuit

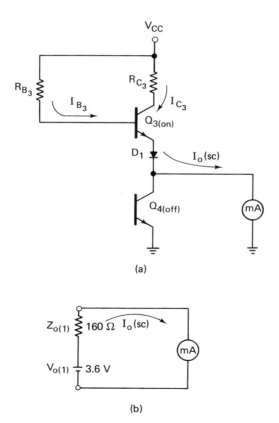

Fig. 8-10 (a) The test circuit for measuring the short-circuited output current. (b) An equivalent of the output circuit.

$$I_{o(sc)} = \frac{V_{o(1)}}{Z_{o(1)}}$$

$$\approx \frac{3.6}{160}$$

$$I_{o(sc)} \approx 22.5 \text{ mA}$$

Noise Immunity. The logic 0-noise immunity $N_{(0)}$ of the TTL *NAND* gate is the difference between the lower acceptance level V_y and the actual output of the driver. The lower acceptance level can be determined from Fig. 8-11(a). If V_{BE_4} equals the cutin voltage $V_\gamma = 0.5$ V, Q_4 is just off and the output will still be at the 1 level. The Q_1 input that will produce $V_{BE_4} = 0.5$ V is the lower acceptance level V_y. From Fig. 8-11(a)

$$V_y = V_{EB_1} + V_{BC_1} + V_{BE_2} + V_{BE_4} \tag{8-4}$$
$$= -0.7 + 0.7 + 0.7 + 0.5$$
$$V_y = 1.2 \text{ V}$$

When the inverter output of the driver is saturated, it applies $V_{CE_{sat}}$ to this circuit. If a worst-case $V_{CE_{sat}} = 0.4$ V is assumed

$$N_{(0)} = V_y - V_{(0)\max} \tag{8-5}$$
$$= 1.2 - 0.4$$
$$N_{(0)} = 0.8 \text{ V}$$

As temperature increases, the base-emitter voltage required to saturate the transistors decreases. Therefore, the lower acceptance level and the 0-noise margin will be less than 0.8 V at higher temperatures.

The logic 1-noise immunity is the difference between the upper acceptance level V_x and the actual output of the driver. The upper acceptance level can be determined from Fig. 8-11(b). Q_1 is just off with $V_{BE_1} = V_\gamma = 0.5$ V and Q_4 is saturated with $V_{BE_4} = 0.7$ V. With Q_4 saturated, the output will be at the 0 level. The Q_1 input that will produce $V_{BE_1} = V_\gamma = 0.5$ V is the upper acceptance level V_x. From Fig. 8-11(b)

$$V_x = V_{EB_1} + V_{BC_1} + V_{BE_2} + V_{BE_4} \tag{8-6}$$
$$= -0.5 + 0.7 + 0.7 + 0.7$$
$$V_x = 1.6 \text{ V}$$

The no-load 1 output of the driver is approximately equal to 3.9 V. However, the actual output is lower with a load connected. Under the worst-

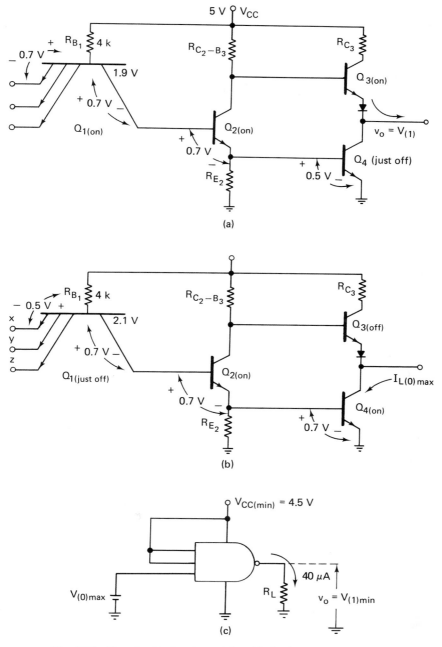

Fig. 8-11 The circuits for determining (a) the lower acceptance level V_y and (b) the upper acceptance level V_x. (c) The worst-case test circuit for measuring the logic 1-output voltage. R_L is the equivalent of a full fan-out of 10 drawing 10 times the 1-input current (40 μA leakage) of each load. (Courtesy Sprague Electric Co.)

case test conditions, shown in Fig. 8-11(c), the logic 1 output is at least 2.4 V. Hence, the 1-level noise margin is

$$N_{(1)} = V_{(1)min} - V_x \qquad (8\text{-}7)$$
$$= 2.4 - 1.6$$
$$N_{(1)} = 0.8 \text{ V}$$

The transfer characteristic for this circuit is shown in Fig. 8-12. At point 1, Q_4 is just off and the output is at the 1 level. At point 2, Q_4 is just saturated and the output is at the 0 level.

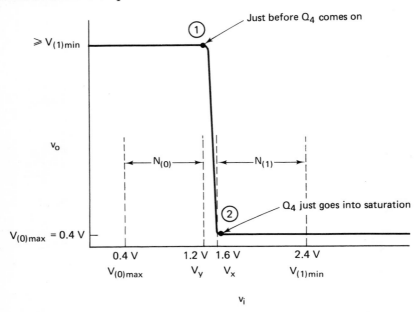

Fig. 8-12 The transfer characteristic for the circuit of Fig. 8-11 showing noise immunity.

8-3 Transient Response of the TTL NAND Gate

In order to see why TTL is faster than DTL, consider first the DTL circuit shown again in Fig. 8-13. When all inputs are at V_H, the *AND* gate diodes are cut off and a current flows from V_{CC} through R_A and the offset diodes D_4 and D_5. It then divides, with part of the current flowing through R_B and the remainder flowing into the base. This sends Q_1 into saturation and produces a 0-level output. The shunt capacitance C_{cs} (the capacitance between the collector that serves as the common anode of the input diodes and the sub-

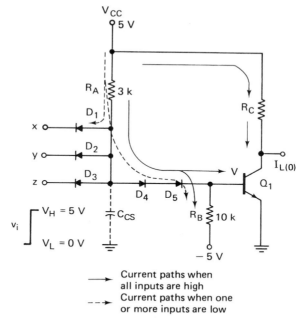

Fig. 8-13 The DTL *NAND* gate of Fig. 8-2(b).

strate) will be charged to $V_{D_4} + V_{D_5} + V_{BE_1} = 2.1$ V. In addition, there will be charge stored in the base of the saturated transistor.

If one of the inputs then goes low, the diode with the low input conducts. The *AND* gate output should go low and the base current to Q_1 should be removed. However, the charge stored in the base and in C_{cs} temporarily holds Q_1 on and slows down the 0-to-1 transition at the output. The current required by the conducting input diode is supplied at first by both V_{CC} and the stored charge. When the excess charge is removed, all of the input current is supplied by V_{CC}.

Now consider the TTL *NAND* gate, shown again in Fig. 8-14. When all inputs are at the 1 level, the three emitters are cut off and a current I_{B_2} flows from V_{CC} through the base-collector junction of Q_1 and into the base of Q_2. This current turns Q_2 on and part of its emitter current flows through R_{E_2}, developing a voltage $I_{E_2} R_{E_2} = 0.7$ V. A base current $I_{B_4} = I_{E_2} - I_{R_{E(2)}}$ saturates Q_4 and sends the output to the 0 level. The total shunt capacitance C_T, which consists of C_{cs} and the input capacitance C_i of Q_2, charges to $V_{BE_2} + V_{BE_4} = 1.4$ V.

If one or more of the emitter inputs then go low, Q_1 should go into saturation and its output should drop to $V_{CE_{sat}} = 0.3$ V. This will cut off Q_2 and remove the base current to Q_4, cutting it off and sending the output to the 1 level. However, it takes time to remove the charge stored in the base of

224 / Transistor-Transistor Logic TTL OR T²L with Output Circuit Variations

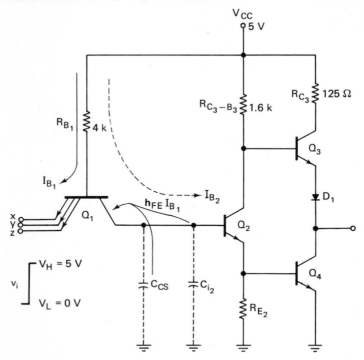

Fig. 8-14 The TTL *NAND* gate. I_{B_2} (in broken lines) flows only when all inputs are high. The current $h_{FE} I_{B_1}$ flows only during the time Q_2 is in the process of turning off.

Q_2 and to discharge C_T from 1.4 V to $V_{CE_{sat}} = 0.3$ V. This charge will temporarily hold Q_2 on, which keeps Q_4 on and the output at the 0 level.

The improvement in turn-off time occurs because Q_1 requires a large collector current $I_{C_1} = \mathbf{h}_{FE} I_{B_1}$ when it saturates, and this current can only be supplied by C_{CS} and the base of Q_2. Although a capacitor can supply a high amount of current, it can do so only for a limited period of time. The low resistance presented by Q_1 (the collector-emitter saturation resistance $R_{CE_{sat}}$ is typically 5–20 Ω) will quickly remove the base charge and discharge C_{CS} to 0.3 V. The result is a rapid turn-off of Q_2. The charge stored in the base of Q_4 is quickly removed by R_{E_2} and a fast 0-to-1 trensition is effected. Since Q_1 removes the charge from the base of Q_2, there is no need for a base pull-down resistor for Q_2. A typical value of turn-off time for Q_2 is calculated in the following example.

Example 8-5: Given typical values of total shunt capacitance $C_T = C_{CS} + C_{i_2} = 50$ pF, $\mathbf{h}_{FE_1} = 20$, $V_{CE_{sat}} = 0.3$ V, and $V_{BE_{sat}} = 0.7$ V, calculate the turn-off time of Q_2 in Fig. 8-14. Use $V_i = V_L = 0$ V.

Transient Response of the TTL NAND Gate | 225

Solution: The base current of Q_1 is supplied by V_{CC} and equals

$$I_{B_1} = \frac{V_{CC} - (V_{BE_{sat}} + V_i)}{R_{B_1}}$$

$$= \frac{5 - (0.7 + 0)}{4 \text{ k}}$$

$$I_{B_1} = 1.075 \text{ mA}$$

Hence, the collector of Q_1 can draw

$$I_{C_1} = h_{FE}I_{B_1}$$
$$= 20 \times 1.075$$
$$I_{C_1} = 21.5 \text{ mA}$$

Since h_{FE} and I_{B_1} are fixed, this is a constant current. It can be supplied only for a limited time by C_{CS} and C_{i_2}. From $I_{C_T} = C_T(\Delta V_{C_T}/\Delta t)$

$$\Delta t = C_T \frac{\Delta V_{C_T}}{I_{C_T}} \qquad (8\text{-}8)$$

where Δt is the turn-off time of Q_2, $I_{C_T} = I_{C_1} = h_{FE}I_{B_1}$, and ΔV_{C_T} is the change in V_{C_T} from $V_{BE_2} + V_{BE_4}$ when Q_1 is off to $V_{CE_{sat}}$ when Q_1 is on.
Therefore,

$$t_{off} = 5 \times 10^{-11} \frac{1.4 - 0.3}{2.15 \times 10^{-2}}$$

$$t_{off} = 2.56 \text{ ns}$$

Higher values of h_{FE} increase I_{C_1} and speed up switching.

When C_{CS} is discharged to $V_{CE_{sat}} = 0.3$ V and Q_2 is cut off, I_{C_1} will be only the minute leakage current I_{CBO_2}. The input emitter current is then supplied entirely by V_{CC} and equals I_{B_1}. If more than one input is at the 0 level, the input current divides among these emitters.

Because of the action of the multi-emitter transistor, the phase splitter Q_2 does not appreciably add to the circuit delay. And, as explained in Sec. 8-2, the totem-pole output provides a low impedance output to both sink and source loads to produce very short RC time constants for both the 0-to-1 and 1-to-0 transitions (see also Sec. 8-9).

The values of R_{B_1}, $R_{C_2\text{-}B_3}$, and R_{E_2} affect switching speeds. Lower values of R_{B_1} increase I_{B_1} and I_{C_1} and permit Q_1 to sweep the stored charge out of the base of Q_2 more quickly. The higher I_{B_1} also increases the unit load current to the driver which decreases its fan-out. When Q_i is off, the value of R_{B_3} determines the base current to Q_3. A low value of R_{B_3} increases the base drive and lowers the turn-on time of Q_3. Q_2, when it comes on, draws current away from the base of Q_3, and also quickly removes the charge stored in its base, resulting in a fast turn-off of the emitter follower. When Q_2 is on, a low value of R_{C_2} increases the current through R_{E_2}, which reduces the time it takes for the turn-on V_{BE} of Q_4 to be reached. A high value of R_{E_2} has the same effect. However, since R_{E_2} also serves as the path for removal of the charged stored in the base of Q_4, a large value of R_{E_2} will increase the Q_4 turn-off time. In short, lower values of resistance increase the overall speed. However, this improvement is not without limitations; the higher currents result in greater power consumption.

While the phase splitter and the totem-pole output are also used in DTL power gates and buffers, their use, in conjunction with the multi-emitter transistor, produces the very high speeds of TTL.

Input Clamping Diodes. Many TTL gates are designed with input clamping diodes. This network, shown in Fig. 8-15(a), prevents excessive *ringing* at the input. The clamping diodes are cut off in the steady state because only positive inputs are applied ($V_{i(0)} = V_{CE_{sat}} \approx 0.2$ V, $V_{i(1)} \geq 2.4$ V). However, during the very rapid rise and fall times they conduct to damp out unwanted oscillations at the input. The problem is more severe on the 1-to-0 transition where an undershoot of more than 2 V is possible. Figure 8-15(b) shows an unclamped input with an unused input connected to V_{CC}, as recommended. The other input has undergone a 1-to-0 transition and has a -2 V ringing pulse on it. The transistor turns on and V_B goes to -1.3 V. This makes the voltage across the reverse-biased junction equal 6.3 V, which can cause that emitter to break down. The excessive current that will be drawn by this emitter creates undesired noise.

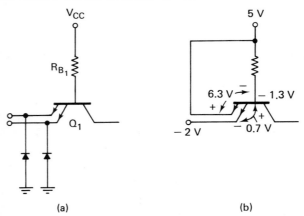

Fig. 8-15 (a) The input clamping network used with TTL gates. (b) An unclamped TTL gate with an unused input tied to V_{CC} and -2 V on the input.

Another problem occurs on the second half-cycle of oscillation. The overshoot that follows the negative excursion can turn the transistor off, which sends the output momentarily to the 0 level.

The ringing diodes prevent breakdown of any reverse-biased emitters by limiting the amplitude of the undershoot. If the input tries to go below -0.7 V, the diode conducts and clamps the input at this level. When on, they draw energy from the line, which limits the amplitude of the following overshoot to a level that cannot turn Q_1 off.

8-4 The TTL NOR Gate

A two-input TTL *NOR* gate is shown in Fig. 8-16(a) and its state chart is given in Fig. 8-16(b). A totem-pole output is used with two phase-splitter transistors Q_{2A} and Q_{2B}. If either Q_{2A} or Q_{2B} is turned on, the base current to Q_3 is removed. The emitter currents of the phase splitters flow through R_4 and raise the base voltage of Q_4 to $V_{BE_{sat}}$. The output then goes to the 0 level. Base current can flow into Q_3 only when both Q_{2A} and Q_{2B} are cut off. This also removes the forward bias for the base-emitter junction of Q_4, which cuts it off. The output then goes to the 1 level.

As shown in the state chart, the two phase splitters are cut off only when both inputs A and B are low. Then Q_{1A} and Q_{1B} both conduct and remove the base currents to Q_{2A} and Q_{2B}. The positive-logic truth table is shown in Fig. 8-16(c). Since the output is high only when both inputs are low, $fz = \bar{A}\bar{B} = \overline{A + B}$.

The number of inputs is increased by adding other input transistors and phase splitters. The collectors of all the phase splitters are tied together as are the emitters. The circuit currents, for both the 1 and 0 states, are the same as those calculated for the *NAND* gate in Exs. 8-1 and 8-2. The fan-out also remains the same, but the power consumption is higher. In the *NAND* gate circuit shown in Fig. 8-4, a current $I_{E_1} = 1.075$ mA flows with one input low and the others high. If other inputs go low, this current doesn't change; it simply divides, with each low emitter drawing an equal share. In the TTL *NOR* gate an additional $I_{E_1} = V_{CC} - (V_{BE_1} + V_i)/R_1$ flows when both inputs are at V_L. This current was calculated in Ex. 8-4 to be 1.075 mA. The total 1-state supply current of the *NOR* gate is

$$I_{CC(1)} = I_{E_{1A}} + I_{E_{1B}} + I_{B_3} + I_{C_3}$$

The values of $I_{B_3} = 25$ μA and I_{C_3} equal to 0.375 mA were also calculated in Ex. 8-4. Hence,

$$I_{CC(1)} = 1.075 + 1.075 + 0.025 + 0.375$$
$$I_{CC(1)} = 2.55 \text{ mA}$$

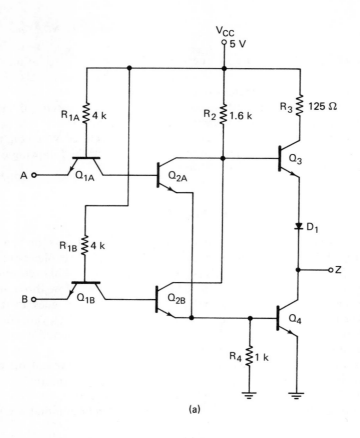

B	A	Q_{1A}	Q_{1B}	Q_{2A}	Q_{2B}	Q_3	Q_4	Z
V_L	V_L	On	On	Off	Off	On	Off	$V_{(1)}$
V_L	V_H	Off	On	On	Off	Off	On	$V_{(0)}$
V_H	V_L	On	Off	Off	On	Off	On	$V_{(0)}$
V_H	V_H	Off	Off	On	On	Off	On	$V_{(0)}$

(b)

B	A	Z
0	0	1
0	1	0
1	0	0
1	1	0

(c)

Fig. 8-16 (a) A two-input positive-logic TTL NOR gate with (b) its state chart and (c) its truth table. $f_z = \overline{AB} = \overline{A+B}$.

and the 1-state power dissipation is

$$P_{CC(1)} = I_{CC(1)}V_{CC}$$
$$= (2.55 \text{ mA})(5)$$
$$P_{CC(1)} = 12.75 \text{ mW}$$

The 0-state supply current doesn't change. From Ex. 8-4, $I_{CC(0)} = 3.225$ mA, so the 0-state power dissipation is still 16.125 mW. Therefore, the average power dissipation of the *NOR* gate is

$$P_{CC(av)} = \frac{P_{CC(1)} + P_{CC(0)}}{2}$$
$$= \frac{12.75 + 16.125}{2}$$
$$I_{CC(av)} = 14.44 \text{ mW}$$

The average power dissipation of the *NAND* gate is 11.75 mW.

8-5 The TTL Inverter

The TTL inverter circuit shown in Fig. 8-17 performs the *NOT* function. The circuit is the same as the TTL *NAND* gate except for the number of inputs to Q_1. The inverter has only one input, which is inverted at the output. The propagation delay, average power dissipation, and fan-out are the same as in the *NAND* gate.

If the Q_1 input is at the 0 level, Q_1 saturates and removes the base drive to Q_2. When Q_2 goes off, the base drive to Q_4 is removed, so it cuts off. A current flows from V_{CC} through R_2 and into the base of Q_3, turning it on. With Q_3 on and Q_4 off, the output is at the logic 1-voltage level.

If the Q_1 input is at the 1 level, the emitter of Q_1 cuts off and a current flows from V_{CC} through R_1 and the *B-C* junction of Q_1 and into the base of Q_2. Q_2 conducts and removes the base drive to Q_3, cutting it off. At the same time, the emitter current of Q_2 develops a voltage across R_4 that turns Q_4 on. With Q_3 off and Q_4 on, the output is at the logic 0-voltage level.

Since the output is low when the input is high and it is high when the input is low, the *NOT* function is performed.

The emitter-follower portion of the totem-pole output is omitted in some inverters. The open-collector output increases the 0-to-1 transition time. As a result, the propagation delay increases to 15 ns. The inverter with the open collector can be used in wired-logic applications (see Sec. 7-5).

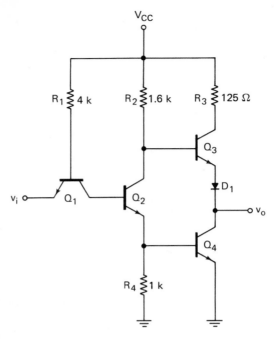

Fig. 8-17 The TTL inverter circuit. (Courtesy Sprague Electric Co.)

8-6 The TTL AND Gate

A two-input positive-logic *AND* gate is shown in Fig. 8-18(a). The circuit is basically the same as the *NAND* gate except for the addition of an internal inverter and the omission of the emitter-follower half of the output circuit. The open-collector output permits its use in wired logic. Another version of the TTL *AND* gate has a totem-pole output with a Darlington amplifier used in the emitter-follower half of the output. This output configuration is one of the circuits explained in Sec. 8-9.

The state chart for the *AND* gate is given in Fig. 8-18(b). If one or both inputs are low, Q_1 conducts and removes the base drive to Q_2. This cuts Q_2 off and removes the forward bias for Q_3 (developed across R_3 by I_{E_2}), which turns it off. At the same time, current flows from V_{CC} through R_2 and D_1 and into the base of Q_4, turning it on. The emitter current of Q_4 develops a voltage across R_5 that saturates Q_5 and sends the output to the 0 level.

When both inputs are high, Q_1 cuts off and a current flows from V_{CC} through R_1 and the B-C junction of Q_1 and into the base of Q_2. When Q_2 conducts, the base drive to Q_4 is removed, cutting it off. This removes the base drive to Q_5, which cuts it off and sends the output to the 1 level. When

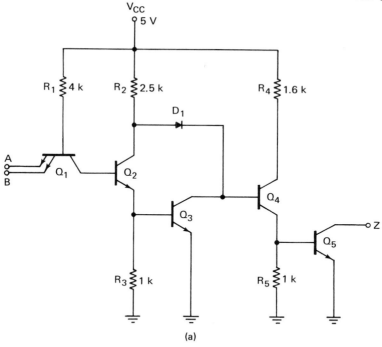

Fig. 8-18 (a) A two-input positive-logic TTL AND gate with (b) its state chart and (c) its truth table.

B	A	Q_1	Q_2	Q_3	Q_4	Q_5	Z
V_L	V_L	On	Off	Off	On	On	$V_{(0)}$
V_L	V_H	On	Off	Off	On	On	$V_{(0)}$
V_H	V_L	On	Off	Off	On	On	$V_{(0)}$
V_H	V_H	Off	On	On	Off	Off	$V_{(1)}$

(b)

B	A	Z
0	0	0
0	1	0
1	0	0
1	1	1

(c)

Q_2 comes on, a voltage is developed across R_3 that turns Q_3 on. The low resistance of Q_3 quickly removes the charge stored in the base of Q_4 and this causes it to turn off rapidly. The diode D_1 is used to guarantee that Q_4 and Q_5 remain off when Q_2 is on. The voltage at the collector of $Q_2(V_{C_2} = V_{CE_2} +$

V_{BE_5}) will always be less than the voltage $(V_{D_1} + V_{BE_4} + V_{BE_5} \approx 2.1 \text{ V})$ required to turn Q_5 on.

The truth table for the AND gate is given in Fig. 8-18(c). The output is high only when both inputs are high simultaneously; thus, $f_Z = AB$.

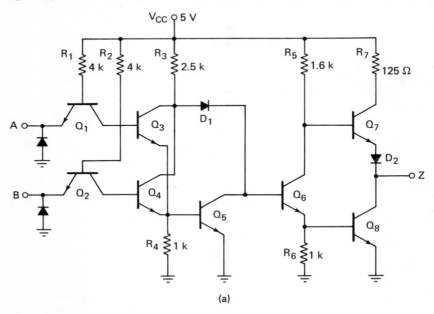

B	A	Q_1	Q_2	Q_3	Q_4	Q_5	Q_6	Q_7	Q_8	Z
V_L	V_L	On	On	Off	Off	Off	On	Off	On	$V_{(0)}$
V_L	V_H	Off	On	On	Off	On	Off	On	Off	$V_{(1)}$
V_H	V_L	On	Off	Off	On	On	Off	On	Off	$V_{(1)}$
V_H	V_H	Off	Off	On	On	On	Off	On	Off	$V_{(1)}$

(b)

B	A	Z
0	0	0
0	1	1
1	0	1
1	1	1

(c)

Fig. 8-19 (a) A two-input positive-logic TTL OR gate (Courtesy Sprague Electric Co.) with (b) its state chart and (c) its truth table. $f_z = A\bar{B} + \bar{A}B + AB = A + B$.

8-7 The TTL OR Gate

A two-input positive-logic OR gate is shown in Fig. 8-19(a). The circuit is basically the same as the NOR gate except for the addition of an internal inverter.

The state chart for the OR gate is given in Fig. 8-19(b). When both inputs are low, both Q_1 and Q_2 saturate. When Q_1 conducts, it turns off Q_3, and when Q_2 conducts, it turns off Q_4. With both Q_3 and Q_4 cut off, the forward bias for Q_5 is removed (it is developed by $I_{E_{3-4}}$ through R_4). At the same time, current flows from V_{CC} through R_3 and D_1 and into the base of Q_6, turning it on. The emitter current of Q_6 develops a voltage across R_6 that saturates Q_8 and sends the output to the 0 level. When Q_6 comes on, it removes the base drive to Q_7, cutting it off.

When either of the inputs goes high, the emitter with the high inputs cuts off. If A is high, the B-C junction of Q_1 conducts and turns Q_3 on. If B is high, the B-C junction of Q_2 conducts and turns Q_4 on. If the two inputs are high, both B-C junctions conduct, turning Q_3 and Q_4 on simultaneously. In any case, the base drive to Q_6 is removed, cutting it off. At the same time, the emitter current of Q_3 and/or Q_4 through R_4 develops a voltage that saturates Q_5. The low resistance of Q_5 quickly removes the charge stored in the base of Q_6, which causes it to turn off rapidly. When Q_6 cuts off, the base current to Q_8 is removed, so it cuts off and the output goes to the 1 level. A current flows from V_{CC} through R_5 and into the base of Q_7, which turns it on. The offset diodes serve the same purpose as in the AND and $NAND$ gates previously explained. D_1 ensures that the voltage $V_{C_{3-4}}$ cannot turn on Q_6 and Q_8 when Q_3 and/or Q_4 conduct. D_2 guarantees that the voltage V_{C_6} cannot turn Q_7 on when Q_6 conducts.

The truth table for the OR gate is given in Fig. 8-19(c). The output is high if one or the other input is high or if both inputs are high simultaneously. Therefore, the $fz = A\bar{B} + \bar{A}B + AB = A + B$.

8-8 The AND-OR-INVERT Circuit

The circuit shown in Fig. 8-20(a) is an AND-OR-$INVERT$ AOI gate. It is a two-wide, two-input circuit capable of implementing the $EXCLUSIVE$-OR function. Nodes E and C are brought out from the collector and emitter of the phase-splitter stage to allow expansion of the AND input. The AOI gate consists of two two-input AND gates driving a NOR gate. The logic diagram is shown in Fig. 8-20(b). The electrical characteristics are the same as those of the NOR gate: The propagation delay is 10 ns, the average power dissipation is 14 mW, and the fan-out is 10. With the inputs shown in Fig. 8-20, the

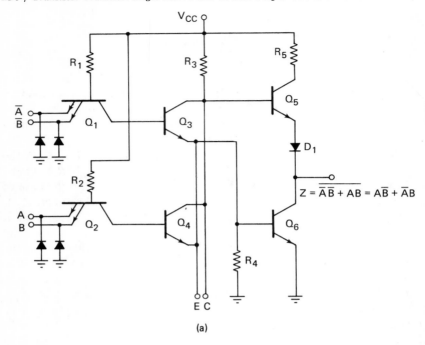

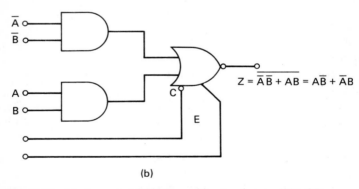

Fig. 8-20 (a) An expandable two-wide, two-input *AND-OR-INVERT AOI* gate with inputs that produce the *EXCLUSIVE-OR* function. (b) The logic diagram of (a). (Courtesy Sprague Electric Co.)

EXCLUSIVE-OR function $A\bar{B} + \bar{A}B$ is produced at z. The operation of the curcuit with these inputs will be explained with the aid of the state chart given in Table 8-1(a).

If either Q_3 or Q_4 conducts, the base current to Q_5 is removed and it cuts off. The emitter current of Q_3 and/or Q_4 develops a voltage across R_4

Table 8-1 (a) The state chart for the *AOI* gate of Fig. 8-20 and (b) its positive-logic truth table.

B	A	Q_1	Q_2	Q_3	Q_4	Q_5	Q_6	Z
V_L	V_L	Off	On	On	Off	Off	On	$V_{(0)}$
V_L	V_H	On	On	Off	Off	On	Off	$V_{(1)}$
V_H	V_L	On	On	Off	Off	On	Off	$V_{(1)}$
V_H	V_H	On	Off	Off	On	Off	On	$V_{(0)}$

(a)

B	A	Z
0	0	0
0	1	1
1	0	1
1	1	0

(b)

that saturates Q_6 and sends the output to the 0 level. Q_5 conducts only if both Q_3 and Q_4 are cut off. Current then flows from V_{CC} through R_3 and into the base of Q_5. At the same time, the voltage across R_4 is removed and Q_6 cuts off. The output then goes to the logic-1 voltage level.

When both *A* and *B* are low, the two emitters of Q_2 conduct. When Q_2 saturates, it removes the base drive to Q_4, cutting it off. Simultaneously, the two inputs to Q_1 are high and Q_1 cuts off. A current then flows from V_{CC} through R_1 and the B-C junction of Q_1 and into the base of Q_3. When Q_3 comes on, it turns Q_5 off and turns Q_6 on, which sends the output to the 0 level.

When both inputs are high, Q_1 saturates because $\bar{A}$ and $\bar{B}$ are low. This removes the base current to Q_3, cutting it off. At the same time, Q_2 cuts off and a current flows from V_{CC} through R_2 and the B-C junction of Q_2 and into the base of Q_4. When Q_4 saturates, Q_5 cuts off. Q_6 comes on, and the output goes to the 0 level.

If only one of the inputs is high and the other is low one input to Q_1 and one input to Q_2 will be low. Both input transistors then saturate and both phase splitters Q_3 and Q_4 cut off. This causes Q_5 to come on and Q_6 to cut off, which sends the output to the 1 level.

The positive-logic truth table is shown in Table 8-1(b). Since *z* is a logic 1 only if one of the inputs is high, $f_z = A\bar{B} + \bar{A}B$. If *A* and $\bar{B}$ are applied to Q_1 and $\bar{A}$ and *B* are applied to Q_2 the *EQUIVALENCE* function $(AB + \bar{A}\bar{B})$ will be implemented.

The AND Expander. A four-input expander of the *AND* input of the *AOI* gate is shown in Fig. 8-21. The *C* and *E* output terminals are connected to the *C* and *E* expander inputs of Fig. 8-20. The expander has a propagation

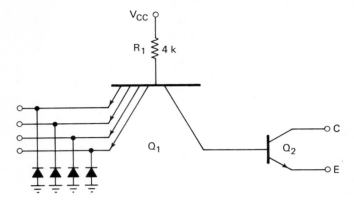

Fig. 8-21 A 4-input *AND* expander

delay of 10 ns and an average power consumption of 4 mW. The number of *AND* expanders permitted is determined by the 0-input threshold voltage considerations. The current sharing in the emitters of the phase splitters reduces the 0-input threshold voltage by approximately 0.02 V per expander. The manufacturer states that as many as four expanders can be used, while still maintaining the dc 0-level noise margin $N_{(0)}$.

8-9 Output Circuit Variations

There are several types of output circuits used in digital circuits. The load on these circuits always includes the input capacitance of the driven circuits. The output circuit serves dual functions: (1) It must act as a current source capable of supplying the high current required to quickly charge the load capacitance. (2) It must act as a current sink capable of sinking the high current that flows when the load capacitance quickly discharges

There are three basic output configurations: the saturated inverter, the emitter follower, and the totem-pole circuit. Only the transient response of these circuits is discussed in this section. A static analysis has already been given for each, whenever they have been encountered.

The Unclamped Saturated Inverter Output. The first output circuit that will be discussed is the simple unclamped saturated inverter, shown in Fig. 8-22 with its input and output waveforms. In order not to complicate the problem by requiring unneccessary calculations, a purely capacitive load is assumed and the transistor delays (see Sec. 2-5) are neglected. With this, the rise and fall times can easily be calculated and the three basic output circuits can be compared.

In Fig. 8-22 assume that the input voltage equals V_L for a long time (greater than 5 τ). The transistor is cut off and C_L has charged through the

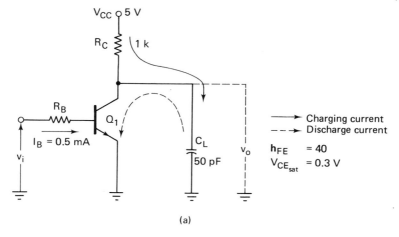

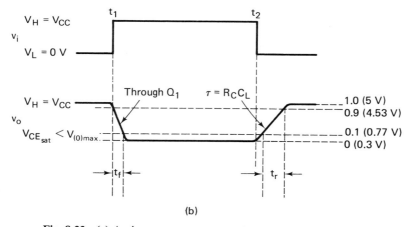

Fig. 8-22 (a) An inverter output stage with a capacitive load and (b) its input and output waveforms. The transistor delays are neglected.

path shown to $V_{CC} = V_H$. If the input now goes high, Q_1 goes into saturation. The capacitor then sees only $V_{CE_{sat}} < V_{(0)max}$. It quickly discharges through the low resistance of the transistor and v_o falls rapidly to $V_{CE_{sat}}$.

When the input again goes low, Q_1 cuts off and C_L must charge back to V_{CC} through R_C. The rise time can be reduced by using lower values of R_C but not without limitations. A low value of R_C causes a high current to flow through Q_1 when it conducts. This limits the amount of 0-output load current that Q_1 can sink and therefore increases fall time and reduces the dc fan-out. It also increases the power consumption of the gate. Typical values of rise and fall times are calculated in the following example.

Example 8-6: Calculate the rise time t_r and fall time t_f of the output signal of Fig. 8-22.

Solution: The fall time is determined first. From Fig. 8-22, v_i rises to V_H at $t_1 +$ and turns Q_1 on. When Q_1 goes into saturation, it acts as a current sink for the discharge current of C_L: The fall time is nearly linear because the npn inverter can sink a constant current equal to $I_{C_{max}}$. The capacitor discharge current

$$I_{C_L} = I_{C_{max}} - I_{R_C}$$

$$I_{C_{max}} = h_{FE}I_B$$

$$= 40 \times 0.5$$

$$I_{C_{max}} = 20 \text{ mA}$$

Part of this current flows through R_C. As shown in Fig. 8-22(b), the output voltage falls almost linearly from 4.53 V to 0.77 V during fall time. Therefore, the average voltage drop across R_C during this time is

$$V_{R_{C(av)}} = \frac{(V_{CC} - 4.53) + (V_{CC} - 0.77)}{2}$$

$$= \frac{0.47 + 4.23}{2}$$

$$V_{R_{C(av)}} = 2.35 \text{ V}$$

and the average current through R_C during fall time is

$$I_{R_{C(av)}} = \frac{V_{R_{C(av)}}}{R_C}$$

$$= \frac{2.35}{1}$$

$$I_{R_{C(av)}} = 2.35 \text{ mA}$$

The rest of the collector current is used to discharge C_L.

$$I_{C_L} = I_{C_{max}} - I_{R_{C(av)}}$$

$$= 20 - 2.35$$

$$I_{C_L} = 17.65 \text{ mA}$$

(*Note:* Don't confuse the collector current $I_{C_{max}}$ with the capacitor current I_{C_L}.)

As long as C_L is discharging, the collector current remains constant. As previously explained, a capacitor can supply any amount of current for a limited time. The greater the collector current, the greater the capacitor current and the faster C_L can discharge. Since the discharge current $I_{C_{max}} - I_{R_C}$ is almost constant, it may be assumed that C_L discharges linearly. Then, from $I = C\,(\Delta v_C/\Delta t)$,

$$\Delta t = C\frac{\Delta v_C}{I}$$

The total change in capacitor voltage is $V_{CC} - V_{CE_{sat}}$, but the fall time is the time that it takes the voltage to fall from the 0.9 to 0.1 levels shown in Fig. 8-22(b). Hence,

$$\Delta v_C = 0.8(V_{CC} - V_{CE_{sat}}) \tag{8-9}$$

and the fall time

$$t_f = C_L \frac{0.8(V_{CC} - V_{CE_{sat}})}{I_{C_L}} \tag{8-10}$$

$$= 5 \times 10^{-11} \frac{0.8(5 - 0.3)}{1.765 \times 10^{-2}}$$

$$t_f = 10.7 \text{ ns}$$

When C_L finishes discharging, V_{C_L} equals $V_{CE_{sat}}$ and remains there until $-t_2$ (just before v_i drops back to V_L).

At t_2+ (just after v_i drops to V_L), Q_1 cuts off and the load capacitance sees $V_{CC} = 5$ V. The actual rise time t_r begins at a point above $V_{CE_{sat}}$ by 10 percent of the difference between $V_H = V_{CC}$ and $V_{CE_{sat}}$. Therefore, the voltage across the load capacitance at the *start* of rise time is

$$^*V_{CS} = V_{CE_{sat}} + 0.1(V_{CC} - V_{CE_{sat}})$$
$$= V_{CE_{sat}} + 0.1 V_{CC} - 0.1 V_{CE_{sat}}$$
$$V_{CS} = 0.9 V_{CE_{sat}} + 0.1 V_{CC}$$

and the voltage across R_C at this time is

$$^*V_{RS} = V_{CC} - V_{CS} \tag{8-11}$$
$$= V_{CC} - (0.9 V_{CE_{sat}} + 0.1 V_{CC})$$
$$V_{RS} = 0.9(V_{CC} - V_{CE_{sat}})$$

By referring again to Fig. 8-22(b), it can be seen that the rise time ends at a point above $V_{CE_{sat}}$ by 90 percent of the difference between $V_H = V_{CC}$

---

*V_{RS} and V_{CS} are used in this text to represent the voltages across R and C at the start S of the particular charge (or discharge) interval.

and $V_{CE_{sat}}$. Therefore, the voltage across load capacitance at the *finish* of rise time is

$$*V_{CF} = V_{CE_{sat}} + 0.9(V_{CC} - V_{CE_{sat}})$$
$$= V_{CE_{sat}} + 0.9V_{CC} - 0.9V_{CE_{sat}}$$
$$V_{CF} = 0.9V_{CC} + 0.1V_{CE_{sat}}$$

and the voltage across the charging resistance R_C at this time is

$$*V_{RF} = V_{CC} - V_{CF}$$
$$= V_{CC} - (0.9V_{CC} + 0.1V_{CE_{sat}}) \quad (8\text{-}12)$$
$$V_{RF} = 0.1(V_{CC} - V_{CE_{sat}})$$

The rise time of the basic inverter can now be calculated by using Eq. (1-7)

$$t = 2.3RC \log \frac{V_{RS}}{V_{RF}} \quad (1\text{-}7)$$

where

$t =$ the rise time t_r;
$R =$ the charging resistance R_C;
$C =$ the load capacitance C_L;
$\frac{V_{RS}}{V_{RF}} =$ the ratio of V_{R_C} at the start of t_r to V_{R_C} at the finish of t_r.

From Eqs. (8-11) and (8-12), the ratio of V_{RS} to V_{RF} equals 9. Hence, the rise time equation can be simplified to

$$t_r = 2.2R_C C_L \quad (8\text{-}13)$$

Substituting the values of R_C and C_L yields

$$t_r = 2.2 \times 10_3 \times 5 \times 10^{-11}$$
$$t_r = 110 \text{ ns}$$

Although the values of t_r and t_f are affected by the transistor delays, the load voltage E_L and the load resistance R_L, the above example shows how the inverter output falls much more rapidly than it rises.

If a *pnp* transistor is used, the rise time is faster than the fall time. As shown in Fig. 8-23, the fall time is caused by C_L charging to $-V_{CC} = V_L$ through R_C, and the rise time is caused by C_L discharging to $V_{CE_{sat}} \approx V_H$.

*V_{RF} and V_{CF} are used in this text to represent the voltages across R and C at the *finish F* of the particular charge (or discharge) interval.

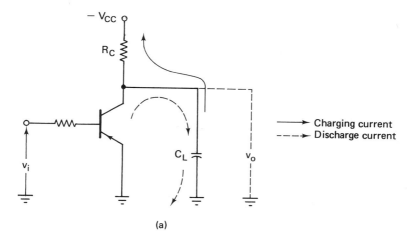

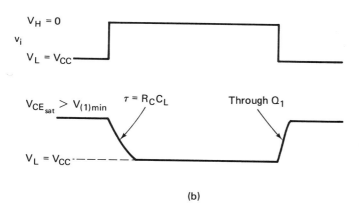

Fig. 8-23 (a) A *pnp* inverter with (b) its input and output waveforms.

through the low resistance of the saturated transistor. However, because of the greater mobility of the majority carriers in *npn* transistors (electrons), they are normally used in high-speed applications.

The Collector-clamped Saturated Inverter. By using a value of V_{CC} greater than V_H (V_L), and a *catching* or *clamping* diode to clamp the collector at V_H, rise (fall) time of the *npn* (*pnp*) saturated inverter can be improved. The circuit of Fig. 8-23 is repeated in Fig. 8-24(a) with the addition of the clamping diode D_1 and clamp voltage $V_{CL} = V_H$.

Its output voltage waveform is shown in Fig. 8-24(b). When the transistor is on, $v_o = V_{CE_{sat}}$ and the diode is reverse biased with $V_D = V_{CE_{sat}} - V_{CL} = -4.7$ V. When the transistor is cut off, the capacitor sees $V_{CC} = 10$ V.

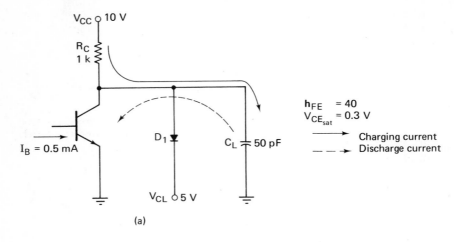

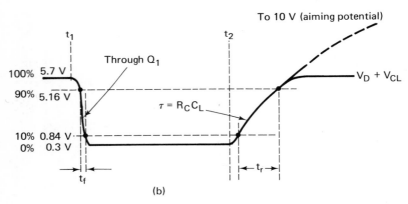

Fig. 8-24 (a) A saturated inverter with a collector clamping diode. (b) Its input and output waveforms.

It begins to charge from $V_{CE_{sat}}$ toward $V_{CC} > V_H$ but, when it exceeds $V_{CL} = V_H$, the diode comes on and clamps v_o at $V_{CL} + V_D = 5.7\text{ V} \approx V_H$.

When the transistor is again turned on, the capacitor quickly discharges through its low resistance to $V_{CE_{sat}}$. The diode cuts off as soon as v_o drops below $V_\gamma + V_{CL} = 5.5$ V. The rise and fall times for this circuit are calculated in the following example.

Example 8-7: Calculate the rise time t_r and the fall time t_f of the output signal shown in Fig. 8-24(b).
Solution: The fall time is determined first. From Fig. 8-24(b), v_o falls from 5.16 V to 0.84 V during the actual fall time. This change in voltage occurs while the capacitor discharges through the saturated transistor. As with the

unclamped inverter, it discharges almost linearly because the *npn* inverter can sink a constant current equal to $I_{C_{max}}$. The capacitor discharge current $I_{C_L} = I_{C_{max}} - I_{R_C}$

$$I_{C_{max}} = h_{FE}I_B$$
$$= 40 \times 0.5$$
$$I_{C_{max}} = 20 \text{ mA}$$

Part of this current flows through R_C. The average voltage across R_C during fall time is

$$V_{R_{C(av)}} = \frac{(V_{CC} - 0.84) + (V_{CC} - 5.16)}{2}$$
$$= \frac{14}{2}$$
$$V_{R_{C(av)}} = 7 \text{ V}$$

Therefore, the average current through R_C during this time is

$$I_{R_{C(av)}} = \frac{V_{R_{C(av)}}}{R_C}$$
$$= \frac{7}{1}$$
$$I_{R_{C(av)}} = 7 \text{ mA}$$

The rest of the collector current is used to discharge C_L

$$I_{C_L} = I_{C_{max}} - I_{R_{C(av)}}$$
$$= 20 - 7$$
$$I_{C_L} = 13 \text{ mA}$$

Hence, from $I = C(\Delta V_C/\Delta t)$,

$$t_f = C_L \frac{0.8[(V_D + V_{CL}) - V_{CE_{sat}}]}{I_{C_L}}$$
$$= 5 \times 10^{-11} \frac{0.8[(0.7 + 5) - 0.3]}{1.3 \times 10^{-2}}$$
$$t_f = 13.3 \text{ ns}$$

When C_L finishes discharging, V_{C_L} equals $V_{CE_{sat}}$ and remains there until $-t_2$ (just before Q_1 is cut off). At t_2+ (just after Q_1 cuts off), the load capacitance sees $V_{CC} = 10$ V, so it begins to charge. The actual rise time t_r begins at a point above $V_{CE_{sat}}$ by 10 percent of the difference between $V_D + V_{CL}$ and $V_{CE_{sat}}$. (*Note*: Don't confuse V_{CL}, the clamp supply voltage with V_{C_L}, the voltage across the load capacitance.) Therefore, the voltage across the C_L at the start of rise time is

$$V_{CS} = V_{CE_{sat}} + 0.1[(V_D + V_{CL}) - V_{CE_{sat}}]$$
$$= V_{CE_{sat}} + 0.1V_D + 0.1V_{CL} - 0.1V_{CE_{sat}}$$
$$V_{CS} = 0.9V_{CE_{sat}} + 0.1V_D + 0.1V_{CL}$$

and the voltage across the charging resistance R_C at this time is

$$V_{RS} = V_{CC} - V_{CS}$$
$$V_{RS} = V_{CC} - (0.9V_{CE_{sat}} + 0.1V_D + 0.1V_{CL}) \qquad (8\text{-}15)$$

By again referring to Fig. 8-24(b) it can be seen that the rise time ends at a point above $V_{CE_{sat}}$ by 90 percent of the difference between $V_D + V_{CL}$ and $V_{CE_{sat}}$. Therefore, the voltage across C_L at the finish of rise time is

$$V_{CF} = V_{CE_{sat}} + 0.9[(V_D + V_{CL}) - V_{CE_{sat}}]$$
$$= V_{CE_{sat}} + 0.9V_D + 0.9V_{CL} - 0.9V_{CE_{sat}}$$
$$V_{CF} = 0.1V_{CE_{sat}} + 0.9V_D + 0.9V_{CL}$$

and the voltage across R_C at this time is

$$V_{RF} = V_{CC} - V_{CF}$$
$$V_{RF} = V_{CC} - (0.1V_{CE_{sat}} + 0.9V_D + 0.9V_{CL}) \qquad (8\text{-}16)$$

The rise time of the collector-clamped inverter can now be calculated by using Eq. (1-7)

$$t = 2.3RC \log \frac{V_{RS}}{V_{RF}} \qquad (1\text{-}7)$$

where

t = the rise time t_r;
R = the charging resistance R_C;
C = the load capacitance C_L;
$\dfrac{V_{RS}}{V_{RF}}$ = the ratio of V_{R_C} at the start of t_r to V_{R_C} at the finish of t_r.

Hence,

$$t_r = 2.3 R_C C_L \log \frac{V_{CC} - (0.9 V_{CE_{sat}} + 0.1 V_D + 0.1 V_{CL})}{V_{CC} - (0.1 V_{CE_{sat}} + 0.9 V_D + 0.9 V_{CL})} \quad (8\text{-}17)$$

$$= 2.3 \times 10^3 \times 5 \times 10^{-11} \log \frac{10 - (0.27 + 0.07 + 0.5)}{10 - (0.03 + 0.63 + 4.5)}$$

$$= 1.15 \times 10^{-7} \log \frac{10 - 0.84}{10 - 5.16}$$

$$= 1.15 \times 10^{-7} \log 1.89$$

$$t_r = 32 \text{ ns}$$

Notice that the ratio of V_{RS} to V_{RF} is reduced from 9 in the unclamped inverter to 1.89 in this clamped inverter. This causes a reduction of over 70 percent in the rise time. By using higher values of V_{CC}, rise time can be further improved. Note also, however, that the higher value of V_{CC} increases the average current through R_C during fall time. This reduces the current that can be used to discharge C_L which results in an increase in fall time compared to the unclamped inverter.

The Emitter-follower Output. The next output circuit to be discussed is the emitter follower. A basic circuit is shown in Fig. 8-25 with its input and output waveforms. Notice that it has characteristics opposite of those of the inverter. The rise time is faster than the fall time. Again, a purely capacitive load is assumed and the transister delays are neglected.

Note in Fig. 8-25(b) that $v_{i(0)}$ is greater than V_L and $v_{i(1)}$ is greater than V_H. This is fairly common in the emitter follower. One use of the emitter follower is to produce a negative level shift to cancel unwanted positive level shifts produced in previous circuits. The voltages 0.7 V and 5.7 V shown on the input waveform could be the outputs of an *AND* gate with 0 V and 5 V input levels. As shown in Fig. 8-25(b), the emitter follower shifts the levels back very close to 0 V and 5 V.

Assume that the input has been high for a long time. The transistor has been on in the active region and C_L has charged to $(5.7 \text{ V} - V_{BE}) \approx V_H = 5 \text{ V}$. Then when v_i goes low, $V_{BE} = v_i - V_{C_L} = 0.7 \text{ V} - 5 \text{ V} = -4.3 \text{ V}$. The transistor cuts off and C_L begins to discharge through R_E. When V_E decreases to the point where $V_{BE} = V_\gamma$, Q_1 comes on. Assuming that $V_\gamma = 0.5 \text{ V}$, this occurs when V_{C_L} discharges to approximately 0.2 V. As soon as Q_1 comes on, however, it charges C_L to more than 0.2 V and Q_1 cuts off again. This permits C_L to again discharge to 0.2 V. The net result is that the emitter is clamped at 0.2 V and V_{BE} is clamped at $V_\gamma = 0.5 \text{ V}$. The small differences between $V_H = 5 \text{ V}$ and the actual high output $(5.7 - V_{BE})$ and the small difference between $V_L = 0 \text{ V}$ and the actual low output $(0.7 \text{ V} - V_\gamma)$ are

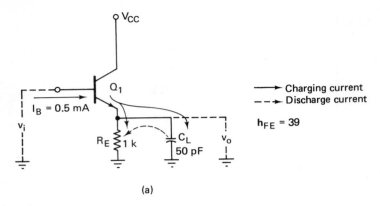

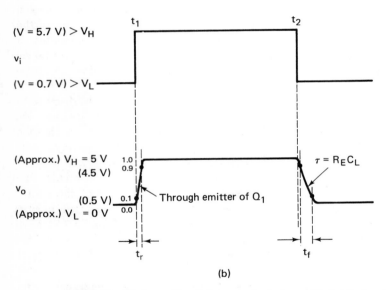

Fig. 8-25 (a) An emitter-follower output stage with (b) its input and output waveforms.

neglected. This provides easier numbers with which to work. Typical values of t_r and t_f can then easily be determined. An example follows.

Example 8-8: Calculate the rise and fall times of the emitter-follower output signal shown in Fig. 8-25(b).
Solution: The rise time is calculated first. From Fig. 8-25(b), the actual rise time is the time it takes the load capacitance to charge from 0.5 V to 4.5 V. This is a nearly linear rise because an almost constant current equal to $I_E - I_{R_E}$ charges C_L. The emitter current is constant at

$$I_E = (1 + h_{FE})I_B$$
$$= 40 \times 0.5$$
$$I_E = 20 \text{ mA}$$

and the current through R_E during rise time is relatively small. Since the average voltage across R_E during rise time is

$$V_{R_{E(\text{av})}} = \frac{[V_L + 0.1(V_H - V_L)] + [V_L + 0.9(V_H - V_L)]}{2}$$
$$= \frac{0.5 + 4.5}{2}$$
$$V_{R_{E(\text{av})}} = 2.5 \text{ V}$$

the average current through R_E during rise time is

$$I_{R_{E(\text{av})}} = \frac{V_{R_{E(\text{av})}}}{R_E}$$
$$= \frac{2.5}{1}$$
$$I_{R_{E(\text{av})}} = 2.5 \text{ mA}$$

The rest of the emitter current is used to charge C_L

$$I_{C_L} = I_E - I_{R_{E(\text{av})}}$$
$$= 20 - 2.5$$
$$I_{C_L} = 17.5 \text{ mA}$$

Hence, from $I = C(\Delta V_C/\Delta t)$,

$$t_r = C_L \frac{0.8(V_H - V_L)}{I_{C_L}}$$
$$= 5 \times 10^{-11} \frac{0.8(5 - 0)}{1.75 \times 10^{-2}}$$
$$t_r = 11.4 \text{ ns}$$

When C_L finishes charging, V_{C_L} equals $V_H = 5$ V and remains there until $-t_2$ (just before Q_1 cuts off). At t_2+ (just after Q_1 cuts off), C_L sees only R_E and begins to discharge. The actual fall time is the time it takes C_L to discharge from 4.5 V to 0.5 V. With Q_1 off, $V_{R_E} = V_{C_L}$; thus,

$$V_{RS} = V_{CS} = 0.9(V_H - V_L)$$
$$= 0.9(5 - 0)$$
$$V_{RS} = V_{CS} = 4.5 \text{ V}$$

and
$$V_{RF} = V_{CF} = 0.1(V_H - V_L)$$
$$= 0.1(5 - 0)$$
$$V_{RF} = V_{CF} = 0.5$$

Therefore,
$$t_f = 2.3 R_E C_L \log \frac{V_{RS}}{V_{RF}} \qquad (8\text{-}18)$$
$$= 2.3 R_E C_L \log \frac{4.5}{0.5}$$
$$t_f = 2.2 R_E C_L \qquad (8\text{-}18a)$$

Substituting the values of R_E and C_L yields
$$t_f = 2.2 \times 10^3 \times 5 \times 10^{-11}$$
$$t_f = 110 \text{ ns}$$

Note that while the rise time of the emitter follower is short, the fall time is the same as the rise time of the unclamped inverter calculated in Ex. 8-6.

The Darlington Amplifier. The rise time of the emitter follower can be improved by using a special configuration known as the *Darlington* amplifier. The circuit is shown in Fig. 8-26(a). An input current I_{B_1} that turns Q_1 on is applied. This causes a current I_{E_1} that serves as the base current of Q_2. I_{B_2} turns Q_2 on and produces an emitter current I_{E_2}. Figure 8-26(b) shows the two transistors replaced by a *composite* or equivalent transistor that has an $\mathbf{h}_{FE}$ equal to

$$\mathbf{h}_{FE(eq)} = \frac{I_{E(eq)}}{I_{B(eq)}} = \frac{I_{E_2}}{I_{B_1}} \qquad (8\text{-}19)$$

The equivalent $\mathbf{h}_{FE}$ can be expressed in terms of the individual $\mathbf{h}_{FE}$ of each transistor. By referring to Fig. 8-26(a), it can be seen that

$$I_{E_2} = I_{B_2} + I_{C_2}$$
$$= I_{B_2} + \mathbf{h}_{FE_2} I_{B_2}$$
$$I_{E_2} = (1 + \mathbf{h}_{FE_2}) I_{B_2} \qquad (8\text{-}20)$$

But
$$I_{B_2} = I_{E_1}$$

and
$$I_{E_1} = I_{B_1} + I_{C_1}$$
$$= I_{B_1} + \mathbf{h}_{FE_1} I_{B_1}$$
$$I_{E_1} = (1 + \mathbf{h}_{FE_1}) I_{B_1} \qquad (8\text{-}21)$$

Output Circuit Variations / 249

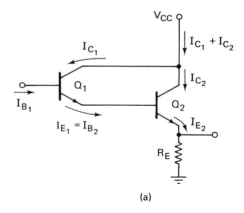

(a)

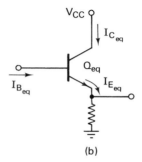

(b)

Fig. 8-26 (a) The Darlington amplifier. (b) An equivalent circuit showing the two transistors replaced by a composite transistor.

Substituting Eq. (8-21) for I_{B_2} in Eq. (8-20) yields

$$I_{E_2} = (1 + \mathbf{h}_{FE_2})(1 + \mathbf{h}_{FE_1})I_{B_1} \tag{8-22}$$

Finally, by substituting Eq. (8-22) in Eq. (8-19), the equivalent $\mathbf{h}_{FE}$ is

$$\mathbf{h}_{FE(eq)} = \frac{(1 + \mathbf{h}_{FE_2})(1 + \mathbf{h}_{FE_1})I_{B_1}}{I_{B_1}}$$

which reduces to

$$\mathbf{h}_{FE(eq)} = (1 + \mathbf{h}_{FE_2})(1 + \mathbf{h}_{FE_1})$$

If each transistor has the same $\mathbf{h}_{FE}$,

$$\mathbf{h}_{FE(eq)} = (1 + \mathbf{h}_{FE})^2 \tag{8-23}$$

The higher output current gives the Darlington configuration a lower output resistance and permits the load capacitance to charge more quickly than it can in the basic emitter-follower circuit. This improvement is at the expense of higher power consumption.

The Totem-pole Output. By combining the emitter follower with the saturated inverter, in what is known as the *totem-pole* circuit, very fast rise and fall times are obtained. A basic totem-pole circuit is shown with its input and output waveforms in Fig. 8-27. A detailed static analysis of the

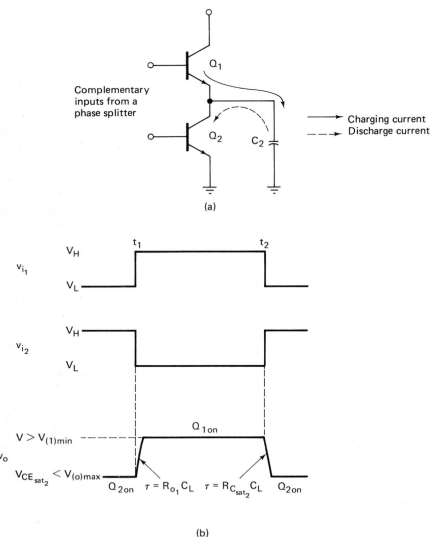

Fig. 8-27 (a) The totem-pole output. (b) Its input and output signals.

totem-pole circuit is given in Sec. 8-2. Its switching characteristics are discussed in this section.

In Fig. 8-27 it can be seen that at $-t_1$, v_{i_1} has been low for a long time while v_{i_2} has been high. Therefore, Q_1 has been off and Q_2 has been saturated for a long time and the output voltage v_o equals the $V_{CE_{sat}}$ of Q_2 (which is less than $V_{(0)max}$). When v_{i_1} goes high and v_{i_2} goes low, Q_2 cuts off and Q_1 comes on in the active region. The circuit is now the same as Fig. 8-25(a), except for R_E, which is now the cut off inverter Q_2. This permits all of the high output current of the emitter follower to be used to quickly charge C_L to $V_{o(1)}$, which is greater than $V_{(1)min}$.

At t_2+, v_{i_1} goes low and v_{i_2} goes high again. This causes Q_1 to cut off and Q_2 to saturate. The circuit is now the same as Fig. 8-22(a), except for R_C, which is now the cut off emitter follower Q_1. This allows all of the high current capacity of Q_2 to be used for sinking a large discharge current from C_L. Thus, the output quickly falls to $V_{CE_{sat}}$. Note that in this simplified circuit, with a purely capacitive load (and leakage current neglected) V_{CE} would fall to 0 after C_L finishes discharging. The dc load (not shown) would, however, keep a current flowing through Q_1 and hold V_{CE} at $V_{CE_{sat}}$.

A totem-pole circuit that utilizes the Darlington connection is shown in Fig. 8-28. This is a two-input gate which is the same as Fig. 8-18(a) except for the addition of the Darlington pull-up in the output. Q_4 now acts as a

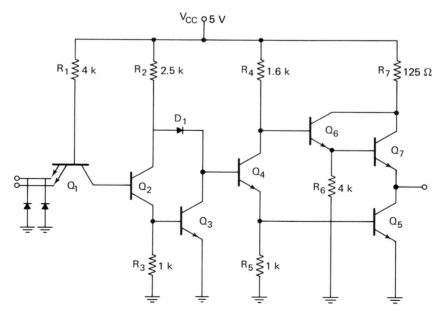

Fig. 8-28 A two-input *AND* gate with a Darlington pull-up in a totem-pole output. (Courtesy Sprague Electric Co.)

phase splitter. When it is cut off, the base drive to Q_5 is removed and a current flows from V_{CC} through R_4 and into the base of Q_6. Q_6 comes on in the active region and its emitter current $I_{E_6} = (1 + \mathbf{h}_{FE_6})I_{B_6}$ divides into the base of Q_7 and the base pull-down resistor R_6. This turns Q_7 on, producing an output current $I_{E_7} = (1 + \mathbf{h}_{FE_7})I_{B_7}$, which quickly charges the load capacitance to the 1-level output voltage.

When Q_4 comes on, it quickly removes the charge stored in the base of Q_6, turning it off. The base pull-down resistor R_6 removes the charge stored in the base of Q_7 to turn off that transistor. At the same time the emitter current of Q_4 develops a voltage across R_5 that turns Q_5 on. The load capacitance quickly discharges to the 0 level through the saturated inverter.

The propagation delay of this AND gate is 15 ns; the AND gate with the open-collector output has a 20-ns delay. However, the average power consumption of the circuit with the Darlington pull-up is higher (20 mW compared to 18 mW). Power dissipation can be reduced by returning R_6 to the output terminal instead of to ground. With either connection, it dissipates no dc power when Q_6 and Q_7 are off. When R_6 is returned to ground, $V_{R_6} = V_{BE_7} + V_{o(1)}$. It draws some of the emitter current of Q_6 away from the base of Q_7 and dissipates power. If returned to the output V_{R_6} is limited to V_{BE_7} reducing the current drawn by R_6 and the power dissipated by that resistor.

The offset diode in the emitter of other totem-pole circuits (e.g., D_1 in Fig. 8-3) is not needed because of the additional emitter-base junction in the Darlington circuit.

Conclusion: By combining the emitter follower with the saturated inverter in such a way that the emitter follower acts as the collector pull-up resistance of the inverter and the inverter acts as the emitter pull-down resistance of the emitter follower, the rise time and fall time can both be made very fast. If the emitter follower utilizes the Darlington connection, the rise time is further improved.

8-10 Summary

Transistor-transistor logic TTL or T²L employs a multi-emitter transistor. Each emitter-base junction serves the same function as the diodes in diode-transistor logic DTL. The main advantage of TTL is its higher speed with good noise immunity. When one of the EB junctions is forward biased, the multi-emitter transistor saturates and presents a very low resistance, through which the charge stored in the base of the following transistor and in its own collector-substrate capacitance C_{CS} is quickly removed. Switching speeds as low as 2 ns are possible with TTL.

Transistor-transistor logic circuits utilize both the open-collector and totem-pole outputs. The open-collector output permits the use of wired logic.

Some totem-pole outputs employ a Darlington amplifier pull-up which permits even faster 0-to-1 transitions than the basic totem-pole circuit.

The basic building blocks of TTL are the *NAND, NOR, OR, AND, AND-OR-INVERT AOI*, and the *INVERTER*. An *AND* expander is also available to increase system flexibility.

Problems

8-1. Set up a chart showing the conduction state of each transistor for every combination of inputs to the circuit of Fig. 8-29. Use $V_L = 0$ and $V_H = V_{CC}$ as inputs. Also include the output $V_{(0)}$ or $V_{(1)}$ for each input combination and determine the positive-logic function.

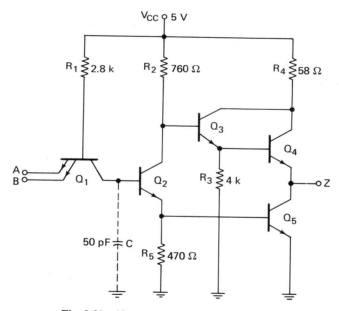

Fig. 8-29 (Courtesy Sprague Electric Co.)

8-2. Calculate $h_{FE_{min}}$ for Q_2 in the circuit of Fig. 8-3. Use $V_{BE_{sat}} = 0.7$ V, $V_{CE_{sat}} = 0.3$ V, $V_{BC_1} = 0.7$ V, and $V_i = V_H = V_{CC}$ at all inputs. Neglect leakage currents.

8-3. Calculate the steady-state currents I_{E_1}, I_{B_3}, I_{R_3}, and I_{B_4} for the circuit shown in Fig. 8-29. Assume an output voltage equal to 3.56 V with A at $V_L = 0$ V and B at $V_H = V_{CC}$. Use $h_{FE_3} = 29$, $h_{FE_4} = 14$, $V_{BE_{sat}} = 0.7$ V, and $V_{CE_{sat}} = 0.3$ V. (Note: I_{C_3}, I_{C_4}, and I_{E_4} depend on the load current. In the steady state this is only the 1-input leakage current of

8-4. Calculate the dc fan-out of the circuit in Fig. 8-29, if Q_5 has an $h_{FE_{min}}$ equal to 5. Use $V_{CE_{sat}} = 0.3$ V and $V_{BE_{sat}} = 0.7$ V. Neglect leakage currents. Assume that the driver and loads are identical to Fig. 8-29. (Note: Q_5 will actually have a minimum $h_{FE} > 5$ so that it will be able to sink a high discharge current from the load capacitance and bring about a fast 1-to-0 transition.)

8-5. Calculate the 0-level power consumption of the circuit shown in Fig. 8-29. Use $V_{BE_{sat}} = 0.7$ V and $V_{CE_{sat}} = 0.3$ V. Assume that this gate drives and is driven by others exactly like itself.

8-6. Determine the 0-level noise immunity $N_{(0)}$ for the circuit of Fig. 8-29. Assume a worst-case 0-input voltage $V_{(0)max} = 0.4$ V, a B-E cutin voltage $V_\gamma = 0.5$ V, and $V_{BE_{sat}} = 0.7$ V.

8-7. Determine the 1-level noise immunity $N_{(1)}$ for the circuit of Fig. 8-29. Assume a worst-case 1-input voltage $V_{(1)min} = 2.4$ V, $V_{BE_{sat}} = 0.7$ V, and a B-E cut in voltage $V_\gamma = 0.5$ V.

8-8. If one of the inputs of Fig. 8-29 goes to 0.3 V after both inputs have been high for a long time, Q_1 turns on and Q_2 turns off. Calculate the turn-off time of Q_2. Use $V_{BE_{sat}} = 0.7$, $V_{CE_{sat}} = 0.3$ V, h_{FE} of $Q_1 = 30$, and $C_T = 50$ pF. (Note: $C_T = C_{cs} + C_{is}$, the collector-substrate capacitance of Q_1 and the input capacitance of Q_2.)

8-9. Calculate the 1-level dc power consumption of the AND gate of Fig. 8-18(a). Use $V_{BC_1} = 0.7$ V, $V_{BE_{sat}} = 0.7$ V, $V_D = 0.7$ V, and $V_{CE_{sat}} = 0.3$ V.

8-10. (a) Calculate the 0-level dc power consumption for the AND gate of Fig. 8-18(a). Use $V_{BE_{sat}} = 0.7$ V, $V_D = 0.7$ V, $V_{CE_{sat}} = 0.3$ V, and $V_{i(0)} = 0.3$ V.

(b) Using the answers from part (a) and Prob. 8-9, calculate the average power consumption for a 50 percent duty cycle.

8-11. Calculate $h_{FE_{min}}$ for Q_4 and I_{B_5} in Fig. 8-18(a). Use $V_{BE_{sat}} = 0.7$ V, $V_{CE_{sat}} = 0.3$ V, and $V_D = 0.7$ V.

8-12. Calculate the power consumption of the OR gate of Fig. 8-19 when both inputs equal 0.3 V. Use $V_{BE_{sat}} = 0.7$ V, $V_D = 0.7$ V, and $V_{CE_{sat}} = 0.3$ V.

8-13. Calculate the power consumption of the OR gate of Fig. 8-19
(a) When one input is at 0.3 V and the other is high.
(b) When both inputs are high.
Use $V_{BE_{sat}} = 0.7$ V, $V_D = 0.7$ V, $V_{CE_{sat}} = 0.3$ V, $V_{BE} = 0.7$ V, a 1-input current to each load equal to 40 μA, a fan-out $FO = 10$, and a 1-level output voltage $V_{o(1)} = 3.56$ V.

8-14. If the dc fan-out is 10 and the maximum 0-input current of each load is 1.6 mA, what is the minimum value of h_{FE} for Q_8 in Fig. 8-19?

8-15. Calculate the rise and fall times of the circuit of Fig. 8-22(a). Use $V_{CE_{sat}} = 0.3$ V, $R_C = 2$ k, $V_{CC} = 5$ V, $C_L = 40$ pF, $I_B = 0.4$ mA, and $h_{FE} = 50$.

8-16. Calculate the rise and fall times of the circuit of Fig. 8-23(a). Use $R_C = 2$ k, $V_{CE_{sat}} = -0.3$ V, $V_{CC} = -10$ V, $C_L = 50$ pF, $I_B = 0.4$ mA and $h_{FE} = 30$.

8-17. Calculate the rise and fall times of the circuit of Fig. 8-24(a). Use $V_{CC} = 12$ V, $R_C = 2$ k, $V_{CL} = 6$ V, $C_L = 40$ pF, $I_B = 0.4$ mA, $V_{CE_{sat}} = 0.3$ V, and $h_{FE} = 50$.

8-18. Calculate the rise and fall times of the circuit of Fig. 8-25(a). Use $R_E = 2$ k, $C_L = 40$ pF, $I_B = 0.4$ mA, $h_{FE} = 49$. $V_{BE} = 0.7$ V, $V_{i(1)} = 6.7$ V, and $V_{i(0)} = 0.7$ V.

9

ASTABLE

AND

MONOSTABLE MULTIVIBRATORS

There are three types of multivibrator MV circuits: *the astable, the monostable*, and *the bistable*. Each is basically two amplifiers connected as shown in Fig. 9-1. The output of one amplifier drives the other.

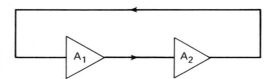

Fig. 9-1 Two amplifiers with the output of one driving the other.

    The astable MV is a free-running oscillator that requires input signals only to synchronize its oscillating frequency. It has two states that continuously alternate from A_1 on and A_2 off to A_2 on and A_1 off, and back again. The frequency of oscillation depends (when $V_{BB} = V_{CC}$ and operation is outside the active region) primarily on the RC circuits that couple the output of one amplifier to the input of the other. The astable MV is widely used as a generator of rectangular waves and it is also sometimes used as a clock pulse generator.

    The monostable or one-shot multivibrator has one stable state. One amplifier is *normally on* and the other is *normally off*. A trigger pulse sends it into its *quasi-stable* state, where the normally on circuit goes off and the normally off circuit comes on. It automatically returns to its stable state after a time that depends on an RC circuit that couples the normally off amplifier to the normally on amplifier. The monostable is used to shape waves, adjust pulse widths, and delay pulses.

The bistable MV (*Eccles-Jordan circuit flip-flop FF*) has two stable states. If A_1 is on and A_2 is off, it remains in that state unless a trigger pulse is applied. It then remains in the new state until another trigger pulse is applied. The flip-flop circuit is widely used in counters and memory registers of digital computers. There are several types of bistable multivibrators encountered in computers. They are discussed in detail in Chap. 10. Only the astable and monostable multivibrators are explained in this chapter.

7-1 The Collector-base Coupled Astable Multivibrator

The circuit shown in Fig. 9-2 is known as the collector-base coupled (or simply the collector-coupled) astable MV.* The waveforms for this circuit are shown in Fig. 9-3. The charge path for $C_{C(2)}$ and the discharge path for $C_{C(1)}$ are shown on the circuit. During the alternate half-cycle of oscillation $C_{C(1)}$ charges through R_{C1} and the base-emitter junction of Q_2, and $C_{C(2)}$ discharges through Q_2, the V_{BB} supply, and R_2.

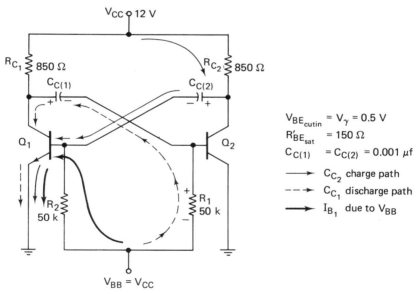

Fig. 9-2 An astable multivibrator circuit.

*There are several other astable multivibrators that can be found in pulse circuit texts. The collector-base coupled astable MV, which is the most frequently used circuit, is the only one discussed in this text.

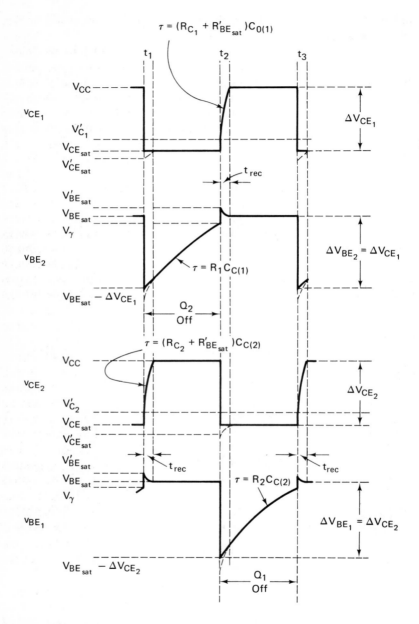

Fig. 9-3 The collector and base waveforms of the astable MV shown in Fig. 9-2.

The circuit is self-starting. When power is applied, a slight imbalance that always exists, even in symmetrical circuits, is quickly amplified until one transistor is cut off and the other is saturated. Operation in the saturation region is preferred because this makes the circuit action independent of the ac parameters of the transistors. Only the dc parameter $h_{FE_{min}}$ is important. It must be large enough to keep the ON transistor in saturation after the capacitor in its base circuit finishes charging. For example, as shown in Fig. 9-2, when $C_{C(2)}$ charges, its charging current adds to the current from the V_{BB} supply and thus provides a high overdrive base current. After $C_{C(2)}$ is fully charged, I_{B_1} is reduced to $(V_{BB} - V_{BE_{sat}})/R_2$. Q_1 must have a value of $h_{FE_{min}}$ that permits it to remain in saturation with the lower I_B, while it sinks the discharge current of $C_{C(1)}$ plus the current that flows from V_{CC} through R_{C_1}.

Circuit Operation. When power is first applied, both transistors conduct and both capacitors begin to charge. Because of a slight imbalance (which always exists even in a symmetrical MV), one transistor begins to conduct harder than the other. Assuming that it is Q_1, v_{CE_1} will decrease slightly and the drop will be coupled through $C_{C(1)}$ to base B_2. Since this now makes Q_2 conduct even less, v_{CE_2} rises. The rise is coupled through $C_{C(2)}$ to base B_1. While $C_{C(2)}$ charges, there is an overdrive voltage and current at the base of Q_1 which makes it conduct so hard that v_{CE_1} drops to $V_{CE_{sat}}$. Since all the above action is almost instantaneous, the entire drop in v_{CE_1} is coupled to B_2 and cuts Q_2 off.

The collector of Q_2 rises to V_{CC} as $C_{C(2)}$ charges through the path shown in Fig. 9-2. The output curves of the transistors are shown in Fig. 9-4(a). When a transistor is cut off, its collector voltage goes to V_{CC}. When it comes on, v_{CE} decreases to $V_{CE_{sat}}$. The small undershoot at each collector, seen on the waveforms of Fig. 9-3, can be accounted for by examining Fig. 9-4(b), which shows the saturation region expanded. While the charging current flows into the base, a large overdrive current $I'_{B_{sat}} \gg I_{B_{sat}}$ flows. (Note: Values with prime marks are for the intervals during which the capacitor charges and overdrive is present at the base.) This reduces $R'_{CE_{sat}}$ to less than 20 Ω and $V'_{CE_{sat}}$ to less than 0.3 V. However, because the charging current continuously decreases, $I'_{B_{sat}}$, $R'_{CE_{sat}}$, and $V'_{CE_{sat}}$ continuously change. The undershoot is small and is neglected in the analysis that follows in Sec. 9-2. Trying to find the value of this voltage only makes the analysis more complex than is necessary. From Fig. 9-4(b) it is obvious that $V'_{CE_{sat}}$ is in the order of 0.1 V regardless of circuit component values.

After $C_{C(2)}$ finishes charging, $v_{CE_2} = V_{CC}$, $v_{BE_1} = V_{BE_{sat}}$, and $v_{C_{C(2)}} = V_{CC} - V_{BE_{sat}}$. The transistor Q_2 remains off and Q_1 remains on as $C_{C(1)}$ discharges through the path shown in Fig. 9-2. As $v_{C_{C(1)}}$ decreases, v_{R_1} decreases and v_{BE_2} reaches its cutin value $V_\gamma = 0.5$ V. This permits Q_2 to come on, which causes its collector voltage to decrease. The drop in v_{CE_2} is coupled through $C_{C(2)}$ to B_1 and causes Q_1 to conduct less. This makes V_{CE_1} rise and

260 / *Astable and Monostable Multivibrators*

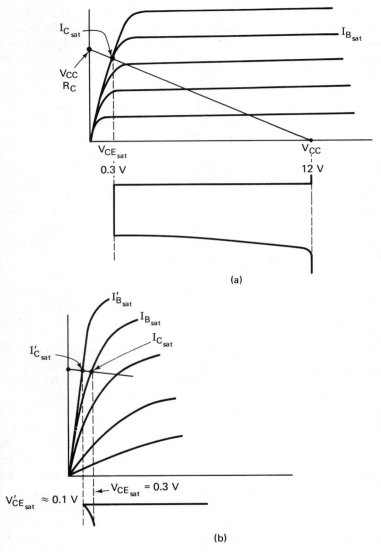

Fig. 9-4 (a) The output characteristics showing the V_{CE} swing from V_{CC} to $V_{CE_{sat}}$. (b) The saturation region expanded to show $V'_{CE_{sat}}, V_{CE_{sat}}$.

the rise is coupled through $C_{C(1)}$ to B_2. While $C_{C(1)}$ charges from V_{CC} through R_{C_1} and the *B-E* junction of Q_2, there is an overdrive current at the base of Q_2. The resulting heavy conduction of Q_2 reduces v_{CE_2} to $V'_{CE_{sat}} < V_{CE_{sat}}$. The drop is coupled to B_1 and cuts off Q_1.

When $C_{C(1)}$ finishes charging, $v_{CE_1} = V_{CC}$, $v_{BE_2} = V_{BE_{sat}}$, and $v_{C_{C(1)}} = V_{CC} - V_{BE_{sat}}$. The transistor Q_1 remains off and Q_2 remains on as $C_{C(2)}$ discharges through Q_2, V_{BB}, and R_2. As $v_{C_{C(2)}}$ decreases, v_{R_2} decreases and v_{BE_2} reaches its cutin value $V_\gamma = 0.5$ V. This permits Q_1 to conduct again, which decreases v_{CE_1}, and the cycle repeats.

9-2 Circuit Analysis

An astable MV is shown in Fig. 9-5(a). The waveforms for C_1, the collector of Q_1, and B_2, the base of Q_2, are shown in Fig. 9-5(b). The C_2 and B_1 waveforms are the same but displaced one-half cycle with respect to the waveforms for C_1 and B_2. (See Fig. 9-3.) The equivalent circuit that used to produce the C_1 and B_2 waveforms is shown in Fig. 9-6. The development of this equivalent circuit is explained before beginning the analysis. By referring to Fig. 9-5(a) it can be seen that there are two ground-return paths from the left side of $C_{C(1)}$: through the saturated transistor Q_1 and through R_{C_1} in series with V_{CC}. There are also two ground-return paths from the right side of $C_{C(1)}$: through R_1 in series with V_{BB} and through the base-emitter junction of Q_1.

As shown in Fig. 9-4, V_{CE} is, for all practical purposes, constant in the saturation region. The transistor is therefore represented by a constant-voltage source $V_{CE_{sat}}$ in the equivalent circuit of Fig. 9-6. As shown in Fig. 9-7(a), V_{BE} is relatively constant once it exceeds V_γ. Regardless of the values of R_1 and V_{BB}, which determine I_B after $C_{C(1)}$ finishes charging, the B-E junction (in the steady state) can be represented by a constant-voltage source $V_{BE_{sat}}$. This is shown in the equivalent circuit of Fig. 9-6.

At the instant Q_2 comes on and cuts off Q_1, $C_{C(1)}$ begins to charge and a high overdrive current flows into the base of Q_2. This current, which is several milliamperes, is many times larger than the steady-state base current

$$I_{B_{sat}} = \frac{(V_{BB} - V_{BE_{sat}})}{R_1}$$

$$= \frac{10 - 0.7}{40 \text{ k}}$$

$$I_{B_{sat}} = 0.2325 \text{ mA}$$

The base current may be as much as 50 times greater when $C_{C(1)}$ begins to charge. It therefore is not available on the input characteristics. The B-E junction is represented by a small dc resistance $R'_{BE_{sat}} = 200\ \Omega$ at the instant $C_{C(1)}$ begins to charge. The operating point as it would appear if the input curve were extended is shown in Fig. 9-7(b).

$R'_{BE_{sat}}$, like $R'_{CE_{sat}}$, is not a constant resistance, but one that continuously increases as $C_{C(1)}$ charges and $I'_{B_{sat}}$ decreases. However, the value of

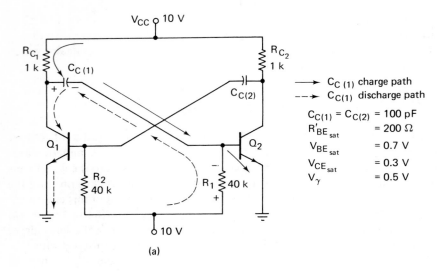
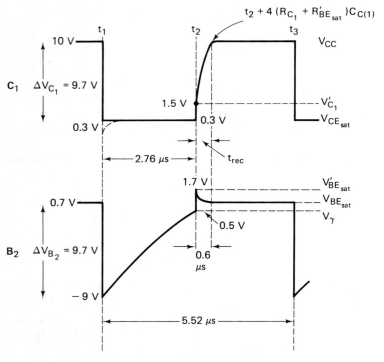

Fig. 9-5 (a) The astable MV for Ex. 9-1 with (b) its waveforms.

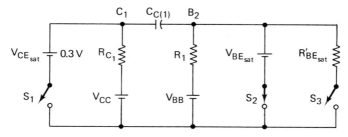

Fig. 9-6 The equivalent circuit used to develop the V_{CE_1} and V_{BE_2} waveforms.

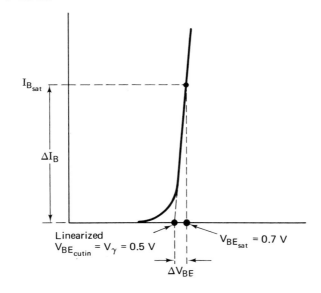

(a)

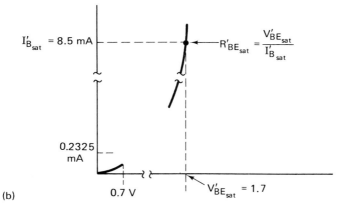

(b)

Fig. 9-7 (a) The input characteristics for the normal operating region including the linearized curve. (b) The operating point at the instant the coupling capacitor begins to charge and applies the high overdrive current.

264 / Astable and Monostable Multivibrators

200 Ω* is typical for the instant Q_2 comes on in this circuit and it permits calculation of the amplitude of the important base overshoot. The voltages and times shown on the waveforms of Fig. 9-5 are calculated in the following example:

Example 9-1: Develop the collector C_1 and base B_2 waveforms for the circuit of Fig. 9-5(a). Also calculate the collector recovery time t_{rec} and the frequency.

Solution: Just before Q_1 comes on $(-t_1)$. $C_{C(1)}$ has fully charged through the path shown in Fig. 9-5(a); so

$$V_{CE_1} = V_{CC} = 10 \text{ V}$$
$$V_{BE_1} = V_{BE_{sat}} = 0.7 \text{ V}$$

and

$$v_{C_{C(1)}} = v_{CE_1} - v_{BE_2}$$
$$= 10 - 0.7$$
$$v_{C_{C(1)}} = 9.3 \text{ V (positive on the } C_1 \text{ side)}$$

Just after Q_1 *comes on* $(t_1 +)$. The collector voltage of C_1 drops from V_{CC} to $V_{CE_{sat}} = 0.3$ V and, since $v_{C_{C(1)}}$ can't change instantaneously, the change in voltage at C_1 is coupled to B_2.

$$\Delta v_{CE_1} = V_{CC} - V_{CE_{sat}}$$
$$= 10 - 0.3$$
$$\Delta v_{CE_1} = 9.7 \text{ V}$$

This is a decrease in voltage, which drives B_2 from $V_{BE_{sat}} = 0.7$ V to -9 V and cuts off Q_2. At the instant Q_2 cuts off,

$$v_{R_1} = V_{BB} - v_{BE_2}$$
$$= (10) - (-9)$$
$$v_{R_1} = 19 \text{ V}$$

This is the voltage v_{RS}, the voltage on R at the *start* S of discharge.

*The value of $R'_{BE_{sat}}$ depends on the amount of charging current at the instant Q_2 comes on.
 The charging current depends primarily on V_{CC} and the size of R_{C_1}. Values of $R'_{BE_{sat}}$ from 100 Ω to 400 Ω are typical.

Summary of the Conditions at t_1+.

$$v_{CE_1} = 0.3 \text{ V} \qquad v_{BE_2} = -9 \text{ V}$$
$$v_{C_{C(1)}} = 9.3 \text{ V} \qquad v_{RS} = 19 \text{ V}$$

Just before Q_2 *comes on* $(-t_2)$. Q_2 remains off as $C_{C(1)}$ discharges and v_{R_1} decreases. When $v_{BE} = V_\gamma = 0.5$ V, Q_2 begins to come on. Just before Q_2 comes on, the voltage across $C_{C(1)}$ is

$$v_{C_{C(1)}} = v_{BE_2} - v_{CE_1}$$
$$= V_\gamma - V_{CE_{sat}}$$
$$= 0.5 - 0.3$$
$$v_{C_{C(1)}} = 0.2 \text{ V (positive on the } B_2 \text{ side)}$$

The voltage on the discharge resistance R at the *finish* F of the discharge period is

$$v_{RF} = V_{BB} - v_{BE_2}$$
$$= 10 - 0.5$$
$$v_{RF} = 9.5 \text{ V}$$

Frequency Calculation. The time that Q_2 is cut off can now be calculated by using Eq. (1-7)

$$t_{Q_2(\text{off})} = 2.3RC \log \frac{v_{RS}}{v_{RF}}$$

where $R = R_1$, $C = C_{C(1)}$, and v_{RS} and v_{RF} are the values calculated at t_1+ and $-t_2$ above.

Note that the ratio $v_{RS}/v_{RF} = 19/9.5$ is 2. It will always be very close to 2 as long as $V_{BB} = V_{CC}$. Thus, the time that Q_2 is off (t_1 to t_2) is determined by

$$t_{Q_2(\text{off})} = 2.3R_1C_{C(1)} \log 2$$
$$= 2.3R_1C_{C(1)}(0.3)$$
$$t_{Q_2(\text{off})} = 0.69R_1C_{C(1)} \qquad (9\text{-}1)$$
$$= 0.69 \times 4 \times 10^4 \times 10^{-10}$$
$$t_{Q_2(\text{off})} = 2.76 \text{ } \mu\text{s}$$

And since the circuit is symmetrical,

$$t_{Q_1(\text{off})} = t_{Q_2(\text{off})} = 2.76 \text{ } \mu\text{s}$$

Hence, the time for a cycle is

$$t_1 \text{ to } t_3 = 2(0.69\, RC) \tag{9-2}$$
$$= 2 \times 2.76\ \mu s$$
$$t_1 \text{ to } t_3 = 5.52\ \mu s$$

and the frequency

$$F = \frac{1}{t_1 \text{ to } t_3} \tag{9-3}$$
$$= \frac{1}{5.52\ \mu s}$$
$$F = 181\ \text{kHz}$$

Summary of the Conditions at $-t_2$.

$$v_{CE_1} = 0.3\ \text{V}$$
$$v_{BE_2} = 0.5\ \text{V}$$
$$v_{C_{C(1)}} = 0.2\ \text{V (positive on the } B_2 \text{ side)}$$

Just after Q_2 comes on (t_2+). When Q_2 comes on, it cuts Q_1 off and C_{C_1} begins to charge. The equivalent circuits of Fig. 9-8 are used to calculate the v'_{BE_2}, v'_{CE_1}, and i'_{B_2} at t_2+. In Fig. 9-8(a) $R_{BTH} \approx R'_{BE_{sat}} = 200\ \Omega$ and

$$v'_{BTH} = \frac{(R'_{BE_{sat}})(V_{BB})}{R_1 + R'_{BE_{sat}}}$$
$$= \frac{(0.2)(10)}{40 + 0.2}$$
$$v'_{BTH} = 0.05\ \text{V} \approx 0\ \text{V}$$

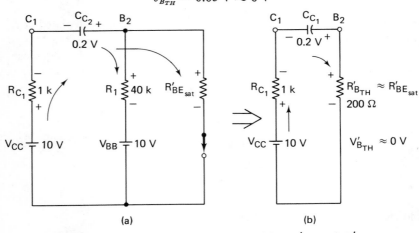

(a)   (b)

Fig. 9-8 The equivalent circuits for determining V'_{CE_1} and V'_{BE_2} at t_2+.

Because $R'_{BE_{sat}} \ll R_1$, $v'_{B_{TH}}$ is always very small. The base-to-ground Thévenin equivalent can therefore be accurately represented by using only $R'_{BE_{sat}}$, as shown in Fig. 9-8(b).

The voltage on the charging resistance $R = R_{C_1} + R'_{BE_{sat}}$ at the *start* of the charging interval is

$$v'_{RS} = V_{CC} + v_{C_{C(1)}}$$
$$= 10 + 0.2$$
$$v'_{RS} = 10.2 \text{ V}$$

Hence,

$$v'_{R_{C(1)}} = \frac{(R_{C_1})(v'_{RS})}{R_{C_1} + R'_{BE_{sat}}}$$
$$= \frac{(1)(10.2)}{1.2}$$
$$v'_{R_{C(1)}} = 8.5 \text{ V}$$

and

$$v'_{R_{BE_{sat}}} = v'_{RS} - v'_{R_{C(1)}}$$
$$= 10.2 - 8.5$$
$$v'_{R_{BE_{sat}}} = 1.7 \text{ V}$$

The polarity of these voltages is shown on the resistors in Fig. 9-8(b).

Summary of the Conditions at t_2+.

$$v'_{CE_1} = v'_{R_{C(1)}} + V_{CC}$$
$$= -8.5 + 10$$
$$v'_{CE} = 1.5 \text{ V}$$

and

$$v'_{BE_2} = v'_{R_{BE_{sat}}} + v'_{B_{TH}}$$
$$\approx 1.7 + 0$$
$$v'_{BE_2} \approx 1.7 \text{ V}$$

Note that the instantaneous changes in v_{BE_2} and v_{CE_1} (because C_1 and B_2 are capacitively coupled) are the same. From Fig. 9-8(b), if $v'_{BE_2} = 1.7$ V,

$$i'_{B_{sat}} = \frac{v'_{BE_{sat}}}{R'_{BE_{sat}}}$$
$$= \frac{1.7}{0.2 \text{ k}}$$
$$i'_{B_{sat}} = 8.5 \text{ mA}$$

268 / Astable and Monostable Multivibrators

This current is approximately 37 times greater than

$$I_{B_{sat}} = \frac{V_{BB} - V_{BE_{sat}}}{R_1} = 0.2325 \text{ mA}$$

the current that keeps Q_2 in saturation after C_{c_1} finishes charging.

Recovery Time t_{rec}. $C_{C(1)}$ charges back to $V_{CC} - V_{BE_{sat}} = 9.3$ V, bringing the circuit to the end of the cycle. As previously explained, $R'_{BE_{sat}}$ continuously increases as the charging current decreases. Thus, $C_{C(1)}$ is always charging through a different resistance toward a different voltage. When it finishes, $v_{CE_1} = V_{CC} = 10$ V, $v_{BE_2} = V_{BE_{sat}} = 0.7$ V, and $v_{C_{C(1)}} = v_{CE_1} - v_{BE_2} = 9.3$ V (positive on the C_1 side). The recovery time can be accurately calculated with

$$t_{rec} = 5(R_{C_1} + R'_{BE_{sat}})C_{C(1)} \tag{9-4}$$
$$= 5 \times 1.2 \times 10^3 \times 10^{-10}$$
$$t_{rec} = 0.6 \text{ } \mu s$$

9-3 A Modified Collector-coupled Astable MV

Both of the astable multivibrator circuits discussed in the preceding sections of this chapter are symmetrical types (mark-to-space ratio $m/s = 1$). However, it is often necessary to have one transistor conduct for only a short time while the other conducts for a long time. In order to decrease the off time of a transistor (without changing the supplies), the time constant of the RC coupling network between its base and the collector of the other transistor must be reduced. For example, if it is desired to have Q_2 in Fig. 9-5 remain off for less than the 2.76 μs calculated in Ex. 9-1, R_1 and/or $C_{C(1)}$ must be made smaller. The amount that they may be reduced is, however, not without limitations. If R_1 is made too small, Q_2 may come on before the collector of Q_1 returns to V_{CC}. If $C_{C(1)}$ is made too small, the stray capacitance of the circuit has a greater influence on the circuit. Both of these effects decrease the stability of the oscillator.

A modified circuit is shown in Fig. 9-9(a). A resistor $R_{C_{1B}}$ is connected between $C_{C(1)}$ and the collector of Q_1. The equivalent circuit used to calculate the C_1 and B_2 waveforms is shown in Fig. 9-9(b). The other half of the circuit, as shown in Fig. 9-9(c), is the same as it is for the basic MV of Fig. 9-5.

All voltages on the C_2 and B_1 waveforms are the same as they are in Fig. 9-5. The Q_1 off time is also unchanged. The most important change is the Q_2 off time (Q_1 on time), which is greatly reduced. This is because $C_{C(1)}$ is tied to point X, the junction of the voltage divider formed by $R_{C_{1A}}$ and $R_{C_{1B}}$, instead of the collector of Q_1. While the collector still drops from $V_{CC} = 10$ V

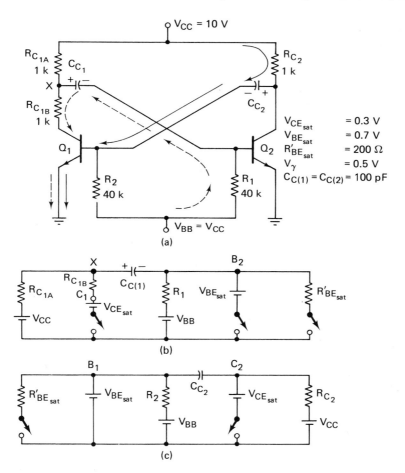

Fig. 9-9 (a) A modified collector-coupled astable. (b) The overall equivalent circuit used to develop the C_1 and B_2 waveforms. (c) The overall equivalent circuit used to develop the C_2 and B_1 waveforms.

to $V_{CE_{sat}} = 0.3$ V, V_x (from Eq. 1-4) goes to 5.15 V, a 4.85 V drop from V_{CC}. This drop is coupled to B_2 sending v_{B_2} from 0.7 V to -4.15 V instead of -9 V as it did in the original circuit. The off time of Q_2 can no longer be calculated by Eq. (9-1). Instead, Eq. (1-7) must be used. The recovery time of each collector is still 0.6 μs. If the output is taken from the collector of Q_2, the waveform is the same as it was for the basic MV of Fig. 9-5 except for the mark-to-space ratio. By making $R_{C_{1B}}$ larger, the time that Q_2 is off can be made even smaller. However, it should not be reduced to less than $t_{rec} = 0.6$ μs.

This technique of adding $R_{C_{1B}}$ to change the off time of a transistor in a

270 / *Astable and Monostable Multivibrators*

MV permits larger values to be used in the *RC* coupling networks and, therefore, it improves stability.

9-4 Synchronization of the Collector-coupled Astable MV and Frequency Division

The waveforms of Fig. 9-10 show how a multivibrator can be made to oscillate precisely at some desired frequency. A positive synchronizing pulse is applied to the base of an *npn* transistor of an astable MV that has a free-running frequency slightly lower than the sync pulse frequency. Figure 9-10(a) and (b) illustrates the triggering action. The base would normally reach V_γ and turn the transistor on some time after t_1, as indicated in Fig. 9-10(b). However, the addition of the positive sync pulse raises the base above V_γ at exactly t_1. The multivibrator immediately goes back to its free-running

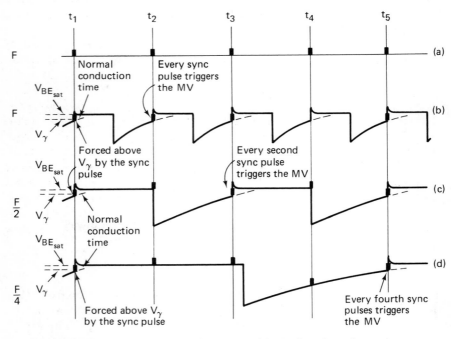

Fig. 9-10 The waveforms showing synchronization of a collector-coupled astable MV. (a) The sync pulse at a frequency F. (b) The base waveform of the MV with a free-running frequency slightly less than F but forced to run at the sync frequency F. (c) The free-running frequency slightly less than $F/2$ but synchronized at $F/2$. (d) The free-running frequency slightly less than $F/4$ but synchronized at $F/4$.

frequency and reaches V_y again a little after t_2. Again, the sync pulse forces it to conduct sooner, this time at t_2.

It may take several cycles for a sync pulse to catch it the first time. But after one of the sync pulses has finally triggered it, each of the following sync pulses arrives at the correct time to trigger it again.

Frequency Division. Figure 9-10(c) and (d) shows how the multivibrator can be made to run at a frequency equal to some submultiple of a particular frequency.

In Fig. 9-10(c) the first sync pulse forces the base above V_y at t_1. The free-running frequency is slightly less than half the sync pulse frequency. Therefore, when the positive sync pulse arrives at t_2, the base is still at $V_{BE_{sat}}$; thus, the transistor remains on. The pulse at t_3 arrives just before the base normally would reach V_y and forces it to conduct at t_3. Each *alternate* sync pulse arrives at the correct time to cause triggering.

In Fig. 9-10(d), every fourth sync pulse arrives at the correct time to cause triggering. The pulse at t_1 forces the transistor into conduction. The pulses at t_2 and t_3 arrive while the base is still at $V_{BE_{sat}}$, and the pulse at t_4 comes in while the base is still too far beyond cutoff.

There is a limit on the rate at which trigger pulses can be applied. If a pulse drives the base above V_y before the recovery time is over, operation becomes unstable. Therefore, it is important, in some circuits, to have the coupling capacitors charge as fast as possible. One way this can be effected is by reducing the size of the collector resistors. This technique, however, has its drawbacks. For one, it increases the power consumption of the circuit. It also causes the transistors to have to sink higher currents. Transistors with higher collector dissipation ratings and higher values of $\mathbf{h}_{FE_{min}}$ will be required.

Another method of improving the recovery time is the use of collector clamping (or catching) diodes. This technique, which was explained in detail in the discussion of the collector-clamped saturated inverter of Sec. 8-9, requires a higher V_{CC} supply. The collector aims for the higher V_{CC}, but it is clamped at a lower clamp supply voltage V_{CL} by the collector diode. The slow portion of the waveform is therefore removed.

9-5 The pnp Collector-coupled Astable MV

When *pnp* transistors are used in the multivibrator, the entire operation is the same as for the *npn* circuit, but the waveforms are inverted. The MV of Ex. 9-1, shown in Fig. 9-5(a) is repeated in Fig. 9-11(a) with *pnp* transistors. The waveforms for the *pnp* circuit are shown in Fig. 9-11(b).

At the instant just before t_1, Q_1 is just about to come on. The capacitor C_{C_1} has fully charged, bringing v_{CE_1} to $V_{CC} = -10$ V and $V_{BE_{sat}} = -0.7$ V. At t_1, Q_1 comes on and its collector voltage goes to $V_{CE_{sat}} = -0.3$ V. This is

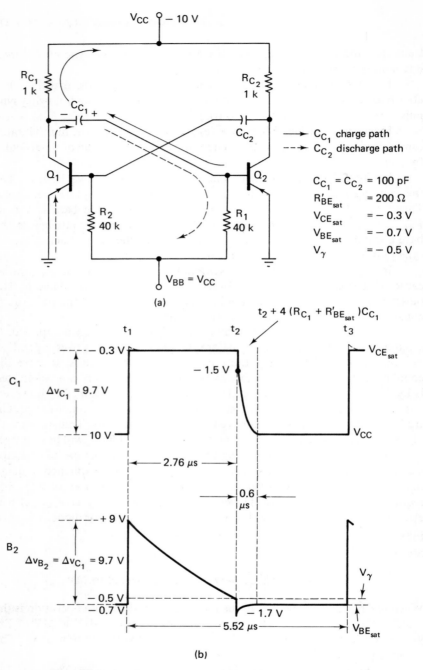

Fig. 9-11 (a) A *pnp* collector-coupled astable MV. (b) The waveforms for the collector of Q_1 and the base of Q_2.

a positive going change of 9.7 V which is coupled to the base of Q_2. The base voltage therefore rises from -0.7 V to $+9$ V and cuts off Q_2.

The rest of the analysis, which is the same as that given in Ex. 9-1, will not be duplicated here. However, it should be noted that negative pulses are needed to synchronize the *pnp* circuit. The free-running frequency must still be slightly lower than the sync frequency.

9-6 The Collector-base Coupled Monostable Multivibrator

The circuit of Fig. 9-12(a) is a collector-base coupled monostable (mono or one-shot) multivibrator.* If this circuit is compared to the astable circuit of Fig. 9-2, two important differences will be observed. Note that the base of Q_1 in the monostable is returned to a negative supply, but in the astable it is returned to a positive supply. The second difference is the resistive coupling between the collector of Q_2 and the base of Q_1. In the astable there is capacitive coupling. Because of these differences, Q_1 is normally off and Q_2 is normally on. This is known as the *stable state*. Only a positive pulse at the trigger input changes the state. It turns Q_1 on, which causes Q_2 to turn off. The circuit remains in this state (known as the *quasi-stable state*) for a time that depends on the time constant of the $R_2 C_C$ coupling circuit. When Q_2 comes on again, it turns Q_1 off and the circuit is returned to its stable state.

Before the following detailed analysis is studied the astable analysis methods given in Sec. 9-2 should be thoroughly understood.

Detailed Analysis. ($-t_1$ *Just before* Q_1 *comes on*). The waveforms for the collector-base coupled monostable MV are shown in Fig. 9-12(b). At the instant just before the trigger pulse is applied ($-t_1$), the circuit is in its stable state. Transistor Q_2 is on, with $v_{B_2} = V_{BE_{sat}} = 0.7$ V, because its base is returned to $V_{BB_2} = V_{CC} = 12$ V. Its collector voltage $v_{C_2} = V_{CE_{sat}} = 0.3$ V. The combination of $v_{C_2} = V_{CE_{sat}} = 0.3$ V, $V_{BB_1} = -3$ V, and the resistive coupling network made up of R_1 and R_3 holds Q_1 off. Since the coupling capacitor C_C has fully charged, $v_{C_1} = V_{CC} = 12$ V.

Example 9-2: Verify that Q_1 is cut off in the stable state and determine the voltage on C_C after it has finished charging.

Solution: The circuit of Fig. 9-13(a) is used to prove that Q_1 is cut off in the stable state. Its base sees R_1 returned to V_{BB_1} and R_3 returned to the fixed

---

*There are several other types of monostable multivibrators which can be found in pulse circuit textbooks. The collector-base coupled *mono*, which is a frequently used circuit, is the only one discussed in this text.

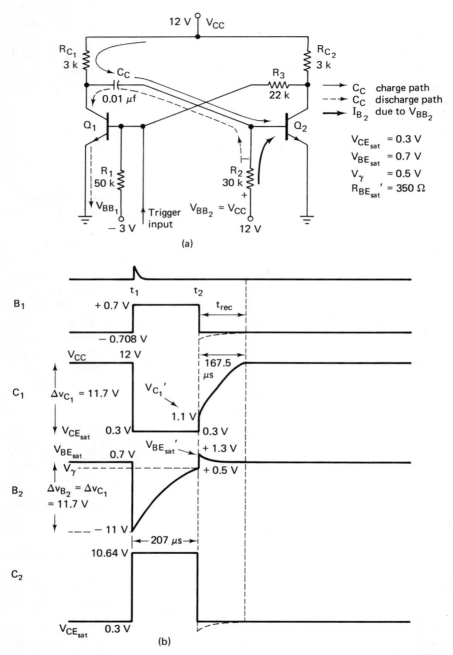

Fig. 9-12 (a) A collector-base-coupled monostable MV with (b) its waveforms.

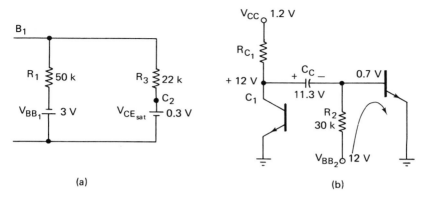

Fig. 9-13 The circuits for Ex. 9-2.

voltage $V_{CE_{sat}}$ at the collector of Q_2. By applying Eq. (1-4), it can be seen that the open-circuited voltage seen by the base of Q_1 is

$$v_{B_1(TH)} = \frac{(V_{BB_1})(R_3) + (V_{CE_{sat}})(R_1)}{R_1 + R_3}$$

$$= \frac{(-3)(22) + (0.3)(50)}{50 + 22}$$

$$= \frac{-66 + 15}{72}$$

$$v_{B_1(TH)} = -0.708 \text{ V}$$

Since the emitter is grounded, $v_{BE_1} = -0.708$ V, and Q_1 is cut off.

The circuit of Fig. 9-13(b) is used to find the voltage on C_C after it is fully charged. With Q_1 cut off and the charging current reduced to zero, there is no current through R_{C_1}. Hence, $v_{R_{C(1)}} = 0$ and $v_{C_1} = V_{CC} = 12$ V.

Q_2 is kept in saturation, with v_{B_2} held at $V_{BE_{sat}} = 0.7$ V by $V_{BB_2} = 12$ V. Therefore, the voltage on C_C at the *start* (S) of discharge is

$$v_{C_C(S)} = v_{C_1} - v_{B_2}$$

$$= V_{CC} - V_{BE_{sat}}$$

$$= 12 - 0.7$$

$$v_{C_C(S)} = 11.3 \text{ V (positive on the } C_1 \text{ side)}$$

Example 9-3: Find the minimum value of h_{FE} that will permit Q_2 to remain saturated after C_C has fully charged and the overdrive that it applies to B_2 is removed.

276 / Astable and Monostable Multivibrators

Solution: The circuit of Fig. 9-14 will be used to calculate $h_{FE_{min}}$. The base current

$$I_{B_2} = \frac{V_{BB_2} - V_{BE_{sat}}}{R_2} \tag{9-5}$$

$$= \frac{12 - 0.7}{30 \text{ k}}$$

$$I_{B_2} = 0.377 \text{ mA}$$

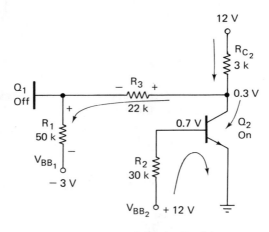

Fig. 9-14 The circuit for Ex. 9-3.

The collector current

$$I_{C_2} = I_{R_{C(2)}} - I_{R_3} \tag{9-6}$$

where

$$I_{R_{C(2)}} = \frac{V_{CC} - V_{CE_{sat}}}{R_{C_2}}$$

$$= \frac{12 - 0.3}{3 \text{ k}}$$

$$I_{R_{C(2)}} = 3.9 \text{ mA}$$

and

$$I_{R_3} = I_{R_1} = \frac{V_{CE_{sat}} - V_{BB_1}}{R_1 + R_3}$$

$$= \frac{0.3 - (-3)}{72 \text{ k}}$$

$$I_{R_3} = I_{R_1} = 45.8 \text{ }\mu\text{A}$$

Therefore,
$$I_{C_2} = 3.9 - 0.0458$$
$$I_{C_2} \approx 3.85 \text{ mA}$$

and

$$h_{FE_{min(2)}} = \frac{I_{C_2}}{I_{B_2}}$$

$$= \frac{3.85}{0.377}$$

$$h_{FE_{min(2)}} = 10.2$$

Just after the trigger pulse is applied (t_1+). A positive pulse applied to the trigger input turns Q_1 on. Its collector voltage drops from $V_{CC} = 12$ V to $V_{CE_{sat}} = 0.3$ V, and the 11.7 V drop is coupled through C_C to the base of Q_2. The conditions in the C_1-to-B_2 circuit just before Q_1 comes on are shown in Fig. 9-15(a). $v_{C_1} = V_{CC} = 12$ V, $v_{B_2} = V_{BE_{sat}} = 0.7$ V, $v_{C_C} = 11.3$ V, and $v_{R_2} = 11.3$ V with the polarity shown.

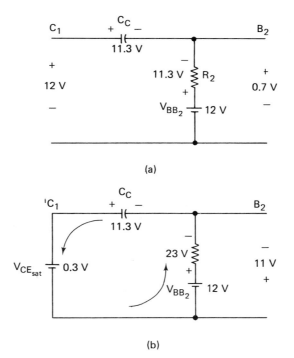

Fig. 9-15 The C_1-to-B_2 circuit (a) just before Q_1 comes on and (b) just after Q_1 comes on.

The conditions that exist in this circuit just after Q_1 comes on are shown in Fig. 9-15(b). The voltage at the collector of Q_1 drops to $V_{CE_{sat}} = 0.3$ V, but v_{C_C} remains the same. Therefore, the voltage at the base of Q_2 at the start of discharge is

$$v_{B_2} = v_{C_C(S)} + v_{C_1} \qquad (9\text{-}7)$$
$$= -11.3 + 0.3$$
$$v_{B_2} = -11 \text{ V}$$

The coupling capacitor begins to discharge through the path shown in Fig. 9-15(b). The voltage on the discharge resistance R_2 at the start of discharge is

$$v_{RS} = V_{BB_2} - v_{B_2} \qquad (9\text{-}8)$$
$$= 12 - (-11)$$
$$v_{RS} = 23 \text{ V}$$

Example 9-4: Calculate the minimum value of h_{FE} that will permit Q_1 to go into saturation just after t_1. This is the time when it must sink the highest discharge current from C_C.

Solution: The circuits of Fig. 9-16 are used to calculate $h_{FE_{min}}$. From Fig. 9-16(a),

$$I_{C_1} = I_{R_{C(1)}} + I_{C_C(S)} \qquad (9\text{-}9)$$

where

$$I_{R_{C(1)}} = \frac{V_{CC} - V_{CE_{sat}}}{R_{C_1}}$$
$$= \frac{12 - 0.3}{3 \text{ k}} \qquad (9\text{-}10)$$
$$I_{R_{C(1)}} = 3.9 \text{ mA}$$

and I_{C_C} at the *start* of discharge is

$$I_{C_C(S)} = \frac{v_{RS}}{R_2} \qquad (9\text{-}11)$$
$$= \frac{23}{30 \text{ k}}$$
$$I_{C_C(S)} = 0.77 \text{ mA}$$

Hence, at t_1+, the collector of Q_1 must sink

$$I_{C_1} = 3.9 \text{ mA} + 0.7 \text{ mA}$$
$$I_{C_1} = 4.67 \text{ mA}$$

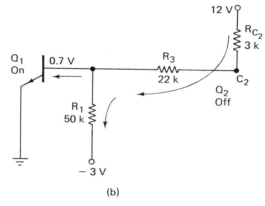

Fig. 9-16 The circuits for Ex. 9-4.

From Fig. 9-16(b), the base current of Q_1 is

$$I_{B_1} = I_{R_3} - I_{R_1} \tag{9-12}$$

where

$$I_{R_3} = I_{RC(2)} = \frac{V_{CC} - V_{BE_{sat}}}{R_3 + R_{C_1}} \tag{9-13}$$

$$= \frac{12 - 0.7}{25 \text{ k}}$$

$$I_{R_3} = I_{RC(2)} = 0.452 \text{ mA}$$

and

$$I_{R_1} = \frac{V_{BE_{sat}} - V_{BB_1}}{R_1} \tag{9-14}$$

$$= \frac{0.7 - (-3)}{50 \text{ k}}$$

$$I_{R_1} = 0.074 \text{ mA}$$

Therefore,
$$I_{B_1} = 0.452 - 0.074$$
$$I_{B_1} = 0.378 \text{ mA}$$

and
$$h_{FE_{\min(1)}} = \frac{I_{C_1}}{I_{B_1}}$$
$$= \frac{4.67}{0.378}$$
$$h_{FE_{\min(1)}} = 12.4$$

Example 9-5: Calculate the voltage that is present at the collector of Q_2 while Q_2 is off.

Solution: The circuit shown in Fig. 9-17 is used to calculate $v_{C_{2(off)}}$. Several techniques may be used to find the voltage at C_2. Since Q_2 is off, the circuit between the two points marked X is open and Thévenin's theorem can be used. By applying Eq. (1-4) to Fig. 9-17, it can be seen that

$$v_{C_{2(off)}} = \frac{(V_{BE_{sat}})(R_{C_2}) + (V_{CC})(R_3)}{R_{C_2} + R_3}$$
$$= \frac{(0.7)(3) + (12)(22)}{25}$$
$$= \frac{2.1 + 264}{25}$$
$$v_{C_{2(off)}} = 10.64 \text{ V}$$

Just before Q_2 comes on $(-t_2)$. The coupling capacitor continues to

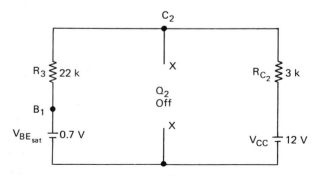

Fig. 9-17 The circuit for Ex. 9-5.

discharge through the path shown in Fig. 9-15(b) [also Fig. 9-12(a)] until v_{BE_2} reaches its cutin value $V_\gamma = 0.5$ V. The circuit at the instant before Q_2 comes on is shown in Fig. 9-18. As it is indicated on the diagram, C_c (because of the value of V_γ chosen for this circuit) actually begins to charge to an opposite polarity voltage equal to $V_{BB_2} - V_{CE_{sat}} = 11.7$ V.

When Q_2 comes on, it causes Q_1 to cut off. The capacitor must then charge back to 11.3 V, positive on the C_1 side (see Ex. 9-2). From Fig. 9-18, since $v_{BE} = V_\gamma = 0.5$ V and Q_1 is still on, with $v_{C_1} = V_{CE_{sat}} = 0.3$ V, the capacitor voltage at the instant before Q_2 comes on is

$$v_{C_C} = V_\gamma - V_{CE_{sat}}$$
$$= 0.5 - 0.3$$
$$v_{C_C} = 0.2 \text{ V (positive on the } B_2 \text{ side)}$$

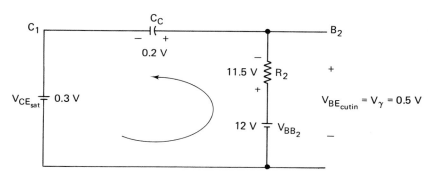

Fig. 9-18 The C_1-to-B_2 circuit at the time $-t_2$ (just before Q_2 comes on).

The voltage on the discharge resistance R_2 when C_C finishes discharging is

$$v_{RF} = V_{BB} - V_\gamma \quad (9\text{-}15)$$
$$= 12 - 0.5$$
$$v_{RF} = 11.5 \text{ V}$$

The pulse width PW can now be calculated.

Example 9-6: Calculate the pulse width PW.
Solution: By using Eq. (1-7), it can be seen that

$$t = 2.3RC \log \frac{v_{RS}}{v_{RF}}$$

282 / *Astable and Monostable Multivibrators*

where

$t = PW$;

$R = R_2$;

$C = C_2$;

$v_{RS} = 23$ V from Eq. (9-8);

$v_{RF} = 11.5$ V from Eq. (9-15).

Note, however, that $v_{RS}/v_{RF} = 2$. The half of the monostable that includes C_C is the same as an astable multivibrator. Therefore, if $V_{BB_2} = V_{CC}$,

$$PW = 0.69 \, R_2 C_C \qquad (9\text{-}16)$$
$$= 6.9 \times 10^{-1} \times 3 \times 10^4 \times 10^{-8}$$
$$PW = 207 \; \mu s$$

Just after Q_2 *comes on* (t_2+). When Q_2 comes on, it cuts off Q_1 and C_C begins to discharge the 0.2 V (positive on the B_2 side) and charge back to 11.3 V (positive on the C_1 side). The charging current provides a high overdrive base current for Q_2, which causes it to quickly go into saturation. The equivalent C_1-to-B_2 circuit at t_2+ is shown in Fig. 9-19(a). The base-

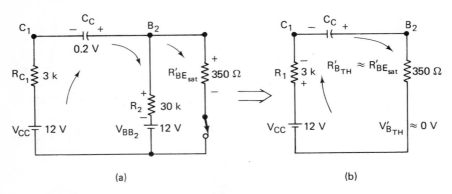

Fig. 9-19 The C_1-to-B_2 circuit just before Q_2 comes on (t_2+). For Ex. 9-7.

emitter junction is represented by a *dc* resistance $R'_{BE_{sa}} = 350 \, \Omega$ (the prime mark indicates that overdrive is present because C_C is charging).

Example 9-7: Calculate the voltages at the base of Q_2 and collector of Q_1 (v'_{B_2} and v'_{C_1}) at t_2+.

Solution: The Thévenin equivalent of the B_2-to-ground circuit is shown in Fig. 9-19(b). $R'_{B_{TH}} \approx R'_{BE_{sat}} = 350 \, \Omega$ and

$$v'_{B_{TH}} = \frac{(R'_{BE_{sat}})(V_{BB_2})}{R'_{BE_{sat}} + R_2}$$

$$= \frac{(0.35)(12)}{30.35}$$

$$v'_{B_{TH}} = 0.14 \, \text{V} \approx 0 \, \text{V}$$

Since $R'_{BE_{sat}}$ is always very much less than R_2, the B_2-to-ground circuit may be represented by $R'_{BE_{sat}}$ as shown in Fig. 9-19(b).

The voltage on the resistance at the start of the charging interval is

$$v'_{RS} = V_{CC} + v_{C_C}$$

$$= 12 + 0.2$$

$$v'_{RS} = 12.2 \, \text{V}$$

The collector voltage is

$$v'_{C_1} = v'_{R_{C(1)}} + v_{CC} \qquad (9\text{-}17)$$

where

$$v'_{R_{C(1)}} = \frac{(R_{C_1})(v'_{RS})}{R_{C_1} + R'_{BE_{sat}}}$$

$$= \frac{(3)(12.2)}{3.35}$$

$$v'_{R_{C(1)}} = 10.9 \, \text{V}$$

The polarity of this voltage is shown in Fig. 9-19(b). Hence,

$$v'_{C_1} = -10.9 + 12$$

$$v'_{C_1} = 1.1 \, \text{V}$$

At the same time, if it is assumed that $V'_{B_{TH}} = 0$

$$v'_{B_2} = v'_{R_{BE(sat)}} \qquad (9\text{-}18)$$

where

$$v'_{R_{BE(sat)}} = v'_{RS} - v'_{R_{C(1)}}$$

$$= 12.2 - 10.9$$

$$v'_{R_{BE(sat)}} = 1.3 \, \text{V}$$

The polarity of this voltage is also shown in Fig. 9-19(b). Therefore,

$$v'_{B_2} = v'_{R_{BE(\text{sat})}}$$
$$v'_{B_2} = 1.3 \text{ V}$$

Note that the changes in v_{C_1} and v_{B_2} at t_2+ are equal. This is to be expected because C_1 and B_2 are connected by a capacitor.

From Fig. 12-19(b), the current into the base of Q_2 at the instant C_C begins to charge is

$$I'_{B_{\text{sat}}} = \frac{v'_{BE_{\text{sat}}}}{R'_{BE_{\text{sat}}}}$$
$$= \frac{1.3}{0.35 \text{ k}}$$
$$I'_{B_{\text{sat}}} = 3.71 \text{ mA}$$

This current is approximately ten times greater than $I_{B_{\text{sat}}}$, the current that keeps the transistor in saturation after C_C finishes charging.

Recovery Time t_{rec}. As explained in the analysis of the astable MV (see Recovery Time in Sec. 9-2), as the coupling capacitor charges, the overdrive current continuously decreases and $R'_{BE_{\text{sat}}}$ increases. Thus, C_C is always charging through a different resistance toward a different voltage. The value of $R'_{BE_{\text{sat}}} = 350 \text{ }\Omega$ is only for the instant just after Q_2 comes on. It permits the base overshoot and v'_{C_1} at t_2+ to be calculated. As explained in Sec. 9-2, the recovery time can be approximated with very good accuracy by using this resistance in Eq. (9-4).

$$t_{\text{rec}} = 5(R_{C_1} + R'_{BE_{\text{sat}}})C_C$$
$$= 5 \times 3.35 \times 10^3 \times 10^{-8}$$
$$t_{\text{rec}} = 167.5 \text{ }\mu\text{s}$$

The circuit has now returned to its stable state: Q_2 on and Q_1 off, with $v_{C_2} = V_{CE_{\text{sat}}} = 0.3 \text{ V}$, $v_{B_2} = V_{BE_{\text{sat}}} = 0.7 \text{ V}$, $v_{B_1} = -0.708 \text{ V}$, $v_{C_1} = V_{CC} = 12 \text{ V}$, and $v_{C_C} = 11.3 \text{ V}$. It remains in this state until the next trigger pulse is applied.

9-7 An Application of the Monostable MV

One important use of the monostable multivibrator is in delay circuits. Figure 9-20(a) shows a pulse (*A*) being used to trigger an *npn* mono that has a pulse width equal to 10 μs. Its output, from the collector of the normally on transistor (v_{C_2} in Fig. 9-12), is the rectangular wave (*B*).

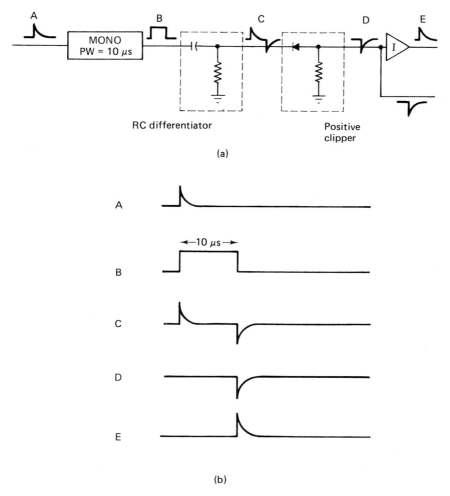

Fig. 9-20 (a) A delay circuit using a monostable MV. (b) The waveforms in a ladder diagram to show the time delay.

The mono output is differentiated (waveform C) and then applied to a positive clipper. The output of the clipper, which is simply a half-wave rectifier, is the negative spike (D) of the differentiator output. If the positive polarity of the input pulse must be maintained, an inverter circuit can be used. The output wave (E), as shown in the ladder diagram in Fig. 9-20(b), is delayed 10 μs with respect to the time of the input wave (A). Complementary outputs can be obtained by using both waveforms D and E. The amount of delay depends on the PW of the monostable MV, which can be varied as required.

9-8 Summary

The astable multivibrator is a free-running, square-wave generator. It is sometimes used as a clock pulse generator in digital computers. If the collector and base supplies are the same, the frequency of oscillation depends only on the time constants of the RC coupling networks. The frequency can be synchronized by positive pulses applied to one of the (*npn*) bases. If the free-running frequency is slightly lower than the sync pulse frequency, the astable will be forced to oscillate at the sync frequency. If it is slightly lower than $\frac{1}{2}$ (or $\frac{1}{3}$, $\frac{1}{4}$, etc.), the sync frequency, the MV will act as a frequency divider. Its output will be exactly $\frac{1}{2}$ (or $\frac{1}{3}$, $\frac{1}{4}$, etc.) of the sync pulse frequency.

The monostable multivibrator has only one stable state. One transistor is normally on and the other is normally off. Only a trigger pulse temporarily changes the state of the circuit. The width of the output pulse depends on the amount of time the circuit spends in its *quasi-stable* state. If the collector and the base supplies of the normally on transistor are the same, the pulse width $PW = 0.69RC$, where RC is the time constant of the collector-base coupling network. The mono is frequently used in time delay circuits. It is also used to shape pulses and to change the width of a pulse.

Problems

9-1 In Fig. 9-2, $R_{C_1} = R_{C_2} = 1.2$ k, $C_{C(1)} = C_{C(2)} = 200$ pF, $R_1 = R_2 = 100$ k, $V_{BB} = V_{CC} = 12$ V, $V_{BE_{sat}} = 0.7$ V, $V_{CE_{sat}} = 0.3$ V, $V_\gamma = 0.5$ V, and $R'_{BE_{sat}} = 220$ Ω. Using the procedure of Ex. 9-1 (Sec. 9-2), draw the collector and base waveforms to scale. Indicate all important voltages.

9-2 In the circuit of Prob. 9-1
  (a) How much base current $I'_{B_{sat}}$ flows at the instant Q_2 comes on?
  (b) What is the Q_1 collector recovery time t_{rec}?
  (c) Determine the frequency F.

9-3 What changes will occur in the answers to Probs. 9-1 and 9-2 if $C_{C(2)}$ is changed to 100 pF?

9-4 What value of R_C will permit each collector in the circuit of Prob. 9-1 to recover in 5 μs?

9-5 If $V_{CC} = 12$ V but V_{BB} is changed to 6 V, what is the frequency of the circuit of Prob. 9-1?

9-6 The *pnp* astable MV of Fig. 9-11 has the following values: $R_{C_1} = R_{C_2} = 1$ k, $C_{C(1)} = C_{C(2)} = 200$ pF, $R_1 = R_2 = 60$ k, $V_{CC} = V_{BB} = -12$ V,

$V_{BE_{sat}} = -0.7$ V, $V_{CE_{sat}} = -0.3$ V, $V_\gamma = -0.5$ V, and $R'_{BE_{sat}} = 200\,\Omega$. Draw the collector and base waveforms to scale and indicate all important voltages.

9-7 Calculate the value of $I'_{B_{sat}}$ (the base current at the instant Q_2 comes on), the Q_1 collector recovery time t_{rec}, and the frequency of the circuit of Prob. 9-6.

9-8 Given the monostable circuit of Fig. 9-12(a) with $R_{C_1} = R_{C_2} = 4$ k, $R_2 = R_3 = 20$ k, $R_1 = 80$ k, $C_C = 680$ pF, $V_{BB_2} = V_{CC} = 10$ V, $V_{BB_1} = -6$ V, $V_{BE_{sat}} = 0.7$ V, $V_{CE_{sat}} = 0.3$ V, $V_\gamma = 0.5$ V, and $R'_{BE_{sat}} = 250\,\Omega$, determine the voltage on C_C and on each base and collector in the stable state.

9-9 Determine the minimum value of h_{FE} that will permit Q_2 in the circuit of Prob. 9-8 to remain saturated after C_C has fully charged.

9-10 Determine the voltage on each base and collector just after a trigger pulse puts the one-shot of Prob. 9-8 in its quasi-stable state.

9-11 Calculate the minimum value of h_{FE} that will permit Q_1 in the circuit of Prob. 9-8 to saturate just after it comes on and turns Q_2 off. (Note: Q_1 must be able to sink the discharge current of C_C.)

9-12 Calculate the base overshoot and v_{C_1} at the instant just after Q_2 comes back on and turns Q_1 off in the mono of Prob 9-8.

9-13 Calculate the pulse width PW and collector recovery time t_{rec} of the monostable MV of Prob. 9-8.

9-14 Given the one-shot MV of Fig. 9-17(a) with *pnp* transistors, $V_{CC} = V_{BB_2} = -12$ V, $V_{BB_1} = 3$ V, $R_{C_1} = R_{C_2} = 4$ k, $C_C = 0.002\,\mu\text{f}$, $R_1 = 80$ k, $R_2 = R_3 = 40$ k, $V_{CE_{sat}} = -0.3$ V, $V_{BE_{sat}} = -0.7$ V, $V_\gamma = -0.5$ V, and $R'_{BE_{sat}} = 550\,\Omega$, repeat Probs. 9-8 through 9-13.

10

THE BASIC BISTABLE

AND

MODERN LOGIC FLIP-FLOP

As explained in the introduction of Chap. 9, the bistable is used as a memory device in digital computers. It is used in counters, frequency dividers, and shift registers. The flip-flop has two stable states. One transistor is saturated and the other is cut off. Only a pulse of proper polarity changes the state, causing the saturated transistor to cut off and the cut-off transistor to saturate. It remains in the new state until another pulse of proper polarity is applied.

In this chapter the operation of a basic bistable multivibrator is discussed in detail. Some basic logic flip-flops are discussed and various modern logic flip-flops are explained in detail.

10-1 The Basic Flip-flop

The basic flip-flop, as shown in Fig. 10-1, is simply two cross-coupled inverters. The output of one drives the other. When Q_1 is on, its collector applies a low voltage ($V_{CE_{sat}} = 0.3$ V) to R_1, the input resistor of the Q_2 inverter. This cuts off the *npn* transistor and sends V_{C_2} to a high voltage (nearly V_{CC}). This positive voltage is applied to R_2, the input resistor of the Q_1 inverter, and holds the *npn* transistor in saturation.

In the following examples it will be proved that if one transistor is on, it cuts off the other transistor. The minimum $\mathbf{h}_{FE}$ required to keep the on transistor saturated will be determined and all stable-state currents and voltages will be calculated.

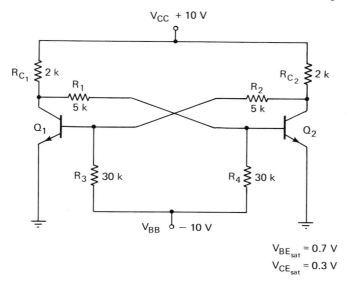

Fig. 10-1 A basic flip-flop.

Example 10-1: Assuming that Q_1 in Fig. 10-1 is on with $V_{BE} = V_{BE_{sat}} = 0.7$ V and $V_{CE} = V_{CE_{sat}} = 0.3$ V, verify that Q_2 is off if Q_1 is on.

Solution: The circuit seen by the base of Q_2 when Q_2 is off and Q_1 is saturated is shown in Fig. 10-2. There are two ground-return paths: (1) through R_4 and V_{BB} and (2) through R_1 and the saturated transistor Q_1. Since the voltage from the collector of Q_1 to ground is fixed at $V_{CE_{sat}} = 0.3$ V, regardless of the values of R_{C_1} and V_{CC}, the path through these components can be omitted from the circuit.

The open-circuit voltage $V_{B_{TH}}$ is

$$V_{B_{TH}} = \frac{(V_{C_1})(R_4) + (V_{BB})(R_1)}{R_1 + R_4} \qquad (1\text{-}4)$$

$$= \frac{(0.3)(30) + (-10)(5)}{5 + 30}$$

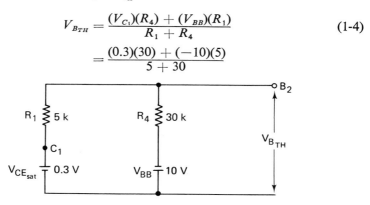

Fig. 10-2 The circuit seen by the base of Q_2 when Q_2 is off and Q_1 is saturated.

$$V_{B_{TH}} = -1.171 \text{ V}$$

Since the emitter is grounded,

$$V_{BE_2} = -1.171 \text{ V}$$

Therefore, Q_2 must be cut off if Q_1 is in saturation.

Example 10-2: Calculate the minimum value of h_{FE} required for Q_1 to remain in saturation with Q_2 cut off.

Solution: The various current paths are shown in Fig. 10-3. The collector current of Q_1 is

$$I_{C_1} = I_{R_{C(1)}} - I_{R_1} \tag{10-1}$$

where

$$I_{R_1} = I_{R_4} = \frac{V_{CE_{sat}} - V_{BB}}{R_1 + R_4}$$

$$= \frac{(0.3) - (-10)}{35 \text{ k}}$$

$$I_{R_1} = I_{R_4} = 0.296 \text{ mA}$$

and

$$I_{R_{C(1)}} = \frac{V_{CC} - V_{CE_{sat}}}{R_{C_1}}$$

$$= \frac{10 - 0.3}{2 \text{ k}}$$

$$I_{R_{C(1)}} = 4.85 \text{ mA}$$

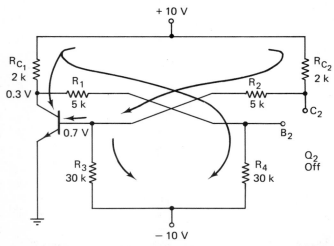

Fig. 10-3 The current paths when Q_1 is saturated and Q_1 is cut off.

Hence,
$$I_{C_1} = I_{R_{C(1)}} - I_{R_1}$$
$$= 4.85 - 0.296$$
$$I_{C_1} = 4.554 \text{ mA}$$

The base current of Q_1 is

$$I_{B_1} = I_{R_2} - I_{R_3} \tag{10-2}$$

where

$$I_{R_2} = I_{R_{C(2)}} = \frac{V_{CC} - V_{BE\text{sat}}}{R_2 + R_{C_2}}$$
$$= \frac{10 - 0.7}{7 \text{ k}}$$
$$I_{R_2} = I_{R_{C(2)}} = 1.329 \text{ mA}$$

and

$$I_{R_3} = \frac{V_{BE\text{sat}} - V_{BB}}{R_3}$$
$$= \frac{(0.7) - (-10)}{30 \text{ k}}$$
$$I_{R_3} = 0.357 \text{ mA}$$

Hence,
$$I_{B_1} = I_{R_2} - I_{R_3}$$
$$= 1.329 - 0.357$$
$$I_{B_1} = 0.972 \text{ mA}$$

The minimum value of h_{FE} can now be calculated.

$$h_{FE\text{min}} = \frac{I_{C\text{sat}}}{I_B}$$
$$= \frac{4.554}{0.972}$$
$$h_{FE\text{min}} = 4.69$$

As long as the transistors have a value of h_{FE} of at least 4.69, the circuit operates as a bistable flip-flop. The actual h_{FE} is higher to permit this circuit to sink current from sink-type loads. A higher h_{FE} also permits the flip-flop to change state more quickly because the circuit will then have overdrive with $I_B = 0.972$ mA.

All currents and voltages except for the voltage on the collector of the off transistor are now known. The value of V_{C_2} is determined in the following example.

Example 10-3: Determine the voltage on the collector of the off transistor V_{C_2}.
Solution: The circuit of Fig. 10-4 shows the two ground-return paths from the collector of Q_2: (1) through R_{C_2} and V_{CC} and (2) through R_2 and the B-E

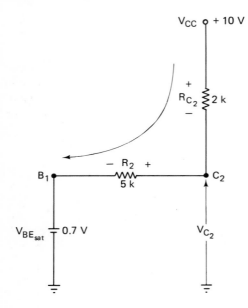

Fig. 10-4 The circuit for Ex. 10-3 showing the two ground-return paths from the collector of Q_2.

junction of Q_1. Since V_{B_1} is fixed at $V_{BE_{sat}} = 0.7$ V, regardless of the values of R_3 and V_{BB}, the path made up of these components can be omitted from the circuit.

From Fig. 10-4,

$$V_{C_2} = V_{R_2} + V_{B_1} \tag{10-3}$$

where

$$V_{R_2} = I_{R_2} R_2$$

The current through R_2 was determined in Ex. 10-2 to be 1.329 mA. Hence,

$$V_{C_2} = (1.329)(5) + 0.7$$
$$= 6.645 + 0.7$$
$$V_{C_2} = 7.345 \text{ V}$$

Summarizing, in the stable state, with Q_1 on and Q_2 off,

$$V_{C_1} = 0.3 \text{ V} \qquad V_{C_2} = 7.345 \text{ V}$$
$$V_{B_1} = 0.7 \text{ V} \qquad V_{B_2} = -1.171 \text{ V}$$

And, in order for the circuit to function properly, h_{FE} must be at least 4.69.

10-2 The Basic Logic Flip-flop

The basic flip-flop circuit of Fig. 10-1 is modified in Fig. 10-5 to form a basic logic flip-flop. The collectors are returned to a 5 V supply through clamping (catching) diodes. The clamping voltage $V_{CL} = V_H$, the high-level voltage of the system. The voltages at the base and collector of the saturated transistor

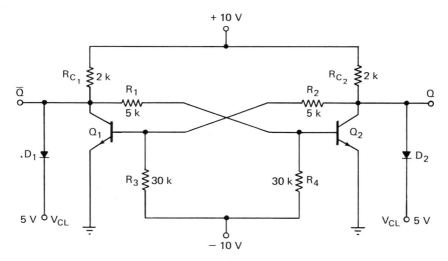

Fig. 10-5 The basic logic flip-flop.

remain the same: $V_C = V_{CE_{sat}} = 0.3$ V and $V_B = V_{BE_{sat}} = 0.7$ V. The voltage at the base of the cut-off transistor also remains the same because that base still sees the same circuit (see Fig. 10-2). However, the voltage at the collector of the off transistor differs from that of the basic flip-flop.

In Ex. 10-3 it was found that the collector of the off transistor of the basic flip-flop goes to 7.345 V. In the logic flip-flop of Fig. 10-5 the off collector *aims* for 7.345 V. But when the collector voltage exceeds V_{CL} by V_γ, the cutin voltage of the diode, the diode comes on and clamps V_C at $V_D + V_{CL}$ = 5.7 V.

Thus, assuming that Q_1 is on and Q_2 is off, the steady-state voltages in the logic flip-flop of Fig. 10-5 are:

$$V_{C_1} = V_{CE\text{sat}} = 0.3\text{ V} \qquad V_{C_2} = V_{D_2} + V_{CL} = 5.7\text{ V}$$
$$V_{B_1} = V_{BE\text{sat}} = 0.7\text{ V} \qquad V_{B_2} = -1.171\text{ V (see Ex. 10-1)}$$

With the Q or 1-output line high and the $\bar{Q}$ or 0-output line low, the flip-flop is said to be in the 1 state. That is, it is storing a logical 1.

It remains in this state until a pulse of proper amplitude and polarity is applied and causes Q_1 to come far enough out of saturation to turn Q_2 on. Regeneration then quickly causes Q_1 to cut off, Q_2 to saturate, and the steady-state voltages to change to:

$$V_{C_1} = V_{D_1} + V_{CL} = 5.7\text{ V} \qquad V_{C_2} = V_{CE\text{sat}} = 0.3\text{ V}$$
$$V_{B_1} = -1.171\text{ V} \qquad V_{B_2} = V_{BE\text{sat}} = 0.7\text{ V}$$

With the $\bar{Q}$ or 0-output line high and the Q or 1-output line low, the flip-flop is said to be in the 0 state. That is, it is storing a logical 0.

10-3 Types of Logic Flip-flops

During the history of computers, five types of logic flip-flops have been developed. They are the *Toggle (T)*, the *Reset-Set (RS)*. The *Reset-Set Toggle (RST)*, the *JK*, and the *D-types*. Today the *JK* and *D-types* are most popular. However, they are usually provided with Direct *PRECLEAR* or *PRERESET* R_D and Direct *PRESET* S_D inputs. They also are frequently connected to act like T-type flip-flops. All operations will therefore be explained.

The Toggle T-type Flip-flop. The logic symbol and state table for a T-type flip-flop are shown in Fig. 10-6(a) and (b). The clock pulse is *AND*ed externally to the T-input control function f_T, that is, $T = f_T \cdot P$. If a clock pulse P is applied while the T input function is high, the state of the flip-flop is complemented. If the T input function is low, the state of the flip-flop remains the same with or without a clock pulse.

The Reset-Set (RS) Flip-flop. The logic symbol and state table for an RS-type flip-flop are shown in Fig. 10-7(a) and (b).

The S_D and R_D inputs are asynchronous direct *PRESET* and *PRE-CLEAR* inputs. They permit the flip-flop to be put in the 0 or 1 state without the presence of a clock pulse. If the flip-flops are used in an up counter, for example, it would be desirable to have all of the flip-flops precleared so the counter can start at zero. The flip-flops in a down counter would be preset (put in the 1 state). The R_D and S_D inputs can be active when at the 1 level or

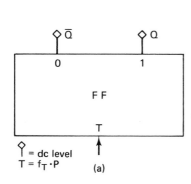
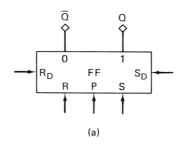

Fig. 10-6 (a) The logic symbol of a Toggle T flip-flop with (b) its state table.

Fig. 10-7 (a) The logic symbol of a Reset-Set RS flip-flop with (b) its state table.

the 0 level, depending on the way the flip-flops are made. See Figs. 10-11, 10-12, and 10-13.

It is assumed here that the R and S inputs are active when at the 1 level. If a clock pulse P is applied while R and S are both low, the state of the flip-flop does not change. If it is applied while R is high, the flip-flop goes to the 0 state regardless of its present state. If S is high, the clock pulse will send the flip-flop to the 1 state regardless of its present state. The R and S inputs must not be made active (1 level here) simultaneously. As shown in the state table of Fig. 10-7(b), if R and S are both high, the next state is indeterminate.

The Reset-Set-Toggle (RST) Flip-flop. The logic symbol and state table for an RST flip-flop are shown in Fig. 10-8. It combines the operations of the RS and T types. As in the RS-type, the R and S inputs must not be made active at the same time. This flip-flop is not made today, but it found frequent

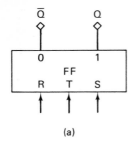

(a)

Q_n*	R	T	S	Q_{n+1}†
0	0	0	0	0
0	0	0	1	1
0	0	1	0	1
0	0	1	1	1
0	1	0	0	0
0	1	0	1	‡
0	1	1	0	0
0	1	1	1	‡
1	0	0	0	1
1	0	0	1	1
1	0	1	0	0
1	0	1	1	1
1	1	0	0	0
1	1	0	1	‡
1	1	1	0	0
1	1	1	1	‡

*Q_n is the state of the flip-flop *now* or before the clock pulse is applied.
†Q_{n+1} is the state of the flip-flop after the clock pulse is applied.
‡Indeterminate.

(b)

Fig. 10-8 (a) The logic symbol of an RST flip-flop with (b) its state table.

use in the recent past and may very likely be encountered in older equipment. Direct Set and Direct Reset inputs were not available and the clock pulse had to be *AND*ed externally to the R, S, and T input functions for synchronous operation. Note that the state table of Fig. 10-8(b) shows that if R and T are high simultaneously, R overrides T, and it shows that if S and T are both high at the same time, S overrides T.

The JK Flip-flop. The logic symbol and state table for a JK flip-flop are shown in Fig. 10-9(a) and (b). It is assumed here that the J and K inputs are active when at the 1 level. If both are low, the state of the flip-flop remains the same. If the clock pulse is applied while J is high, the flip-flop goes to the 1 state regardless of its present state. If K is high, the clock pulse sends the flip-flop to the 0 state regardless of its present state. The important difference between the JK and the RS flip-flops occurs when both inputs are made

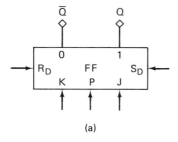

Q_n*	J	K	Q_{n+1}†
0	0	0	0
0	0	1	0
0	1	0	1
0	1	1	1
1	0	0	1
1	0	1	0
1	1	0	1
1	1	1	0

*Q_n is the state of the flip-flop *now* or before the clock pulse is applied.
†Q_{n+1} is the state of the flip-flop after the clock pulse is applied.

(b)

Fig. 10-9 (a) The logic symbol of a JK flip-flop with (b) its state table.

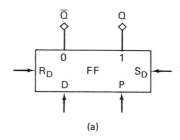

Q_n*	D	Q_{n+1}†
0	0	0
0	1	1
1	0	0
1	1	1

*Q_n is the state of the flip-flop *now* or before the clock pulse is applied.
†Q_{n+1} is the state of the flop-flop after the clock pulse is applied.

(b)

Fig. 10-10 (a) The logic symbol of a D-type flip-flop with (b) its state table.

active simultaneously. While this produces an indeterminate condition in RS flip-flops, the state of the JK flip-flop is complemented. With the absence of this constraint, the JK flip-flop requires less external gating to control its operation in counters. The JK is the most flexible of all types of flip-flops. It is often referred to as the *universal* flip-flop.

The D-type Flip-flop. The logic symbol and state table for a D-type flip-flop are shown in Fig. 10-10(a) and (b). This is a very popular flip-flop today because of the simpler circuitry required to produce it and the fact that one less input is required compared to the JK type. The state table of Fig. 10-10(b) shows that after the clock pulse is applied the state of the flip-flop will be the same as the D input regardless of its present state.

10-4 Practical Logic Flip-flop Circuits

The basic logic flip-flop of Fig. 10-5 consists of two cross-coupled inverters. There are no inputs provided for triggering the circuit. By using *NAND* gates

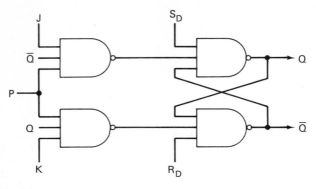

Fig. 10-11 A basic JK flip-flop using *NAND* gates. R_D and S_D are active when at the 0 level. J and K are active when at the 1 level and when P is high.

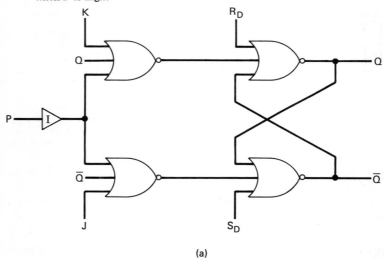

(a)

Q_n	K	J	Q_{n+1}
0	0	0	1
0	0	1	0
0	1	0	1
0	1	1	0
1	0	0	0
1	0	1	0
1	1	0	1
1	1	1	1

(b)

Fig. 10-12 (a) A basic JK flip-flop using *NOR* gates. R_D and S_D are active when at the 1 level. J and K are active when at the 0 level and P is high. (b) The state table of (a).

or *NOR* gates, practical logic flip-flops can be produced. Examples of these circuits are discussed in the following paragraphs.

Basic JK Flip-flops. The circuit of Fig. 10-11 shows a JK flip-flop constructed with *NAND* gates. The state table for this circuit is the same as Fig. 10-9(b). If Q and $\bar{Q}$ are not fed back to the input gates, the circuit will be an RS flip-flop. The J and K inputs become S and R. The state table then is the same as Fig. 10-7(b).

If the Q and $\bar{Q}$ feedback inputs are used and if the J and K inputs are tied together, the circuit becomes a Toggle T-type flip-flop. The common JK input serves as the T input and the state table is the same as Fig. 10-6(b).

The circuit of Fig. 10-12(a) is a JK flip-flop using *NOR* gates. Note that the J and K inputs are active when at the 0 level. The state table is shown in Fig. 10-12(b). If Q and $\bar{Q}$ are not fed back to the input gates, the circuit becomes an RS flip-flop. The J and K inputs become S and R. An indeterminate condition will exist if R and S are brought to the 0 level simultaneously.

If the Q and $\bar{Q}$ feedback inputs are used and if the J and K inputs are tied together, the circuit becomes a Toggle T-type flip-flop. The common JK input then serves as a T input.

Basic D-type Flip-flops. The circuits of Fig. 10-13 are D-type flip-flops. Both circuits operate when the clock pulse P goes high. The state table for these circuits is the same as Fig. 10-10(b).

If the $\bar{Q}$ output is fed back to the D inputs, the circuit becomes a Toggle T-type flip-flop and has the state table of Fig. 10-6(b).

A Modern Integrated Circuit JK Flip-flop. The circuit shown in Fig. 10-14(a) is a modern TTL integrated circuit version of the JK flip-flop. This circuit, used here with the permission of Sprague Electric Company, has multiple J and K inputs, *Direct Set* (S_D) and *Direct Clear* (R_D) inputs for asynchronous operation, and complementary Q and $\bar{Q}$ outputs. The logic symbol is shown in Fig. 10-14(b). There are two J and two K inputs, and there is one $\bar{J}$ and one $\bar{K}$ input. The J and K inputs are enabled by 1-level inputs. The $\bar{J}$ and $\bar{K}$ inputs are enabled by 0-level inputs. Unused J and K inputs should be tied to $V_{CC} = V_H$ or to a used input. Unused $\bar{J}$ and $\bar{K}$ inputs must be tied to ground or a 0-level source.

This circuit responds to the positive edge of the clock pulse. Approximately 10 ns after the pulse reaches its threshold level of 1.5 V, the logic inputs are locked out. This prevents the flip-flop from changing state more than once with each clock pulse (see Sec. 10-5). The clock input is buffered, which gives it a 1 unit-load rating ($I_u = 1.6$ mA). The Q and $\bar{Q}$ outputs are also fully buffered. The fan-out *FO* of this circuit is 10.

The asynchronous set S_D and clear R_D inputs are effective only when the clock pulse is at the 0 level. If the clock is high when either S_D or R_D equals 1, both the Q and $\bar{Q}$ outputs may simultaneously go low. The outputs go to the

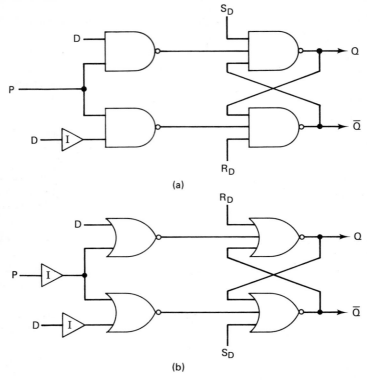

Fig. 10-13 Basic D-type flip-flops using (a) *NAND* gates (R_D and S_D are active when at the 0 level) and (b) *NOR* gates (R_D and S_D are active when at the 1 level).

state called for by the S_D and R_D inputs when the clock goes low. If $S_D = 0$ and $R_D = 0$ simultaneously, Q and $\bar{Q}$ both go low. Unused R_D and S_D inputs should be tied to V_{CC} or a 1-level valtage.

Switching Action. The logic diagram of the dc clocked JK flip-flop is shown in Fig. 10-14(c). The following assumptions are made: the flip-flop is in the 0 state ($\bar{Q}$ high and Q low) before the clock pulse is applied, all J and K inputs are high, the $\bar{J}$ and $\bar{K}$ inputs are low, and the unused R_D and S_D inputs are tied to 1-level voltages.

The unenclosed ones and zeros shown on the diagram are present at those points before the clock goes high. With $\bar{Q}$, J_{1-2} and $R_D = 1$ and $\bar{J} = 0$, all inputs to gate 3 are high, so its output is low. This low is applied to the middle input of gate 4, sending its output to the 1 level.

With $Q = 0$, the output of gate 1 is high, making the middle input of gate 2 high. The top input of gate 2 is a 1 from the output of gate 4, and its bottom (inhibit) input is a 1 from gate 5 ($P = 0$, so $\bar{P} = 1$). These inputs to gate 2 make its output a 1.

With $\bar{Q} = 1$, the output of gate 6 (input to gate 8) must be a 0. This 0 is fed back to the input of gate 7, sending its output to the 1 level. The 1 is inverted by gate 9 to hold Q at the 0 level. The other inputs to gate 7 are high; the middle input is from the gate 4 output and the bottom input is R_D. The 1 out of gate 7 is fed back to the top input of gate 6. $S_D = 1$ and the 1 out of gate 2 make all inputs to gate 6 high, producing the 0 at its output.

The boxed-in ones and zeros shown on the diagram are present just after the clock goes high. When the clock goes high, the output of gate 5 goes low. This zero is applied to the inhibit input of gate 2 and makes the output of gate 2 go low. This low has no effect on the output of gate 4, which would be held at the 1 level by its 0 input from gate 3. And it has no effect on the output of gate 1, which is held at the 1 level by its 0 input fed back from the Q output line. It does, however, make the output of gate 6 go high and thus makes $\bar{Q}$ go low.

The one out of gate 6 now makes all inputs to gate 7 high. This sends the output of gate 7 to the 0 level and the Q output to the 1 level. The state of the flip-flop has been complemented. The conditions after the flip-flop changes state and after the clock drops back to the 0 level are encircled on the diagram. After the flip-flop changes state, the Q input of gate 1 goes to the 1 level. This makes its output and the middle input of gate 2 go low. The output of gate 5, which is the inhibit input of gates 2 and 4, goes high when $P = 0$. This sends the output of gate 2 and middle input of gate 6 to the 1 level. It also maintains the 1 level out of gate 4, which is the middle input of gate 7. Since R_D and S_D are ones, the bottom inputs of gates 6 and 7 remain high. Therefore, the flip-flop is held in the 1 state by the feedback (top) inputs of gates 6 and 7.

10-5 The Timing Problem

The circuit of Fig. 10-15 is an asynchronous (*serial*) (*ripple*) flip-flop counter. Clock pulses are applied only to *FFD*, the 1s flip-flop. If it is assumed that the flip-flops toggle on the leading edge of a *T*-input pulse, *FFC* changes state whenever *FFD* goes from the 1 state to the 0 state. The 4s flip-flop changes state whenever *FFC* goes from the 1 state to the 0 state, and *FFA* changes state whenever *FFB* goes from the 1 state to the 0 state. The 1s circuit changes state once for each clock pulse. It goes from the 0 state to the 1 state after the first pulse and returns to the 0 state after the second pulse. Its 0 output line goes low after the first clock pulse and has no effect on *FFC*. The 0 line goes high after the second pulse, causing *FFC* to go from the 0 state to the 1 state. The 0 output line of *FFC* goes low and has no effect on *FFB*.

In short, the 2s flip-flop changes once for every two clock pulses into T_A, the 4s flip-flop changes once for every four clock pulses into T_A, and the 8s

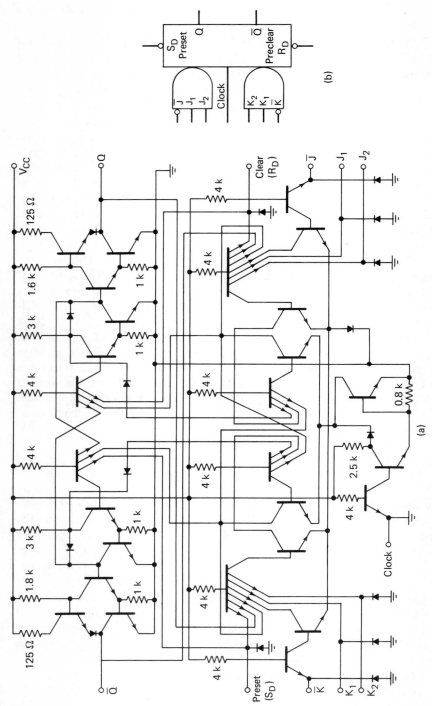

Fig. 10-14 (a) A modern integrated circuit version of a TTL JK flip-flop with (b) its logic symbol (Courtesy Sprague Electric Co.)

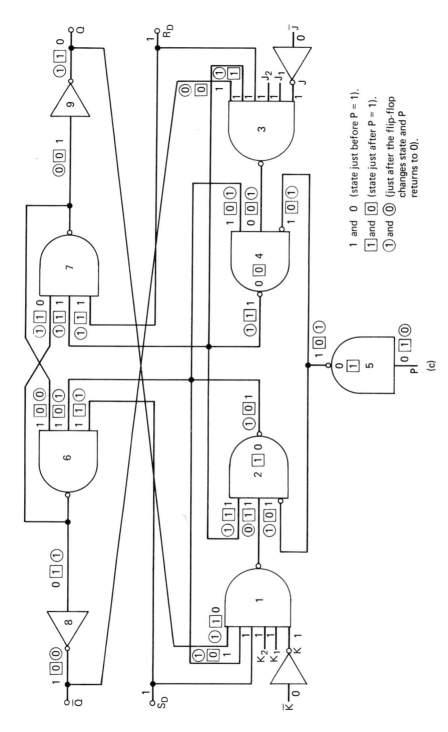

Fig. 10-14 (c) its logic circuit. ($\bar{K}$ and $\bar{J}$ remain low and S_D, R_D, and K_{1-2} remain high throughout the switching action.) (Courtesy Sprague Electric Co.)

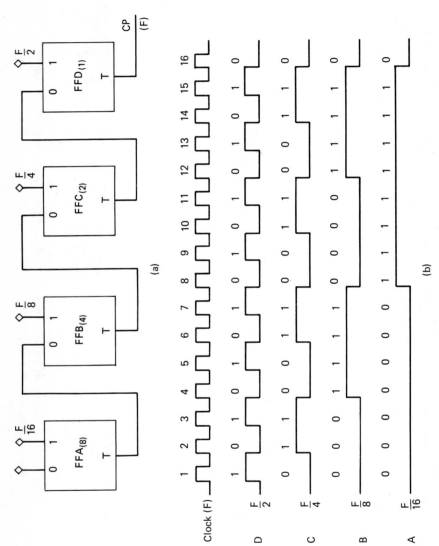

Fig. 10-15 (a) A serial (ripple) flip-flop counter with (b) the pulses seen at the 1-output lines while 16 clock pulses are applied to the T input of *FFD*.

flip-flop changes once for every eight pulses into T_A. A complete pulse is produced at each output after two transitions. As shown in Fig. 10-15(b), $F/2$, $F/4$, $F/8$, and $F/16$ outputs are available. It takes sixteen clock pulses into T_A to effect two changes in the state of FFA. The eighth pulse sends it into the 1 state and the sixteenth pulse returns it to the 0 state.

The timing problem arises from the fact that it takes time for a flip-flop to change state. When the count goes from $0111_2 = 7_{10}$ to $1000_2 = 8_{10}$, each flip-flop changes state. If the propagation time t_p of each flip-flop is 20 ns, there is a total delay of 80 ns before FFA changes state and the count 1000_2 appears. Since it must be allowed to remain long enough to be recognized, the next clock pulse must not be applied until one bit time after FFA changes state. The maximum input frequency is therefore

$$f = \frac{1}{t_p(n+1)} \qquad (10\text{-}4)$$

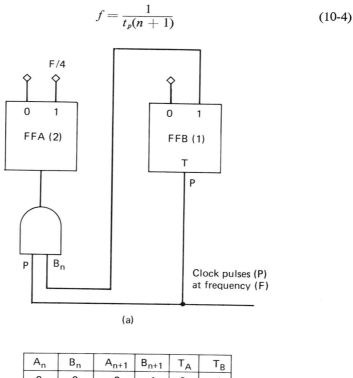

A_n	B_n	A_{n+1}	B_{n+1}	T_A	T_B
0	0	0	1	0	1
0	1	1	0	1	1
1	0	1	1	0	1
1	1	0	0	1	1

(b)

Fig. 10-16 (a) A parallel counter with (b) its state table.

where

t_p = propagation time of each flip-flop;
n = the total number of flip-flops;
$t_p(n + 1)$ = total time between clock pulse leading edges.

Example 10-9: Calculate the maximum clock pulse frequency for the serial counter of Fig. 10-15 if each flip-flop has a propagation time of 20 ns.
Solution: From Eq. (10-14),

$$f = \frac{1}{t_p(n+1)} = \frac{1}{2 \times 10^{-8}(4+1)}$$

$$f = 10 \text{ MHz}$$

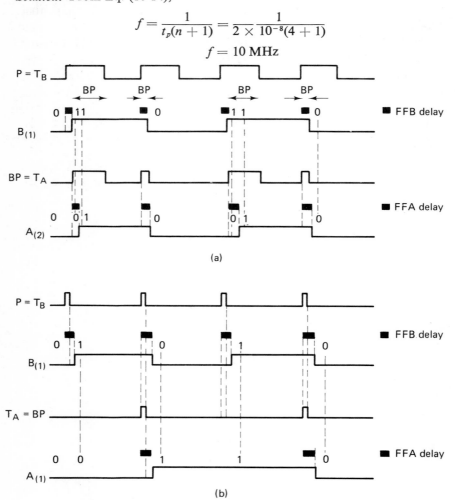

Fig. 10-17 (a) An example of the timing problem that exists in parallel counters. (b) An illustration of how the problem may be solved by applying narrower clock pulses.

Synchronous (Parallel) Counters. One solution to the timing problem is the use of *synchronous* counters, an example of which is shown in Fig. 10-16(a). The clock pulse is applied to all flip-flops simultaneously, allowing the new count to be produced after a delay equal to that of a single flip-flop. Every flip-flop does not have to change state every time, so it is often necessary to use gating circuits to control the clock pulse at the T inputs. These gating circuits add to the delay. The state table for the counter of Fig. 10-16(a) is given in Fig. 10-16(b). Only *FFB* is pulsed when the count goes from 00 to 01 and 10 to 11. Both must be pulsed to send the count from 01 to 10 and 11 to 00. The function $f_{T_A} = 1 \cdot P = P$ while $f_{T_B} = B_N \cdot P$.

A timing problem still exists with parallel counters. It is illustrated in Fig. 10-17(a). If the clock pulse width is too long, it will still be high after *FFB* changes state. This makes $T_A = BP$ high during the first clock pulse and causes *FFA* to change state after its delay. The count then starts at 00 and goes to 01 and 11 although only one clock pulse is applied. It then goes to 00 after the second pulse and again advances two counts 10 and 11 on the third pulse. Since the intermediate counts 01 and 10 won't be recognized by the readout circuit, the sequence will appear to be 0, 3, 0, 3 instead of 0, 1, 2, 3.

One way of solving this problem is by using clock pulses of narrow width. This is shown in Fig. 10-17(b). By the time *FFB* changes state, the first clock pulse is gone, making $T_A = 0$. The correct sequence of counts 0, 1, 2, 3 is produced.

Trailing-edge Logic. Another solution to the timing problem of parallel counters is shown in Fig. 10-18. The T inputs are differentiated and flip-flops

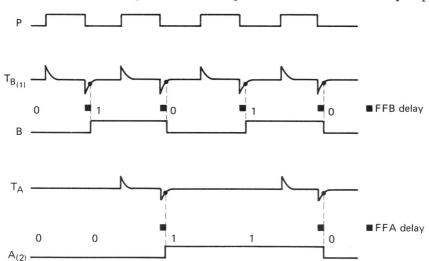

Fig. 10-18 Trailing-edge logic (a second solution to the timing problem in parallel counters).

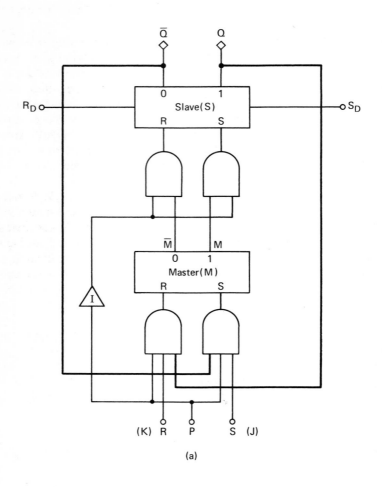
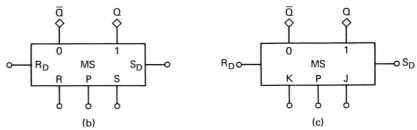

Fig. 10-19 (a) A master-slave RS flip-flop which can be modified by the bold-line connection to act as a master-slave JK flip-flop. (b) The logic symbol for the RS connection. (c) The logic symbol for the JK connection.

that trigger on negative pulses are used. Since the clock pulse must drop back to the 0 level to produce the negative spike, the clock is always gone before the flip-flop changes state. This technique is known as *"trailing-edge logic."*

Master-slave Flip-flops. A third solution to the timing problem is the *master-slave flip-flop*, the logic diagram of which is shown in Fig. 10-19(a). The pulse is applied through AND gates to a control flip-flop known as the master. The Reset and Set input functions are applied to the terminals marked R and S. If the Q and $\bar{Q}$ outputs are fed back, as shown in bold lines, it becomes a JK-type flip-flop. The logic symbols for both are shown in Fig. 10-19(b) and (c).

The clock is also inverted and then applied to the output flip-flop, known as the slave. The operation is best understood by examining Fig. 10-20. The master changes state after P goes high. This sends M high and $\bar{M}$ low. The slave R input equals $\bar{P}\bar{M}$ while its S input equals $\bar{P}M$. Both inputs

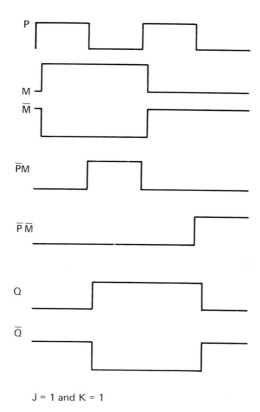

Fig. 10-20 The switching action of the master-slave flip-flop.

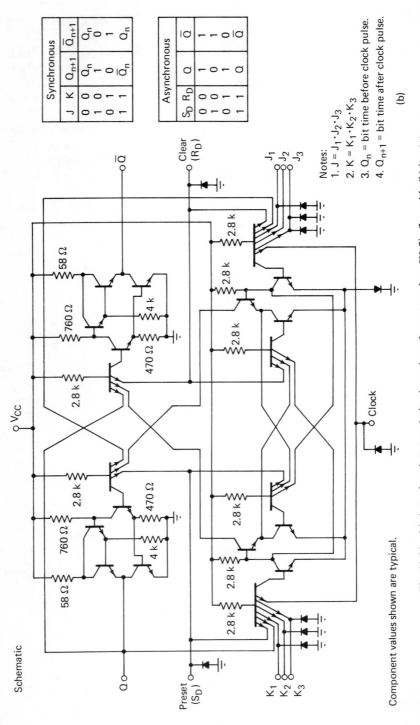

Fig. 10-21 (a) A modern integrated circuit version of a master-slave JK flip-flop with (b) its state tables. (Courtesy Sprague Electric Co.)

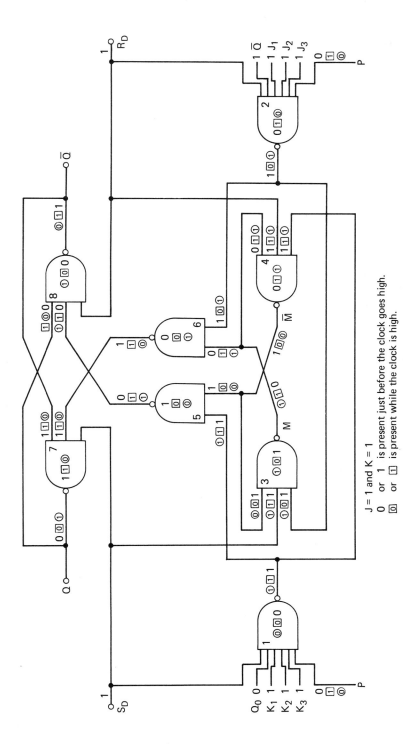

Fig. 10-22 The logic diagram used to explain the switching action of the MS JK flip-flop of Fig. 10-21(a).

will be low as long as P is still high. When the clock returns to the 0 level $\bar{P}M$ is high and the slave, after a delay, is set. The Q output line goes high and the $\bar{Q}$ output line goes low but not until the clock pulse is gone. This is another example of trailing-edge logic.

A Modern Integrated Circuit Version of a Master-slave JK Flip-flop. A schematic of a master-slave JK flip-flop is shown in Fig. 10-21(a). Its state tables are given in Fig. 10-21(b). The asynchronous inputs R_D and S_D must both be at the 1 level when synchronous operation is desired. If R_D is low, $\bar{Q}$ must be high. If S_D is low, Q must be high. If both R_D and S_D are low, both outputs are high.

The logic diagram of this circuit is shown in Fig. 10-22. The flip-flop is assumed to be in the 0 state before the clock is applied. The state of each gate input and output is represented by the unenclosed zeros and ones shown at those points. The conditions that are present while P is high are shown inside the boxes. The $\bar{M}$ output goes low and the M output goes high, but the Q and $\bar{Q}$ outputs remain unchanged. The encircled ones and zeros will be present after the clock returns to the 0 level. The Q output goes high and the $\bar{Q}$ output goes low. The state of the flip-flop has been complemented, but not until the clock pulse is gone.

10-6 A D-type Flip-flop

Figure 10-23(a) shows a very widely used flip-flop. This circuit, known as a D-type flip-flop, contains only D (data) and clock inputs (in addition to the asynchronous R_D and S_D inputs). Data on the D input is transferred to the Q output after the clock pulse goes high.

The logic symbol and truth tables are shown in Fig. 10-23(a) and (b). The synchronous table shows that the Q output will be the same as the D input after a clock pulse is applied. The asynchronous table shows that R_D and S_D are activated by 0-level inputs. If both are low, both outputs go high. If both are high, Q and $\bar{Q}$ remain unchanged.

The operation of this flip-flop is explained with the use of Fig. 10-24. The circuit is assumed to be in the 1 state before the clock goes high. The unenclosed ones and zeros show the condition of each gate input and output for this state. Both asynchronous inputs are inactivated by applying 1-level voltages to them. A 0 level is present on the D input when P goes high. All inputs to gate 3 are high with $P = 1$, so its output goes low. This low sends the $\bar{Q}$ output to the 1 level. The $\bar{Q}$ output is fed into gate 5 making all of its inputs high and sending the Q output to the 0 level. The other gate outputs remain unchanged.

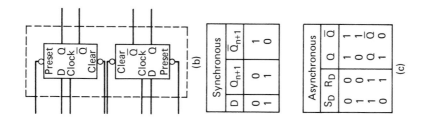

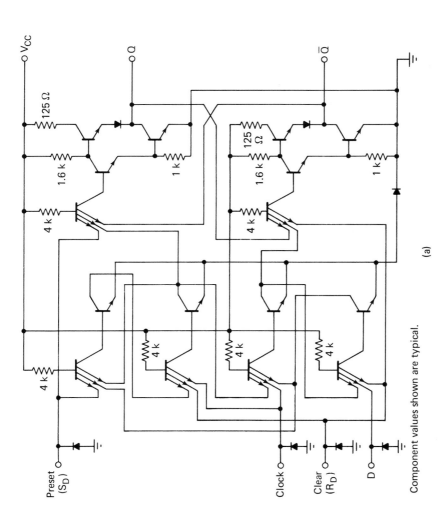

Fig. 10-23 (a) A D-type flip-flop with (b) its logic symbol, and (c) its state table.

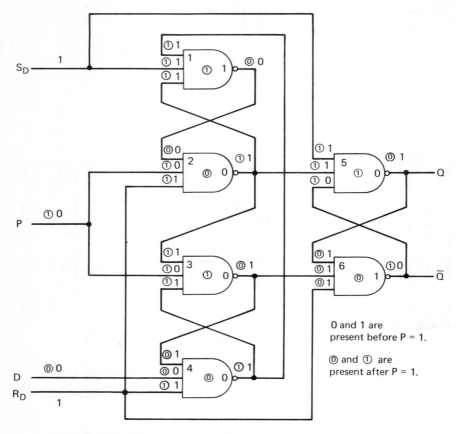

Fig. 10-24 The logic circuit used to describe the switching action of the D-type flip-flop of Fig. 10-23(a).

10-7 The Relationships Between the Various Flip-flops

There are important relationships that exist between the various logic flip-flops. For example, if an RS flip-flop is available, a JK, T, or D-type flip-flop can be produced. The method of conversion is illustrated in Fig. 10-25. The inversion balls shown at the R and S inputs of Fig. 10-25(a) are used to indicate that the *SET* and *RESET* inputs are active with 0-level inputs. A JK flip-flop can be changed to a D or T-type by making the connections shown in Fig. 10-25(b). A D-type flip-flop can be made to operate as a T-type by feeding back the $\bar{Q}$ output to the D input as shown in Fig. 10-25(c).

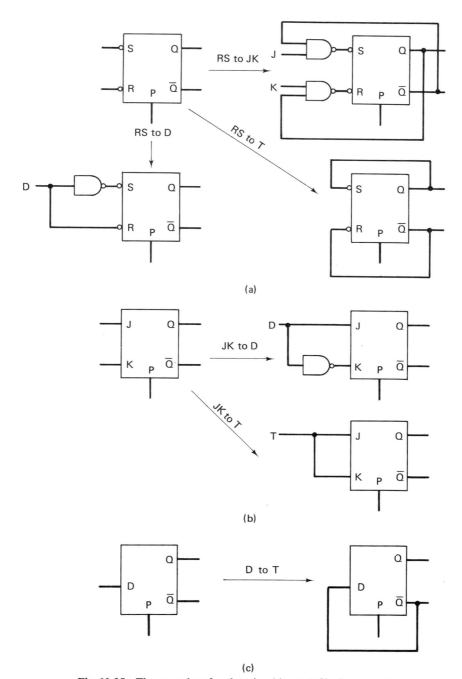

Fig. 10-25 The procedure for changing (a) an RS flip-flop to a JK, T, or D flip-flop, (b) a JK flip-flop to a T or D-type flip-flop, and (c) a D-type flip-flop to a T flip-flop. (Note: The T input of a Toggle type flip-flop is the clock pulse P.)

Problems

10-1 A basic flip-flop such as the one shown in Fig. 10-1 has the following values: $R_{C_1} = R_{C_2} = 3\text{ k}$, $R_1 = R_2 = 17\text{ k}$, $R_3 = R_4 = 80\text{ k}$, $V_{CC} = 12\text{ V}$, and $V_{BB} = -12\text{ V}$. The transistors are *npn* types with $V_{BE_{sat}} = 0.7\text{ V}$ and $V_{CE_{sat}} = 0.3\text{ V}$. Find the minimum h_{FE} for the ON transistor to operate in the saturation region.

10-2 If the circuit of Prob. 10-1 is used to drive the *NAND* gate shown in Fig. 7-4, what $h_{FE_{min}}$ is required for the flip-flop to have a dc fan-out of 10?

10-3 Verify that if one transistor in the flip-flop of Prob. 10-1 is saturated, the other transistor is cut off. Assume that $I_{CBO} = 0$.

10-4 What is the no-load 1-level output voltage of the flip-flop of Prob. 10-1?

10-5 Assume that the flip-flop of Fig. 10-5 has the following values: $R_{C_1} = R_{C_2} = 4\text{ k}$, $R_1 = R_2 = 16\text{ k}$, $R_3 = R_4 = 80\text{ k}$, $V_{BB} = -13\text{ V}$, $V_{CC} = 13\text{ V}$, and $V_{CL} = 5\text{ V}$. What is the $h_{FE_{min}}$ required for the ON transistor to operate in the saturation region?

10-6 Verify that if one transistor in the flip-flop of Prob. 10-5 is saturated, the other transistor must be cut off. Assume that $I_{CBO} = 0$.

10-7 If the flip-flop of Prob. 10-5 is used to drive the *NAND* gate of Fig. 7-4, what $h_{FE_{min}}$ is required for the flip-flop to have a dc fan-out of 10?

10-8 Assume that the diodes in the *NAND* gates of Prob. 10-7 have a maximum leakage current $I_{co} = 40\ \mu\text{A}$. How much current flows through the clamping diode of the flip-flop?

10-9 A serial (ripple) flip-flop divide-by-10 counter uses flip-flops that have a propagation time t_p of 10 ns. What is the maximum clock pulse frequency that can be applied to this counter?

10-10 The flip-flops of Fig. 10-26 toggle on the trailing edge of the clock pulse, but positive going voltages activate the *R* and *S* inputs. Assume

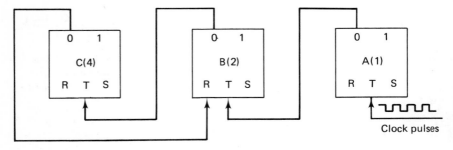

Fig. 10-26 An RTS flip-flop asynchronous counter.

that all flip-flops are initially in the 1 state. What sequence of counts is produced by this asynchronous counter?

10-11 Determine the sequence of counts produced by the synchronous counter of Fig. 10-27.

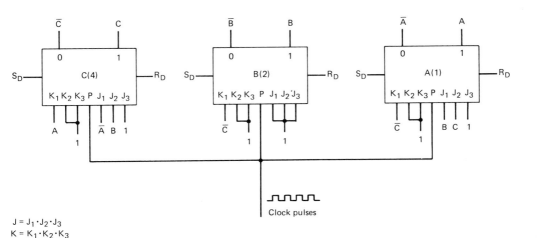

$J = J_1 \cdot J_2 \cdot J_3$
$K = K_1 \cdot K_2 \cdot K_3$

Fig. 10-27 A JK flip-flop synchronous counter.

10-12 There are numbers not included in the normal counting sequence of the counter of Prob. 10-11. Upon application of clock pulses, the numbers outside the normal sequence should jump to one of the numbers in the sequence. With this, the circuit does not have to be preset. Determine the "in" number to which each "out" number jumps. (Note! These are called *ports of entry*.)

10-13 (a) If the inputs to the circuit of Fig. 10-27 are changed to: $f_{J_A} = B$, $f_{J_B} = \bar{A}\bar{C}$, $f_{J_C} = AB$, $f_{K_A} = 1$, $f_{K_B} = AC$, and $f_{K_C} = \bar{B}$, what sequence of counts is produced by the counter?

(b) What are the ports of entry for the numbers not in the normal sequence? (Note: Unused inputs must be tied to a 1 level.)

10-14 (a) Draw an asynchronous divide-by-eight up counter using JK flip-flops that trigger on the trailing edge of the clock pulse.

(b) How is the circuit changed if the flip-flops trigger on the leading edge of the clock?

10-15 Complete the ladder diagram of Fig. 10-28. Note that all flip-flops are in the 0 state before the first clock pulse goes high. The $\bar{A}$, $\bar{B}$, and $\bar{C}$ outputs are high. The flip-flops trigger on the leading edge of the clock pulse. What sequence of counts is produced?

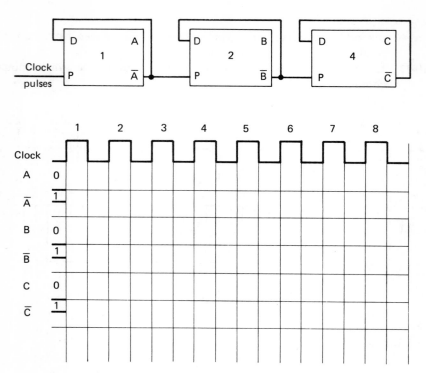

Fig. 10-28 A D-type flip-flop asynchronous counter.

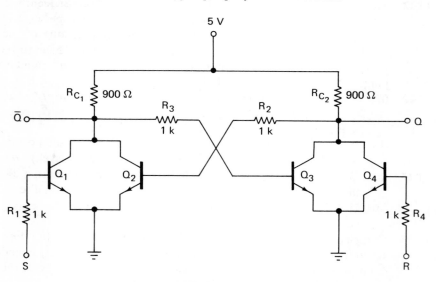

Fig. 10-29 An RS type flip-flop.

10-16 Draw a divide-by-sixteen ripple down counter using D-type flip-flops that operate on the leading edge of the clock pulse.

10-17 Assume that the RTL RS flip-flop of Fig. 10-29 is in the 1 state and has a fan-out of 9. What is the minimum 1-level output $V_{o(1)\min}$?

10-18 Determine $h_{FE\min}$ for the circuit of Prob. 10-17.

11

SPECIAL-PURPOSE CIRCUITS

In addition to the basic building blocks of digital computers discussed in the preceding chapters, there are many *special-purpose circuits*. Some of these are the internal gates of complex arrays; lamp drivers, that indicate when the output is at the 1 or 0 level; MOSFET (see Sec. 2-10) logic circuits, which are useful in low-speed, large-scale applications; crystal-controlled clock pulse generators; relay drivers; transmission line drivers; Schmitt trigger circuits, which can be used to produce square waves from non-square waves; and three-state logic circuits. Examples of these circuits are discussed in this chapter.

11-1 Internal Gates

Digital circuits are available in many different kinds of packages. Some contain only *NAND* gates, *NOR* gates, INVERTERS, or FLIP-FLOPS. Others contain complex arrays. For example, complete 4-bit binary full adders, BCD-to-decimal decoders, binary or BCD decade up and/or down counters, and others are available in single packages (see Chap. 12). The internal gates used in the complex arrays can differ greatly from the input and output gates. The input gates must be able to accept the specified $V_{(0)max}$ and $V_{(1)min}$ of the system. The output gates must provide the correct logic levels and handle the specified maximum load currents of the system. Also, the output gate usually requires an active pull-up because of possible high parasitic load capacitance.

The internal gates do not have these limitations. They may have logic levels that differ from those guaranteed by the system because of lower noise levels. They do not have to handle the high load currents that the output gates must be capable of handling, and they do not require an active pull-up because the parasitic load capacitance will be low. Since they do not have the same limitations, internal gates may be designed with fewer components. Frequently, clamping diodes are added to keep the transistors out of deep saturation and increase switching speed (See Sec. 2-7).

A Simple Resistor-clamped Internal Gate. A frequently used internal gate is shown in Fig. 11-1. A small resistor R_4 is connected between the

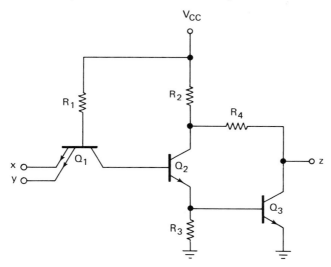

Fig. 11-1 A resistor-clamped internal TTL gate.

collector of the output transistor Q_3 and the collector of the phase-splitter transistor Q_2. This resistor keeps Q_2 out of saturation by keeping its collector positive compared to its base. V_{C_2} can be varied from $V_{CE\mathrm{sat}(3)} = 0.3$ V for $R_4 = \infty$ to $V_{CE\mathrm{sat}(2)} + V_{BE\mathrm{sat}(3)} = 1$ V for $R_4 = 0$. If $V_{C_2} = 1$ V, $V_{CB} = +0.3$ V and Q_3 is out of saturation. The resistance of R_4 is typically around 100 Ω and V_{C_2} is a little less than 1 V.

A Phase Splitter, Inverter-Clamped Internal Gate. An internal gate that clamps the collectors of the phase splitter Q_3 and the output inverter Q_3, keeping both transistors out of saturation, is shown in Fig. 11-2. Q_1 is off with an input $v_i > V_{(1)\min}$ at the y input. The base and second emitter x of Q_1 act as a clamping diode to keep Q_2 out of saturation. The base voltage

$$V_{B_1} = V_{BC_1} + V_{BE_2} + V_{BE_3} = 2.1 \text{ V}$$

This serves as the clamp supply voltage V_{CL} for the collector of Q_2. The emitter x conducts if V_{C_2} drops below V_{B_1} by more than the cutin voltage V_y. It then clamps V_{C_2} at $V_{B_1} - V_{BE(x)} = 2.1 - 0.7 = 1.4$ V. This holds V_{CB_2} at

322 / *Special-Purpose Circuits*

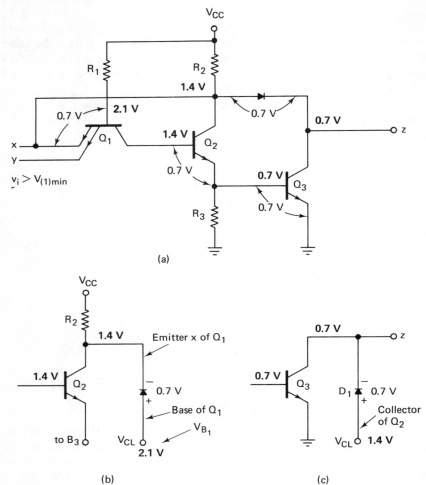

Fig. 11-2 (a) A phase splitter-inverter-clamped TTL gate. (b) A simplified circuit showing how V_{C_2} is clamped at 1.4 V by the $B_1E_{(x)}$ "diode." (c) A simplified circuit showing how D_1 clamps V_{C_3} at 0.7 V.

$V_{C_2} - V_{B_2} = 0$ V and keeps Q_2 out of saturation. The simplified circuit showing this action is given in Fig. 11-2(b).

The diode D_1 in Fig. 11-2(a) conducts when V_{C_3} falls below V_{C_2} by more than its V_γ. As shown in Fig. 11-2(c), it then clamps V_{C_3} at $V_{C_2} - V_{D_1} = 1.4 - 0.7 = 0.7$ V. This holds V_{CB_3} at $V_{C_3} - V_{B_3} = 0.7 - 0.7 = 0$ V, and it keeps Q_3 out of saturation.

An Emitter-clamped Internal Gate. A third technique used to keep the output transistor out of saturation is shown in Fig. 11-3. The phase splitter

Q_2 has a second emitter w which is connected to the collector of Q_3. The base voltage $V_{B_2} = V_{BE_2} + V_{BE_3} = 1.4$ V serves as the clamp supply V_{CL} for the Q_3 collector. When V_{C_3} drops below V_{B_2} by more than V_γ, the second emitter w conducts and clamps V_{C_3} at $V_{B_2} - V_{BE(w)} = 0.7$ V. This holds V_{CB_3} at $V_{C_3} - V_{B_3} = 0.7 - 0.7 = 0$ V and it keeps Q_3 out of saturation.

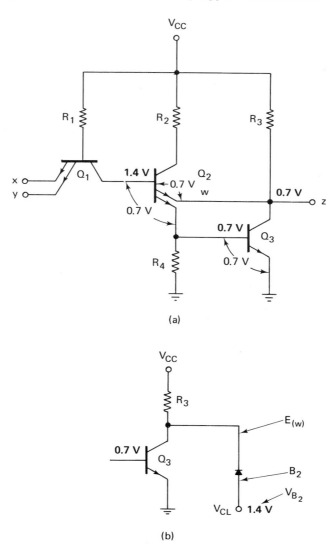

Fig. 11-3 (a) An emitter-clamped TTL gate. (b) A simplified circuit showing how V_{C_3} is clamped at 1.4 V by the $B_2E_{(w)}$ "diode."

11-2 A Schottky-clamped TTL Gate

The circuit of Fig. 11-4(a) employs the Schottky diode clamp. The Schottky diode is produced by using the n-type silicon of the collector as the cathode and a metal as the anode. The contact between the dissimilar materials forms

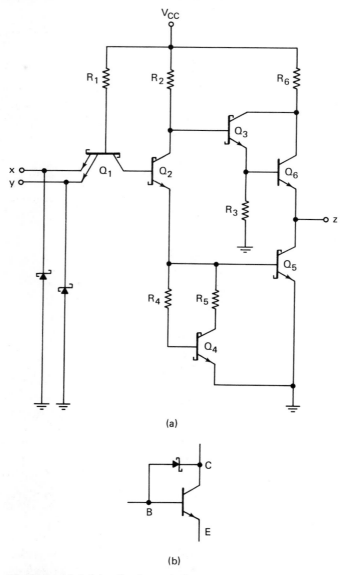

Fig. 11-4 (a) A Schottky-clamped TTL gate. (b) The equivalent of the Schottky transistors $Q_1 - Q_5$.

a diode. The equivalent of the Schottky-clamped transistor is shown in Fig. 11-4(b). Two important characteristics of the Schottky barrier diode (SBD) are: (1) it has a lower forward voltage drop than a silicon *pn* junction and (2) it does not contain minority carriers. The lower forward voltage drop clamps V_{CB} at less than the normal $V_{CB_{sat}} = 0.4$ V and diverts most of the extra base current (see Single-diode Back Clamping in Sec. 2-7). Thus, the transistor does not go into normal saturation. With virtually no charge being stored in the SBD or the transistor, storage time is greatly reduced. This substantially improves the switching time of the transistor.

Until Schottky-clamped TTL, speeds of less than 5 ns were not possible without using current-mode circuitry (see ECL in Sec. 7-2). Unsaturated circuits have the following important disadvantages: (1) relatively high power consumption; (2) expensive power supplies that must have excellent regulation; and (3) system packaging is difficult because of the terminated lines and the higher system cost.

The use of Schottky-clamped transistors reduces the number of components needed to prevent saturation. By being able to use TTL to obtain speeds of less than 5 ns, the disadvantages of ECL are avoided.

Schottky-clamped circuits do have the two following disadvantages: (1) The Schottky clamp becomes less effective as temperature increases and the transistors tend to go into saturation. (2) Because the transistors do not go into deep saturation the 0-level output is higher. These disadvantages are usually outweighed, however, by the faster speeds of Schottky-clamped circuits.

11-3 Simple Lamp Drivers

There are two basic types of lamp drivers: (1) a 1-level indicator driver and (2) a 0-level indicator driver. Examples of both are presented in this section. More complex lamp drivers such as the NBCD-to-Decimal Decoder/Driver and NBCD-to-Seven-Segment Decoder/Driver are discussed in Sec. 12-5.

A 0-Level Lamp Driver. The indicator circuit of Fig. 11-5(a) is useful when brightness is not essential. When $v_i = V_{(1)}$, Q_1 comes on and draws current away from the lamp. Current flows from V_{CC} through R_2 and R_3 and into the collector. When $v_i = V_{(0)}$, Q_1 cuts off and the lamp comes on. Current then flows from V_{CC} through R_2 and the lamp. The 500-Ω resistor limits the brightness of the lamp.

A 1-Level Lamp Driver. The indicator of Fig. 11-5(b) is also useful when brightness is not essential. When $v_i = V_{(0)}$, Q_1 is off and the high resistance of R_3 keeps the lamp off. When $v_i = V_{(1)}$, Q_1 conducts and offers a low resistance return for the lamp current. Current flows from V_{CC} through the lamp, R_2, and Q_1.

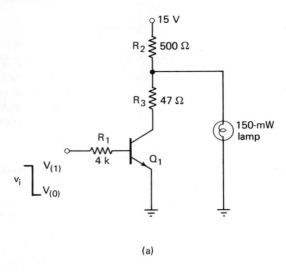

(a)

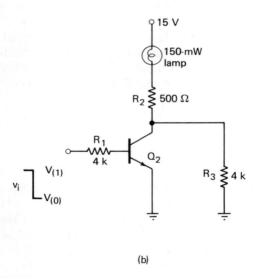

(b)

Fig. 11-5 (a) A 0-level indicator driver. (b) A 1-level indicator driver.

A High-intensity 1-Level Lamp Driver. The circuit of Fig. 11-6 is a 1-level indicator driver that is useful when greater brightness is required. When $v_i = V_{(0)}$, Q_1 is off. The base current required to turn Q_2 on is supplied by the emitter of Q_1. When Q_1 is off, Q_2 is off and the high resistance of R_6 keeps the lamp off.

When $v_i = V_{(1)}$, Q_1 comes on. Part of its emitter current flows into the

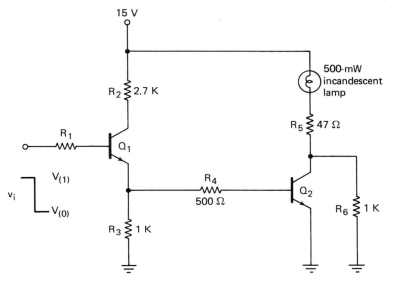

Fig. 11-6 A high-intensity 1-level indicator driver.

base of Q_2, turning that transistor on. The low resistance of Q_2 permits the lamp to light. Current then flows from V_{CC} through the lamp, R_5, and Q_2.

11-4 Relay Drivers

The circuit of Fig. 11-7 is a relay driver. When $v_i = V_{(1)}$, Q_1 conducts and part of its emitter current flows into the base of Q_2. This turns Q_2 on, permitting a high current to flow through the relay coil.

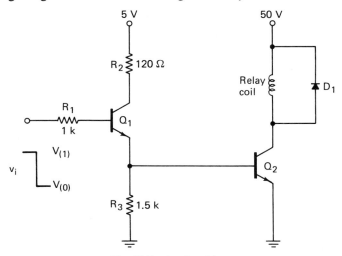

Fig. 11-7 A relay driver.

328 / Special-Purpose Circuits

An important consideration when inductive loads are used is the collector-to-emitter breakdown voltage. When $v_i = V_{(0)}$, Q_1 cuts off, removing the base drive of Q_2. This cuts off Q_2 and opens the return path for the relay current. The rapid change in inductor current produces a large counter EMF that can increase V_{CE} excessively and damage the transistor. The diode provides the suppression that ensures that V_{CE} does not exceed its breakdown value when Q_2 turns off.

11-5 A Transmission Line Driver and Terminator

The circuit of Fig. 11-8 is a transmission-line driver and terminator. The driver and terminator are TTL *NAND* gates with totem-pole outputs.

The characteristic impedance Z_o of the transmission line is 50 Ω. A resister R_x is connected between one input of the terminator and the 5 V source. The combination of R_x and the ON resistance of the terminator gate matches the impedance of the transmission line. The line can be flat cable, coaxial cable, or printed wire.

11-6 A Crystal-controlled Clock Pulse Generator

As mentioned in Chap. 9, astable multivibrators are sometimes used as clock pulse generators. This is permissible when the clock can be periodically adjusted and long-term stability is not necessary. When clock accuracy is more critical, crystal-controlled sine-wave oscillators are used. These oscillators have accuracies of approximately 0.001 percent. The sine-wave output of the clock pulse generator must be converted to a square wave before it can be applied to the digital circuits. A Schmitt trigger (see Sec. 11-7) may be used to change the sine wave to a square wave.

Clock Systems. A computer often requires two-phase (clock and $\overline{\text{clock}}$) pulses at several different frequencies. This can be accomplished with the system shown in Fig. 11-9. The master oscillator is crystal controlled and has a 20 MHz sine-wave output. The sine wave is applied to a Schmitt trigger, which produces a square-wave output at the same frequency as the master oscillator. This square wave is used directly as a 20-MHz clock source of one phase (clock). It is applied to an inverter to produce a $\overline{\text{clock}}$ output. Complementary outputs at lower frequencies may be obtained by applying the Schmitt output to various frequency dividers as shown in Fig. 11-9.

The Schmitt output will not have a 50-percent duty cycle (see Fig. 11-16). If the 20-MHz clock and $\overline{\text{clock}}$ signals must have a 1 : 1 mark-to-space ratio, a 40-MHz master oscillator must be used with the Schmitt output applied to a flip-flop. The $F/2$ output of the flip-flop will be 20 MHz with a 1 : 1 mark-to-

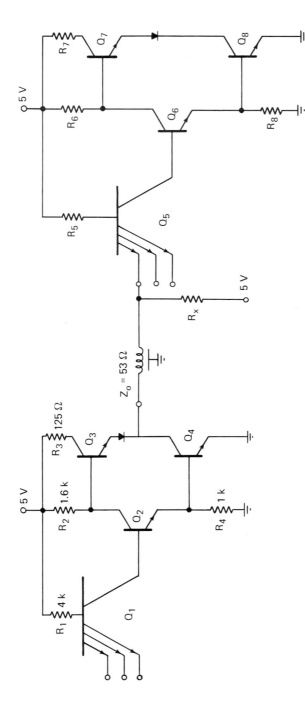

Fig. 11-8 A transmission line driver and terminator.

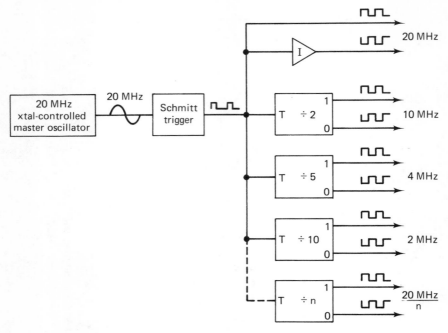

Fig. 11-9 A multifrequency, two-phase clock pulse system.

space ratio. Lower frequency outputs are then obtained by using other appropriate frequency dividers.

11-7 The Schmitt Trigger

The Schmitt trigger is a regenerative double clipper. The circuit is shown in Fig. 11-10. The feedback takes place through the common emitter resistor R_E. The input and output terminals are not in the feedback loop. Therefore, the base voltage v_{B_1} depends only on v_i, and the load capacitance does not affect the regenerative action of the circuit. The load capacitance still affects rise and fall times, however. Lower values of R_{C_2} reduce rise time, but they also reduce the peak-to-peak amplitude of the output pulse while increasing the $h_{FE_{min}}$ requirement of Q_2 and the circuit-power consumption. Fall time remains fast because the load capacitance discharges through the low resistance path consisting of the saturated Q_2 (see Sec. 8-9, The Inverter Output Circuit) and R_E.

As shown on its transfer characteristics in Fig. 11-11, the Schmitt trigger has two stable states. When the output transistor Q_2 is on, the output voltage equals $V_{CE_{sat}} + V_{E_2}$. When it is off, $v_o = V_{CC}$. The curve of Fig. 11-11(a) is a

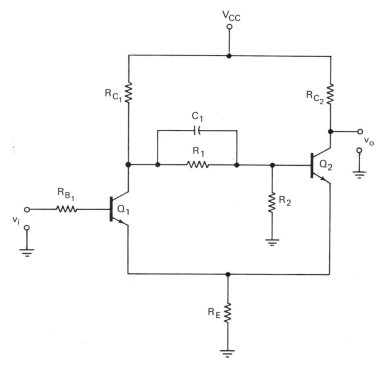

Fig. 11-10 A Schmitt trigger circuit.

plot of v_o versus v_i for a loop gain (in the active region) of less than 1. Regeneration will not occur and there will be a gradual change from one state to the other. If $v_i < V_1$, Q_2 will be on because its base is returned through R_1 and R_{C_1} to V_{CC}. Its emitter voltage $V_{E_2} = I_{E_2} R_E$ is positive and holds Q_1 off. The output voltage then equals $V_{CE_{sat}} + V_{E_2}$. When $v_i > V_1$, Q_1 comes on and, with A_v positive, v_o rises in accordance with v_i. As v_i is increased more and more, Q_1 conducts harder, v_{C_1} decrease and Q_2 cuts off. Then $v_o = V_{CC}$. Beyond this point, v_o again no longer responds to v_i.

The loop gain can be increased by increasing R_{C_1}. With a higher A_v, v_o increases with a steeper slope as v_i is increased above V_1. If R_{C_1} is increased further, the slope becomes steeper and steeper until, at a loop gain = 1, regeneration occurs and the slope becomes infinite. The higher value of R_{C_1} reduces both V_1 and V_2, but the change in V_2 is greater. There is, therefore, a point where they are equal. This occurs when the loop gain equals one. A plot of v_o versus v_i for a loop gain = 1 is given in Fig. 11-11(b). There is an abrupt change from one output level to the other.

If R_{C_1} is increased so that the loop gain is greater than 1, the slope of the transfer curve becomes negative. That is, V_2 becomes less than V_1. From this

332 / *Special-Purpose Circuits*

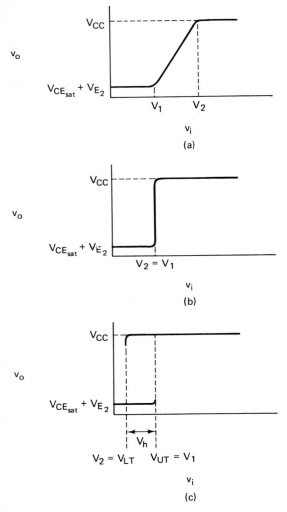

Fig. 11-11 The transfer characteristics of the Schmitt trigger for (a) a loop gain < 1, (b) a loop gain = 1, and (c) a loop gain > 1.

point on, V_1 and V_2 will be referred to as the *upper and lower threshold voltages* V_{UT} and V_{LT} where

$$V_{UT} = V_{E_2} + V_\gamma \tag{11-1}$$

and

$$V_{LT} = V_{E_1} + V_{BE\text{sat}} \tag{11-2}$$

When v_i drops below V_{LT}, Q_1 comes out of saturation, but regeneration doesn't start until Q_2 comes on. Therefore, Q_1 must be brought far enough out of saturation to turn Q_2 on. This will occur if R_{C_1} is large enough. At both threshold levels, I_{B_1} will be small and $V_{R_{B(1)}}$ can be neglected.

The transfer curve for a loop gain > 1 is shown in Fig. 11-11(c). As long as $v_i < V_{UT}$, Q_2 is on and $v_o = V_{CE_{sat}} + V_{E_2}$. For any $v_i > V_{UT}$, Q_1 comes on, Q_2 goes off, and v_o rises straight up to V_{CC}.

If v_i is then decreased, v_o remains at V_{CC} until v_i reaches V_{LT}. Then, for any $v_i < V_{LT}$, Q_2 comes on, Q_1 goes off, and v_o drops straight down to $V_{CE_{sat}} + V_{E_2}$. The transfer curve of Fig. 11-11(c) shows that there will be a *snap-action* change of state. It also shows that the circuit now has *hysteresis*. That is, in order to change v_o from its lower level to its higher level, v_i must be increased beyond V_{LT} to V_{UT}, and in order to change v_o in the opposite direction, v_i must be decreased below V_{UT} to V_{LT}. The hysteresis voltage V_h is the difference between V_{UT} and V_{LT}.

$$V_h = V_{UT} - V_{LT} \qquad (11\text{-}3)$$

Frequently, the size of the hysteresis voltage V_h is not important. However, if the peak-to-peak amplitude of v_i is less than V_h, the circuit will not be able to change state. Reducing R_{C_1} decreases V_h, but it also reduces regeneration and the circuits snap-action. As shown in Fig. 11-11(b), with a loop gain $= 1$, $V_h = 0$ and the abrupt change of state is maintained. However, because of the variation in transistors, if the loop gain is made equal to 1 for the transistor with the lowest gain, it will be greater than 1 for the transistors with a higher gain. Conversely, if the loop gain is made equal to 1 for the transistors with the highest gain, it will be less than 1 for the transistors with less gain. Therefore, although hysteresis may be undesirable, the loop gain is made greater than unity with V_h values of approximately 0.5 V common.

11-8 Analysis of the Schmitt Trigger

The Schmitt trigger will now be analyzed. The circuit is shown with component values in Fig. 11-12. It is assumed here that both transistors operate in the saturation region when they are on. This is permissible because the input supplied by the crystal-controlled clock oscillator will have a constant amplitude and frequency. Operation in the saturation region is not desirable when the input has a varying amplitude. The overdrive provided by the speed-up capacitor C_C, and thus the switching speed of Q_2, will be greater for fast signals than slow signals. Since regeneration cannot begin until Q_2 comes out of saturation, the response of the circuit would vary with the speed of the input waveform.

334 / *Special-Purpose Circuits*

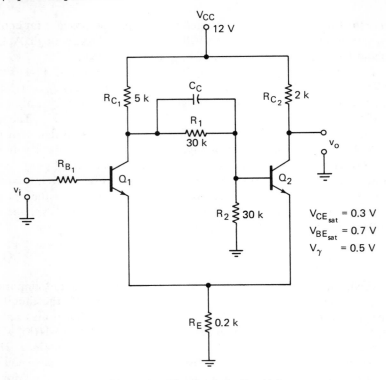

Fig. 11-12 The circuit for Ex. 11-1.

The upper and lower threshold voltages V_{UT} and V_{LT} and the hysteresis voltage V_h will be calculated in the following examples. A simpler and faster method of analysis is given in Sec. 11-9. However, it will be seen from the following analysis that the fast method is limited to specific circuit conditions.

Example 11-1: Calculate the upper threshold voltage V_{UT} for the Schmitt trigger of Fig. 11-12.

Solution: From Eq. (11-1), the emitter voltage V_{E_2} (when Q_2 is on) must be known before V_{UT} can be determined. Thus, V_{E_2} will be calculated first. Begin by assuming that Q_1 is just off with $V_{BE_1} = V_\gamma = 0.5$ V. The circuit is shown with Q_1 removed in Fig. 11-13. The base circuit shown in Fig. 11-13(a) is replaced by its Thévenin equivalent in Fig. 11-13(b).

$$R_{B(TH)} = \frac{R_2(R_1 + R_{C_1})}{R_2 + R_1 + R_{C_1}}$$

$$= \frac{(30)(60)}{90}$$

$$R_{B(TH)} = 20 \text{ k}$$

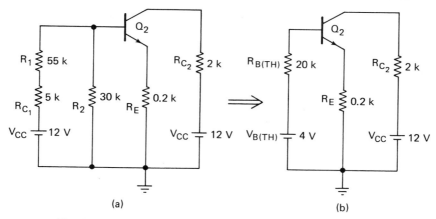

Fig. 11-13 (a) The circuit of Fig. 11-12 with the cut-off transistor Q_1 removed. (b) The circuit of (a) with the base circuit replaced by its Thévenin equivalent.

and

$$V_{B(TH)} = \frac{V_{CC}R_2}{R_2 + R_1 + R_{C_1}}$$

$$= \frac{(12)(30)}{90}$$

$$V_{B(TH)} = 4 \text{ V}$$

The circuit is further simplified in Fig. 11-14(a). The emitter-base junction and the transistor (from emitter to collector) are represented by constant voltage sources $V_{BE_{sat}} = 0.7$ V and $V_{CE_{sat}} = 0.3$ V, respectively. The series opposing sources are combined in Fig. 11-14(b). The emitter voltage V_{E_2} can now be calculated by applying Millman's theorem to Fig. 11-14(b).

$$V_{E_2} = \frac{\dfrac{V_{B(TH)} - V_{BE_{sat}}}{R_{B(TH)}} + \dfrac{0}{R_E} + \dfrac{V_{CC} - V_{CE_{sat}}}{R_{C_2}}}{\dfrac{1}{R_{B(TH)}} + \dfrac{1}{R_E} + \dfrac{1}{R_{C_2}}} \quad (11\text{-}4)$$

$$= \frac{\dfrac{3.3}{20} + \dfrac{0}{0.2} + \dfrac{11.7}{2}}{\dfrac{1}{20} + \dfrac{1}{0.2} + \dfrac{1}{2}}$$

$$= \frac{\dfrac{3.3 + 0 + 117}{20}}{\dfrac{1 + 100 + 10}{20}}$$

$$= \frac{120.3}{111}$$

336 / Special-Purpose Circuits

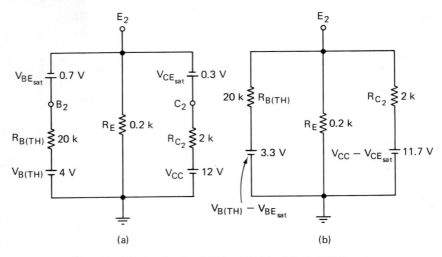

Fig. 11-14 (a) The circuit of Fig. 11-13(b) with the EB junction and the transistor from the emitter to collector represented by constant-voltage sources. (b) The circuit of (a) with the series batteries combined.

$$V_{E_2} = 1.08 \text{ V}$$

Now the upper threshold voltage can be calculated.

$$V_{UT} = V_{E_2} + V_\gamma \quad (11\text{-}1)$$
$$= 1.08 + 0.5$$
$$V_{UT} = 1.58 \text{ V}$$

As long as v_i is less than this value, Q_1 is off, Q_2 is on, and the output voltage is

$$v_o = V_{CE_{sat}} + V_{E_2} \quad (11\text{-}5)$$
$$= 0.3 + 1.08$$
$$v_o = 1.38 \text{ v}$$

When $v_i > V_{UT}$, $V_{BE_1} > V_\gamma$ and Q_1 comes on to begin the regenerative cycle that turns Q_2 off and sends Q_1 into saturation.

Example 11-2: Calculate the lower threshold voltage V_{LT} for the circuit of Fig. 11-12.
Solution: From Eq. (11-2), the emitter voltage V_{E_1} (when Q_1 is on) must be known before V_{LT} can be determined. Therefore, V_{E_1} will be calculated first. Its exact value depends on the input signal. However, if h_{FE_1} is large and R_{B_1} is several times greater than R_E (to minimize changes in V_{E_1} when v_i is very

positive), V_{E_1} can be made almost independent of v_i. The value of V_{E_1} in this analysis represents the voltage developed by I_{E_1} when Q_1 is at the edge of saturation. At this point $V_{R_{B_1}}$ is negligible and $V_{B_1} \approx v_i$.

The simplified circuit with Q_1 on and Q_2 off is given in Fig. 11-15(a). The collector circuit is replaced by its Thévenin equivalent in Fig. 11-15(b).

$$R_{C(TH)} = \frac{R_{C_1}(R_1 + R_2)}{R_{C_1} + R_1 + R_2}$$

$$= \frac{(5)(85)}{90}$$

$$R_{C(TH)} = 4.72 \text{ k}$$

and

$$V_{C(TH)} = \frac{V_{CC}(R_1 + R_2)}{R_1 + R_2 + R_{C_1}}$$

$$= \frac{(12)(85)}{90}$$

$$V_{C(TH)} = 11.3 \text{ V}$$

If h_{FE} is large, $I_{C_1} \approx I_{E_1}$ and, from Fig. 11-15(b),

$$V_{E_1} \approx \frac{R_E(V_{C(TH)} - V_{CE_{\text{sat}}})}{R_E + R_{C(TH)}} \tag{11-6}$$

$$\approx \frac{(0.2)(11)}{4.92}$$

$$V_{E_1} \approx 0.447 \text{ V}$$

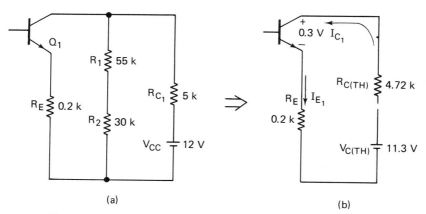

Fig. 11-15 (a) The circuit of Fig. 11-12 with Q_1 on and Q_2 off. (b) The circuit of (a) with the collector circuit replaced by its Thévenin equivalent.

Now the lower threshold voltage can be calculated.

$$V_{LT} = V_{E_1} + V_{BE_{sa}} \tag{11-2}$$
$$= 0.447 + 0.7$$
$$V_{LT} = 1.147 \text{ V}$$

Until v_i drops back below this value, Q_1 is saturated, Q_2 is off, and the output voltage is

$$v_o = V_{CC} = 12 \text{ V} \tag{11-7}$$

When $v_i < V_{LT}$, Q_1 comes far enough out of saturation to begin the regenerative cycle that turns Q_1 off and sends Q_2 into saturation.

Example 11-3: Calculate the hysteresis voltage V_h for the circuit of Fig. 11-12.
Solution: Substituting the values of V_{UT} and V_{LT} determined in Exs. 11-1 and 11-2 in Eq. 11-3 yields

$$V_h = V_{UT} - V_{LT} \tag{11-3}$$
$$= 1.58 - 1.147$$
$$V_h = 0.433 \text{ V}$$

The output waveform for a 10-V *PP* sine wave is shown in Fig. 11-16. As was mentioned in the discussion of clock generators (see Sec. 11-6), the mark-to-space ratio *M/S* of the Schmitt trigger output will be less than one. If a 50-percent duty cycle is required, it can be obtained by operating the sine-wave oscillator at twice the desired frequency and then applying the Schmitt output to a flip-flop. As shown in Fig. 11-17, the flip-flop output will have exactly a 1-to-1 mark-to-space ratio with a frequency of one-half the input frequency. The method of calculating the mark-to-space ratio of the Schmitt trigger is shown in the following example.

Example 11-4: Calculate the mark-to-space ratio *M/S* of the Schmitt trigger of Fig. 11-12 for a 10-V *PP* sine-wave input.
Solution: First, the conduction angle of Q_1 must be determined. The input transistor is off until v_i exceeds the upper threshold voltage $V_{UT} = 1.58$ V. The angle at which this occurs is called the *upper threshold angle* θ_{UT}.

$$\theta_{UT} = \sin^{-1} \frac{V_{UT}}{V_{max}} \tag{11-8}$$
$$= \sin^{-1} \frac{1.58}{5}$$
$$\theta_{UT} = 18.4°$$

As shown in Fig. 11-16, the input transistor is on and the output

Analysis of the Schmitt Trigger / 339

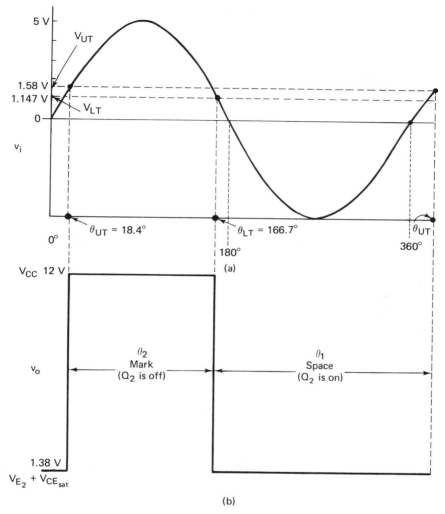

Fig. 11-16 (a) A 10-V *PP* sine-wave input signal with (b) the resulting output of the Schmitt trigger circuit of Fig. 11-12.

transistor is off until v_i drops back to the lower threshold voltage $V_{LT} = 1.047$ V. The angle at which this occurs is called the *lower threshold angle* θ_{LT}.

$$\theta_{LT} = 180° - \sin^{-1} \frac{V_{LT}}{V_{max}} \qquad (11\text{-}9)$$

$$= 180° - \sin^{-1} \frac{1.147}{5}$$

$$= 180° - 13.3°$$

$$\theta_{LT} = 166.7°$$

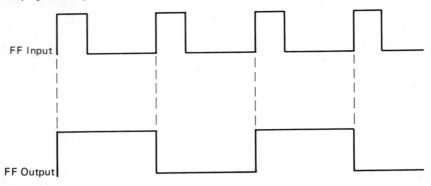

Fig. 11-17 (a) A square wave with any M/S ≠ 1 applied to a flip-flop results in an output (b) with a M/S = 1.

The conduction angle of Q_1, which is called θ_1, can now be calculated.

$$\theta_1 = \theta_{LT} - \theta_{UT} \tag{11-10}$$
$$= 166.7 - 18.4$$
$$\theta_1 = 148.3°$$

The conduction angle of Q_2 is

$$\theta_2 = 360° - \theta_1 \tag{11-11}$$
$$= 360° - 148.3°$$
$$\theta_2 = 211.7°$$

Finally, the mark-to-space ratio is

$$\text{M/S} = \frac{\theta_1}{\theta_2} \tag{11-12}$$
$$= \frac{148.3°}{211.7°}$$
$$\text{M/S} = 0.7$$

The final step in this analysis is the verification that Q_2 is off when Q_1 is saturated. This will be done in the following example.

Example 11-5: Prove that Q_2 is cut off when Q_1 is saturated.
Solution: The circuit of Fig. 11-18 will be used. If Q_1 is saturated, $V_{CE_1} = V_{CE_{sat}} = 0.3$ V and the collector voltage is

$$V_{C_1} = V_{CE_{sat}} + V_{E_1}$$
$$= 0.3 \text{ V} + 0.447$$
$$V_{C_1} = 0.747 \text{ V}$$

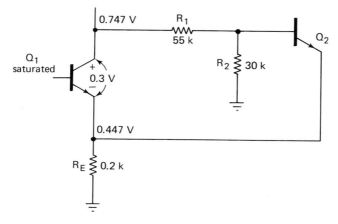

Fig. 11-18 The circuit for Ex. 11-5.

Assuming that Q_2 is off, its base voltage is

$$V_{B_2} = V_{C_1} \frac{R_2}{R_1 + R_2}$$
$$= 0.747 \frac{30}{85}$$
$$V_{B_2} = 0.264 \text{ V}$$

Since $V_E = 0.447$ V when Q_1 is on,

$$V_{BE_2} = V_{B_2} - V_E$$
$$= 0.264 - 0.447$$
$$V_{BE_2} = -0.183 \text{ V}$$

Since its base is negative with respect to its emitter, Q_2 is definitely cut off.

11-9 Fast Analysis of the Schmitt Trigger

When the base current of Q_2 is small, it may be assumed that $I_{E_2} \approx I_{C_2}$. Then the emitter voltage of Q_2 is

$$V_{E_2} \approx (V_{CC} - V_{CE\text{sat}}) \frac{R_E}{R_E + R_{C_2}} \qquad (11\text{-}13)$$
$$\approx (12 - 0.3) \frac{0.2}{2.2}$$
$$V_{E_2} \approx 1.06 \text{ V}$$

This is almost equal to the value calculated in Ex. 11-1. The base current I_{B_2} is small in this circuit because R_1 is large. Therefore, this approximation is permissible.

When the sum of R_1 and R_2 is much greater than R_{C_1}, it may be assumed that $R_{C(TH)}$ in Fig. 11-15 is approximately equal to R_{C_1} and $V_{C(TH)}$ is approximately equal to V_{CC}. Then the emitter voltage of Q_1 is

$$V_{E_1} \approx (V_{CC} - V_{CE\text{sat}}) \frac{R_E}{R_E + R_{C_1}} \qquad (11\text{-}14)$$

$$\approx (12 - 0.3)\frac{(0.2)}{5.2}$$

$$V_{E_1} \approx 0.45 \text{ V}$$

This is almost equal to the value calculated in Ex. 11-2. Since the ratio $(R_1 + R_2)/R_{C_1}$ equals 17 in this circuit, this approximation is permissible. The rest of the analysis is the same as that presented in Sec. 11-8. The high value of R_1 that makes this fast analysis accurate results in a high value of $h_{FE\text{min}}$ required for Q_2 (see Prob. 11-9).

11-10 FET Logic Circuits

As previously explained in Sec. 2-10, the field-effect transistor has both advantages and disadvantages as a switch. Assuming that the input is from gate to source in the FET and from base to emitter in the bipolar transistor, the FET has a much higher *dc* input resistance and therefore a much higher *dc* fan-out. Also, the small size,* low power requirements, and simple fabrication procedure* of the MOSFET make it ideal in *large-scale integration* (LSI). At the present time, however, the FET does not compare favorably with the bipolar transistor as a switching device because of its slower speed. The input capacitance C_{gs}, which is in the order of 10 pF, must discharge through the relatively high output resistance R_{ON} of the driver. The drain-to-source resistance is approximately from 100 Ω to 400 Ω when the FET is on, compared to only a few ohms for $R_{CE\text{sat}}$ in bipolar transistors. In order to improve switching speeds, the fan-out capability must be reduced. Propagation times in the order of 100 ns are typical in the FET. Thus, it may be concluded that the FET is best suited for low-cost, large-scale, low-speed serial configurations. Several FET logic circuits are discussed in the following paragraphs.

*The fabrication process of MOS ICs requires only one-third as many steps as the standard double-diffused bipolar IC. Each MOS transistor requires an area of approximately 1 square mil compared to approximately 50 square mils for the bipolar transistor. Because of this as many as 5000 devices can be included on a 150 × 150 square mils.

The FET *NOR* Gate. A two-input *NOR* gate using *n*-channel enhancement mode MOSFETS is shown in Fig. 11-19(a). By using a MOS device as the pull-up resistor, a high resistance is obtained using a much smaller area on the IC chip. In amplifiers the MOS load resistor has another advantage. The nonlinearity of the MOS load compensates for the nonlinearity of the FET in amplifier operation. The enhancement-type MOSFET is off if $V_{GS} < V_{GST}$ and conducts heavily if $V_{GS} > V_{GS(on)}$ (see Fig. 2-32(b)). In the *n*-channel type the gate-source voltage must be positive in order for the FET to conduct. In the *p*-channel type it must be negative.

In Fig. 11-19(a) if any input is $\geq V_{GS(on)}$ a high current will flow. A high voltage is then dropped across the load, causing the output at Z to drop to a low voltage $< V_{(0)max}$. When all inputs are less than $V_{GST} > 0$ V, all FETs are cut off and only a minute current flows through the load. Since a small current does flow through the load, its V_{GS} must equal V_{GST}. And since the drain is tied to the gate, $V_{DS} = V_{GS} = V_{GST}$. Therefore, the 1-level output at Z equals $(V_{DD} - V_{GST}) > V_{(1)min}$. In the state table of Fig. 11-19(b) it can be seen that the output is high only when both inputs are low. Hence,

$$f_Z = \overline{AB} = \overline{A+B}$$

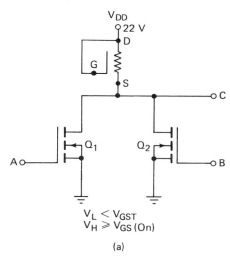

B	A	Q_2	Q_1	Z
V_L	V_L	Off	Off	$> V_{(1)min}$
V_L	V_H	Off	On	$< V_{(0)max}$
V_H	V_L	On	Off	$< V_{(0)max}$
V_H	V_H	On	On	$< V_{(0)max}$

(b)

Fig. 11-19 (a) A *NOR* gate using *n*-channel enhancement mode MOSFETS and a MOS pull-up resistor with (b) its state table.

344 / Special-Purpose Circuits

The *dc* noise margin is the difference between the acceptance level of the circuit and the actual output voltage. That is,

$$N_{(1)} = V_{o_{(1)}} - V_x$$

and

$$N_{(0)} = V_{o_{(0)}} - V_y$$

where V_x is the upper acceptance level and V_y is the lower acceptance level. $V_{(1)\min}$ equals V_x plus the guaranteed minimum 1-level noise margin; $V_{(0)\max}$ equals V_y minus the guaranteed minimum 0-level noise margin. With the parallel arrangement of Fig. 11-19(a), the 0-level noise margin is increased as more devices conduct because the parallel resistance decreases. The 1-level noise margin is reduced as more devices are parallelled because the leakage currents add through the common pull-up resistor.

The FET *NAND* Gate. The circuit of Fig. 11-20(a) is a *NAND* gate. The state table is given in Fig. 11-20(b). When both inputs are low, both FETs

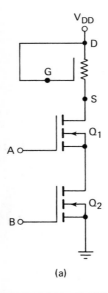

B	A	Q_1	Q_2	Z
V_L	V_L	Off	Off	$> V_{(1)\min}$
V_L	V_H	Off	On	$> V_{(1)\min}$
V_H	V_L	On	Off	$> V_{(1)\min}$
V_H	V_H	On	On	$< V_{(0)\max}$

(a)

(b)

Fig. 11-20 (a) A *NAND* gate using *n*-channel enhancement mode MOSFETS and a MOS pull-up resistor with (b) its state table.

are off and, as in the *NOR* gate, $V_Z = (V_{DD} - V_{GST}) > V_{(1)\text{min}}$. When one input is low and the other is high, one FET conducts, but the current is limited by the nonconducting device. Thus, the output is still high. Only when all inputs are high can a high current flow. Then all FETs offer a low resistance and the output drops to $< V_{(0)\text{max}}$. An important factor to be considered in the cascade arrangement is the higher resistance as more devices are connected in series. This results in a higher 0-level output voltage and reduces the 0-level noise margin. The output can be made approximately the same as for the *NOR* gate by making the FET channels wider in the *NAND* gate, thus reducing R_{ON}.

Other FET Logic Circuits. Several other FET logic circuits are given in Figs. 11-21, 11-22, and 11-23. The circuit of Fig. 11-21(a) is an inverting buffer. The circuit of Fig. 11-21(b) is a non-inverting buffer. The *EXCLUSIVE-OR* and *EQUIVALENCE* functions are produced by the circuits of Fig. 11-22(a) and (b), respectively. An RS flip-flop is shown in Fig. 11-23. When the *FF* is in the 1 state, Q_2 is off and Q_3 is on. The reset and set input transistors are off unless a trigger pulse is applied. If a pulse is applied to the *R* input, Q_1 comes on and sends the Q output to the 0 level. This turns off Q_3 and sends the $\bar{Q}$ output to the 1 level.

If *p*-channel enhancement mode MOSFETS are used, FET logic circuits can be made to be compatible with *npn* bipolar circuits.

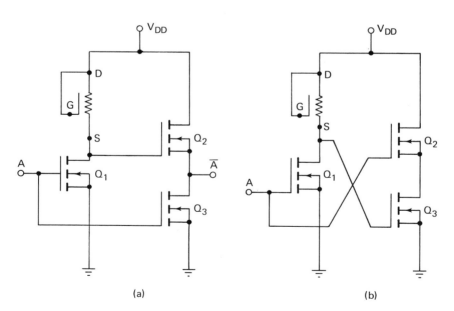

Fig. 11-21 (a) An inverting buffer. (b) A non-inverting buffer.

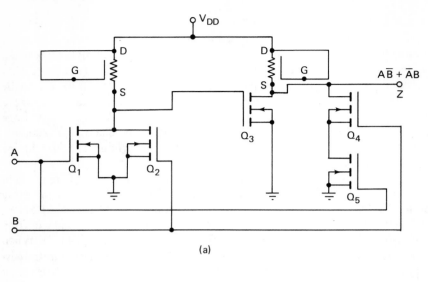

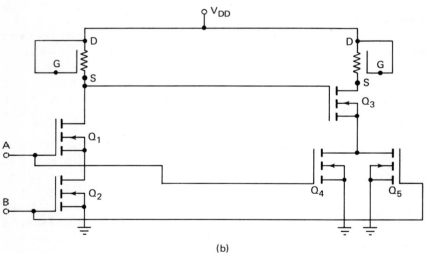

Fig. 11-22 (a) A FET *EXCLUSIVE-OR* circuit. (b) A FET *EQUIVALENCE* circuit.

11-11 CMOS Logic

By using complementary n and p channel transistors a series of MOS logic circuits that requires only a single 5 V supply can be produced. These circuits fall into the category known as *complementary MOS* (CMOS) or *comple-*

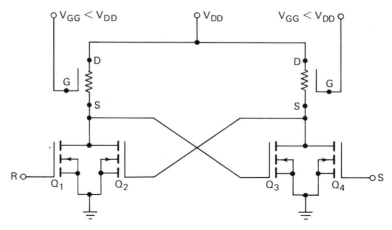

Fig. 11-23 An RS type flip-flop using *n*-channel MOSFETS.

mentary symmetry MOS (COS/MOS) *Logic*. The basic gates of this family are discussed in this section.

The CMOS Inverter. A CMOS inverter is shown in Fig. 11-24(a). When the input voltage V_i is low, V_{GS} of Q_1 is negative turning that transistor on. At the same time $V_{GS} = 0$ V for Q_2 turning off that transistor. The result is a very high resistance path from the output to ground and a low resistance path to $V_{DD} = 5$ V. When the input is high, the *p*-channel transistor Q_1 is cut off and the *n*-channel transistor Q_2 is turned on. This produces a very high resistance path from the output to V_{DD} and a low resistance path to ground. When the input is low the output is high and when the input is high the output is low; hence, the *NOT* function is produced.

Since both active pull-up and pull-down are used, power consumption is extremely low. Values in the order of a few nanowatts are typical. The transfer characteristic curve for the basic CMOS inverter is shown in Fig. 11-24(b). When the input is below the lower acceptance level $V_y = 2$ V, Q_2 is off and Q_1 is on. The quiescent drain current I_{D_1} is limited to the leakage current of Q_2. The loads, because of their high input resistance (in the order of 10^{12} ohms), draw no current. The output voltage V_o is therefore approximately equal to V_{DD} and it is virtually independent of the number of loads being driven.

When the input voltage reaches the upper acceptance level $V_x = 3$ V, Q_2 comes on and Q_1 cuts off. The drain current I_{D_2} is now limited to the leakage current of Q_1. Because of the low ON resistance of Q_2, the output voltage falls to approximately 0 V.

The dc noise margins are:

$$N_{(0)} = V_y - V_{o(0)}$$
$$= 2 - 0$$

348 / *Special-Purpose Circuits*

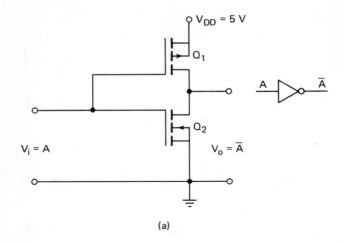

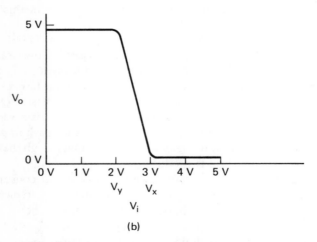

Fig. 11-24 (a) A CMOS inverter. (b) The transfer characteristic curve for the basic CMOS inverter.

$$N_{(0)} = 2 \text{ V}$$

and

$$N_{(1)} = V_{o(1)} - V_x$$
$$= 5 - 3$$
$$N_{(1)} = 2 \text{ V}$$

Note that in addition to the very low power consumption and high dc fan-out, CMOS circuits also have excellent noise immunity. The noise margin is approximately 40% of V_{DD}. It is increased by increasing V_{DD}. Values of V_{DD} as high as 18 V are used. Although the output resistance is higher with the MOS transistor than it is with bipolar transistors, there is no storage time problem in the MOS type. Therefore, the speeds of CMOS circuits can

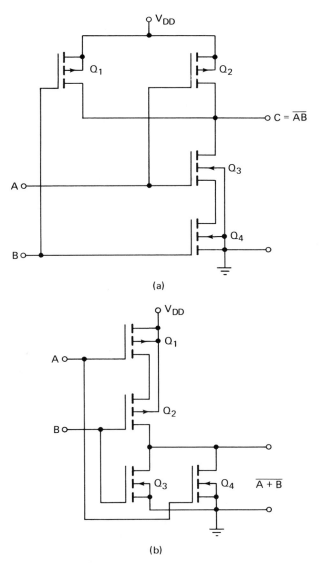

Fig. 11-25 (a) A two-input *CMOS NAND* gate. (b) A two-input *CMOS NOR* gate.

compete with circuits that use saturated bipolar transistors. The highest speeds are still obtained by using bypolar transistors and some non-saturating techniques. CMOS circuits are generally used when low power consumption and high noise immunity are desired.

The CMOS NAND Gate. The circuit of Fig. 11-25 (a) is a two-input *CMOS NAND* gate. When both inputs are high, the *p*-channel pull-up transistors Q_1 and Q_2 are cut off and the *n*-channel pull-down transistors Q_3 and Q_4 are on. This sends the output at C to the 0 level. When the input at A (and/or B) goes low, Q_1 (and/or Q_2) comes on, and Q_3 (and/or Q_4) cuts off. Since Q_3 and Q_4 are in series, only one of them has to be off to produce a high resistance path from the output to ground. And since the pull-up transistors are in parallel, only one of them has to be on to produce a low resistance path from the output to V_{DD}. Therefore, if any input is low the output is high. Only when both inputs are high does the output go low. Hence, the *NAND* function $\overline{AB}$ is implemented.

The CMOS NOR Gate. The circuit of Fig. 11-25 (b) is a two-input *CMOS NOR* gate. Note that the *p*-channel pull-up transistors Q_1 and Q_2 are now in series and the *n*-channel pull-down transistors Q_3 and Q_4 are in parallel. For the output to see a low resistance path to V_{DD}, both inputs must be low to turn on both Q_1 and Q_2, and to simultaneously turn off Q_3 and Q_4. If either input is high, one of the series pull-up transistors is cut off, producing a high-resistance path to V_{DD} and one of the paralleled pull-down transistors is on to produce a low-resistance path to ground. Since the output is high only when both inputs are low, the *NOR* function $\overline{AB} = \overline{A + B}$ is implemented.

11-12 Three-state Logic TSL

Three-state logic TSL differs from standard TTL in that it has a third output state. In addition to the two normal (high and low voltage) outputs of the other circuits discussed in this text, a TSL gate, when disabled, presents an impedance that effectively disconnects it from its buss. This greatly simplifies data bussing because it permits up to 128 TSL gates to be connected on a single buss. The number of gates that can be connected on a single line is limited by the total leakage currents of disabled gates.

Three-state logic circuits are available from National Semiconductor (which introduced it), Texas Instruments, and Signetics.

A TSL Circuit and its Operation. A TSL gate is shown in Fig. 11-26. If the *ENABLE* line is low, Q_8 comes on, turning off Q_9 and Q_6. Then, depending on the *INPUT* level, either Q_4 or Q_5 conducts. These are the two normal states of TTL.

If the *ENABLE* line is high, Q_8 cuts off, allowing Q_9 and Q_6 to conduct.

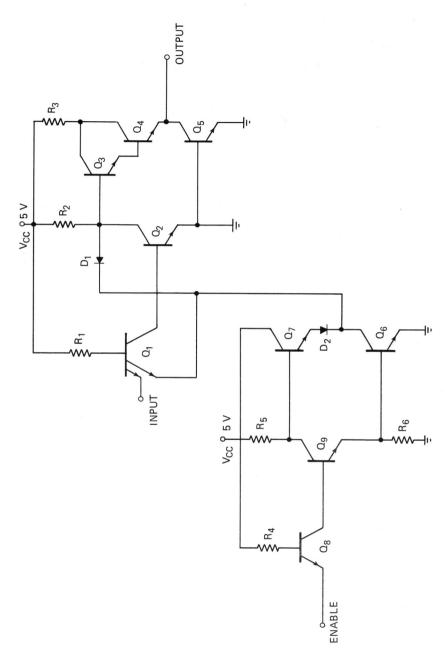

Fig. 11-26 A three-state logic circuit.

The low voltage at the collector of Q_6 turns on Q_1 which turns off Q_2 and Q_5. In TTL, Q_3 and Q_4 would come on when Q_2 cuts off, but in the TSL gate they are also off. The low V_{C_6} causes D_1 to conduct, reducing V_{B_3} to the $V_{CE_{sat}}$ of Q_6 plus V_{D_1} (approximately 1 V) which is not enough to turn on Q_3 and Q_4. Since both output transistors Q_4 and Q_5 are off, the output is in its third (high Z) state.

The propagation delay t_p depends on whether the gate is being enabled or disabled. When the gate is being disabled (*ENABLE* input raised to the 1 level), Q_6 actively turns on Q_1 and pulls down V_{B_3}. When Q_1 comes on, it actively pulls down V_{B_2}. With Q_2 and Q_3 off, Q_4 and Q_5 cut off and the output is disabled.

When the gate is being enabled (*ENABLE* input sent to the 0 level), Q_8 conducts to actively pull down V_{B_9} and cut off Q_9. This cuts off Q_6 and allows Q_7 to conduct. The cathode of D_1 and the emitter of Q_1 are pulled up by Q_7 until they cut off. Then the base capacitance of Q_2 is charged through R_1 if the *INPUT* is high or the base capacitances of Q_3 and Q_4 are charged through R_2 if the *INPUT* is low.

11-13 Interface Elements

Interface elements are used to provide translation between high-voltage logic levels and the standard logic level of 5 V.

The 8T18 2-Input *NAND* Gate Interface Element. Figure 11-27 shows a 2-input *NAND* high-to-low-voltage interface gate (Signetics 8T18). It is used to provide translation between logic levels of up to 30 V to the standard logic level of 5 V. It operates from two power supplies: the input section from a high-voltage supply of from 20 V to 30 V and the output section from a standard 5 V supply.

The input supply accepts voltage swings from 8 V to 30 V. It has a 50 V reverse breakdown rating. The output stage provides active pull-up and pull-down for fast 1-to-0 and 0-to-1 transitions even when driving heavy capacitance loads. The threshold voltage of the 8T18 is independent of temperature because the internal junctions are equal in number but opposite in polarity. It therefore can be used as an excellent high-level threshold detector.

An application of the 8T18 is shown in Fig. 11-29. It should be studied after the 8T80 and 8T90 interface elements (that follow immediately) are understood.

The 8T80 Two-input *NAND* Gate and 8T90 Inverter Interface Elements. The 8T80 circuit is shown in Fig. 11-28. It is a 2-input *NAND* gate. The 8T90 circuit is the same except for the input transistor. It has only one emitter making the circuit an inverter. These circuits are used to provide translation

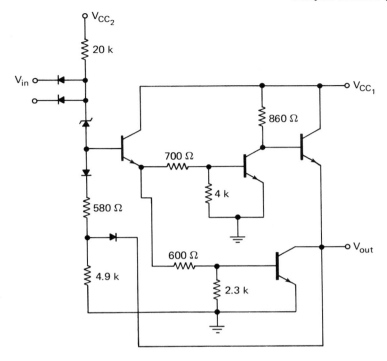

Fig. 11-27 A two-input *NAND* high-to-low voltage interface gate. (Courtesy Signetics Corporation.)

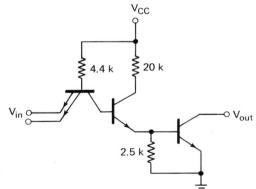

Fig. 11-28 A two-input *NAND* gate interface element. (Courtesy Signetics Corporation.)

from the standard 5 V logic levels to high-voltage logic levels of up to 30 V. Each features a high-voltage-transistor open-collector output. This allows logic swings of up to 30 V and permits the circuits use in wired-logic applications (see Sec. 7-3).

An Application of the 8T18 and 8T80/90 Interface Elements. A typical application of the 8T18 and 8T80/90 interface circuits is shown in Fig. 11-29.

354 / Special-Purpose Circuits

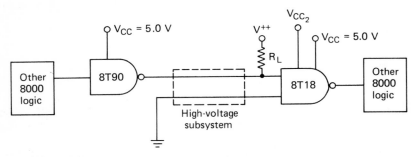

Fig. 11-29 An application of the 8T18 and 8T80/90 interface elements of Figs. 11-27 and 11-28. (Courtesy Signetics Corporation.)

A low-voltage gate like those discussed in Chaps. 7 and 8 is connected to a high-voltage subsystem through the low-to-high interface 8T90 *NAND* gate. The subsystem is then connected to another low-voltage gate through the high-to-low interface 8T18. The V^{++} supply ranges from 20 V to 30 V. If $V^{++} > 30$ V, either a series resistor must be used to limit current to less than 2 mA or a Zener diode of from 20 V to 30 V rating must be connected in shunt.

11-14 Summary

There are circuits that can be categorized as *special-purpose* circuits. For example, the internal gates of complex arrays do not have the limitations of the basic building blocks discussed in the earlier chapters of this book and therefore can be modified to improve speed.

Other circuits perform the special function of driving lamps, relays, or transmission lines. Crystal-controlled oscillators are used to generate accurate clock pulses. The Schmitt trigger circuit is used to reshape pulses that have been distorted after passing through other circuits or to change the sine-wave output of the clock oscillator to a square wave.

The field-effect transistor FET is used to produce the various circuits found in digital computers when high speed is not essential. The FET is ideal in low-speed, large-scale serial applications.

Three-state logic TSL gates have a third, high output impedance state that allows as many as 128 devices to be placed on a single buss.

Interface elements are used between high- and low-voltage logic systems.

Problems

11-1 The circuit of Fig. 11-10 has the following values: $V_{CC} = 10$ V, $R_{C_1} = 2$ k, $R_1 = 8$ k, $R_2 = 10$ k, $R_{C_2} = 1$ k and $R_E = 200$ Ω. Calculate the upper threshold voltage V_{UT}. Use the method presented in Ex. 11-1.

11-2 Calculate (a) the lower threshold voltage V_{LT} and (b) the hysteresis voltage V_h for the circuit of Prob. 11-1. Use the method presented in Ex. 11-2.

11-3 Determine the mark-to-space ratio of the circuit of Prob. 11-1 for a sine-wave input of 10 V PP.

11-4 Verify that Q_2 in the circuit of Prob. 11-1 is cut off when Q_1 is saturated.

11-5 Calculate the minimum h_{FE} required to keep Q_2 in saturation when Q_1 is off in the circuit of Prob. 11-1.

11-6 Calculate V_{E_2} for the circuit of Prob. 11-1 by using Eq. (11-13). Compare this answer to the one found in Prob. 11-1.

11-7 Calculate V_{E_1} for the circuit of Prob. 11-1 by using Eq. (11-14). Compare this answer to the one found in Prob. 11-2.

11-8 What will the mark-to-space ratio M/S be if the amplitude of the sine-wave input to the circuit of Fig. 11-12 is increased to 20 V PP? Compare this answer to the one calculated in Ex. 11-4.

11-9 Calculate the value of $h_{FE\min}$ required to keep Q_2 saturated in Fig. 11-12.

12

PACKAGING
AND APPLICATIONS
OF DIGITAL CIRCUITS

In this chapter some applications of digital circuits are given. Also included are examples of the way these circuits are packaged. Knowledge of the way in which the circuits are packaged is useful because it influences the design of the system. Only Transistor-Transistor Logic TTL packages are described. Diode Transistor Logic DTL and Emitter-Coupled Logic ECL circuits are produced in packages similar to those of TTL. Included are packages containing *NAND* gates, *NOR* gates, *INVERTERS*, *OR* gates, *AND* gates, *FLIP-FLOPS*, and complex arrays. Applications of the individual gates are seen in the complex arrays. Some of these applications appear in Divide-by-N Counters, NBCD-to-Decimal Decoder/Drivers, XS3-to-Decimal Decoders, NBCD-to-Seven Segment Decoder/Drivers, Full Adders, Digital Multiplexers, and Shift Registers.

12-1 TTL Packages

The TTL circuits discussed in the preceding chapters come in various package types, examples of which are discussed in this chapter. The integrated circuit chips are mounted on several different kinds of packages, three of which are shown in Figs. 12-1 and 12-2. Figure 12-1 shows 14-pin and 16-pin dual-in-line packages. A 14-pin flat-pack is shown in Fig. 12-2.

The Sprague Series 5400/7400 integrated circuits consist of TTL circuits. The 5400 series is designed for the military temperature range of $-55°C$ to $+125°C$. The less expensive 7400 series is guaranteed over the smaller temperature range of 0°C to +70°C. Both are available in the 14-pin flat-pack and the 14- or 16-pin dual-in-line packages. They are fabricated using the planar epitaxial process. All of the input stages employ small-geometry, high-speed multi-emitter transistors with input-clamping diodes to prevent ringing problems. Some of the flip-flops use a clamping transistor instead of a diode. The output stages utilize either the totem-pole or open-collector configuration.

The high-speed series 54H00/74H00 are available in similar packages. They use smaller resistances to increase speed. The resulting high currents increase power consumption.

TTL *NAND* Gate Packages. The circuit of Fig. 12-3(a) is 5400/7400 2-input *NAND* gate. As shown in Fig. 12-3(b), in addition to the quad 2-input 5400/7400 package, it is also available in triple 3-input 5410/7410, dual 4-input 5420/7420, and single 8-input 5430/7430 packages.

The 5401/7401 and 5403/7403 quad 2-input *NAND* gates employ the same circuit except for Q_3 and R_3 which are omitted. The open-collector output permits their use in wired-logic applications (see Sec. 7-3) or when it is necessary to interface with discrete components. An asterisk is placed near the output lead of the logic symbol to indicate an open-collector output.

The 5440/7440 dual 4-input *NAND* buffer gate shown in Fig. 12-4 is designed to drive clock lines and high-capacitance loads. It can sink 48 mA in the 0 state and supply 1.2 mA in the 1 state. The fan-out $FO = 30$, where a unit load $I_u = 1.6$ mA is the 0-input current of the 5400/7400 gates. The output stage employs a Darlington pull-up to provide a very low impedance in the 1 state.

TTL *NOR* Gate Packages. The 2-input *NOR* gate of Fig. 8-16 is repeated in Fig. 12-5. It is available in the 5402/7402 in quad 2-input packages. It gives an overall speed improvement, lower power consumption, and better packaging efficiency when compared to obtaining the *NOR* function by proper connection of *NAND* gates.

TTL *AND* Gate Packages. TTL *AND* gates are available in the packages shown in Fig. 12-6(a). The 5409/7409 open-collector circuit is shown in Fig. 8-18(a). The 5408/7408 and 5411/7411 circuits are the same except for the addition of a Darlington pull-up in the output. The circuit for these packages is shown in Fig. 8-28. The 5411/7411 circuit has three inputs.

TTL *OR* Gate Packages. TTL *OR* gates are available in the packages shown in Fig. 12-6(b). The circuit of the 5432/7432 packages is shown in Fig. 8-19. The 5418/7418 circuit is the same except for the number of inputs.

TTL *AOI* Gate Packages. TTL *AND-OR-INVERT AOI* gates are available in the expandable dual 2-wide, 2-input 5450/7450; dual 2-wide, 2-input 5451/7451; dual 2-wide, 2-3-input 5459/7459; expandable 4-wide, 2-input 5453/7453; and 4-wide, 2-input 5454/7454 packages shown in Fig. 12-7(a). All of these gates may be used to implement the *EXCLUSIVE-OR*

358 / *Packaging and Applications of Digital Circuits*

function. The circuit for each is basically the same as the one shown in Fig. 8-20. The 5460/7460 dual 4-input expander, shown in Fig. 12-7(b), is designed for expanding the *AND* function of the 5450/7450 and 5453/7453 gates. The expander circuit is shown in Fig. 8-21.

TTL *Inverter*. The TTL inverter is available in packages of six. The circuit of the 5405/7405 packages is shown in Fig. 12-8. An inverter that has

Package data
  A package:
    1. The integrated circuit chips are mounted on a 14-lead or 16-lead frame and encapsulated with a plastic compound which adheres tightly over the lead frame, preventing moisture intrusion.
    2. The plastic package will withstand soldering temperatures of 230°C for one minute with no deformation of circuit performance characteristics.
    3. Lead spacing is designed for insertion in mounting hole rows, on 0.300" centers. To facilitate the use of automatic insertion equipment, the leads are compressed to 0.300" separation and sufficient tension is provided to secure the package within a Barnes carrier.
    4. Once inserted in the printed circuit board, sufficient tension is provided to secure the package in the boards during soldering.
    5. Pin shoulders support the package above the printed circuit board, allowing free air circulation and preventing the formation of moisture pockets.
    6. Packages shipped in tubes have a lead separation of 0.375" maximum.

Outline—Package A
Plastic 14-pin dual in-line package

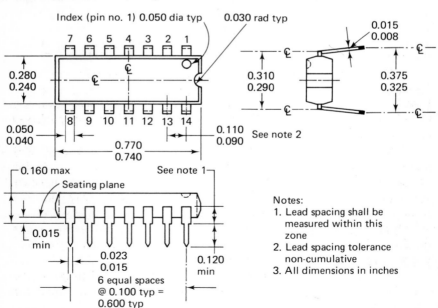

Fig. 12-1 14-pin and 16-pin dual-in-line silicon packages. (Courtesy Sprague Electric Co.)

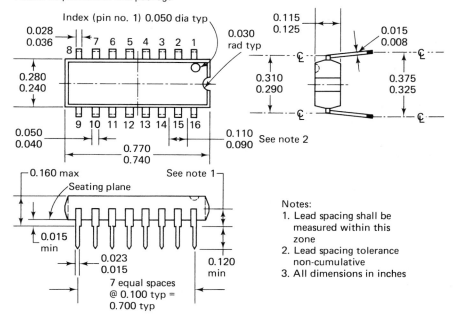

Thermal package characteristics, 'A' package.
Thermal resistance from junction to still air, θ_{JA} = 0.12°C/mW
Thermal resistance from junction to case, θ_{JC} = 0.07°C/mW

Fig. 12-1 (*cont.*)

a totem-pole output is also available. This circuit, which is used in the 5404/7404 packages, is shown in Fig. 8-17. Because they all come in packages of six, they are referred to as *hex-inverters*.

TTL Flip-flop Packages. The 54107/74107 package of Fig. 12-9(a) contains two JK master-slave flip-flops. Each has separate clock inputs for synchronous operation and separate direct reset (or clear) R_D inputs. In order to set the flip-flop (send Q output to the 1 level), a 0-level voltage must be applied to K while $J = 1, J = V_{cc}$, or J is left open and a clock pulse is applied. A 0-level voltage to R_D holds Q at the 0 level. Clock pulses have no effect while $R_D = 0$. The 54107/74107 is available in the 14-pin dual in-line "A" package.

The 5476/7476 package shown in Fig. 12-9(b) is exactly the same as the 54107/74107 except that it is also provided with a direct set S_D input for increased versatility. It is available in the 16-pin dual in-line "A" package.

The 54H78/74H78 packages shown in Fig. 12-9(c) contain two high-speed JK master-slave flip-flops. It has common clock and common clear

Package data (continued)
J package:
1. The integrated circuit chips are mounted in a standard 14-lead (TO-88) flat-pack constructed with a hermetic glass-to-metal seal.
2. The standard 0.050" center-to-center lead spacing and thin rectangular configuration allow for high density circuit board layout where size and weight are important.
3. All devices are shipped in Barnes carriers which simplify handling and reduce costs in incoming inspection, storage, and assembly operation.
4. In applications where flat-packs are mounted on circuit boards with the conductor passing under the package, insulators are available to prevent the metallic base of the package from shorting the conductor. The insulator is a pressure-sensitive plastic tape, 0.355" long x 0.265" wide x 0.003" thick.
5. For mounting convenience, the flat-pack may be ordered with formed and/or tinned leads. Devices supplied with formed leads are shipped in suitable Barnes-type carriers.
6. For convenience of automatic handling, Barnes carriers may be ordered with a magazine-type cartridge.

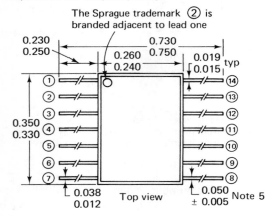

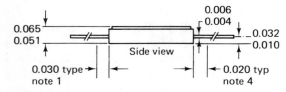

Notes:
(1) Recommended minimum offset before lead bend
(2) All leads weldable and solderable
(3) All dimensions in inches
(4) Lead spacing dimensions apply to this area only
(5) Spacing tolerance non-cumulative

Fig. 12-2 A 14-pin flat-pack. (Courtesy Sprague Electric Co.)

Outline—Package J
14-pin formed leads hermetic flat-pack package

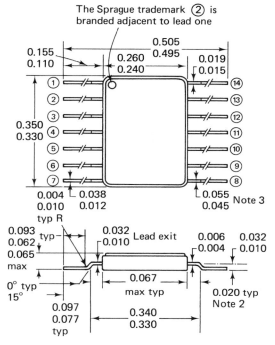

Notes:
(1) Boxed dimensions are inspected after bend
(2) All dimensions in inches
(3) Lead spacing dimensions apply to this area only
(4) Spacing tolerances non-cumulative

Thermal package characteristics, 'J' package
  Thermal resistance from junction to still air, $\theta_{JA} = 0.27°C/mW$
  Thermal resistance from junction to case, $\theta_{JC} = 0.20°C/mW$

Fig. 12-2 (*cont.*)

inputs, which affect all flip-flops in the package. They are available in both the "J" and "A" package types.

The 5472/7472 packages shown in Fig. 12-10(a) contain one JK master-slave flip-flop with multiple J and K inputs for synchronous operation and direct set S_D and reset R_D inputs for asynchronous operation. The multiple J and K inputs reduce the number of external gates required to implement the J and K input functions. Because of the added inputs, each package can only hold one flip-flop.

The 54H71/74H71 packages shown in Fig. 12-10(b) contain one JK flip-flop with 2-wide, 2-input *AND-OR J* and K inputs. The flip-flop has only

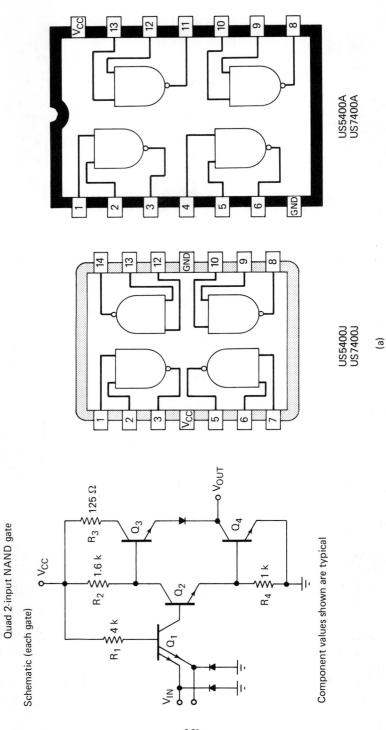

Fig. 12-3 (a) A quad 2-input *NAND* gate. (Courtesy Sprague Electric Co.)

NAND gates:

5400/7400 Quad 2-input
5410/7410 Triple 3-input

5420/7420 Dual 4-input
5430/7430 Single 8-input

These NAND gates are identical in performance and design, with the only difference being the number of inputs per gate and the number of gates per package.

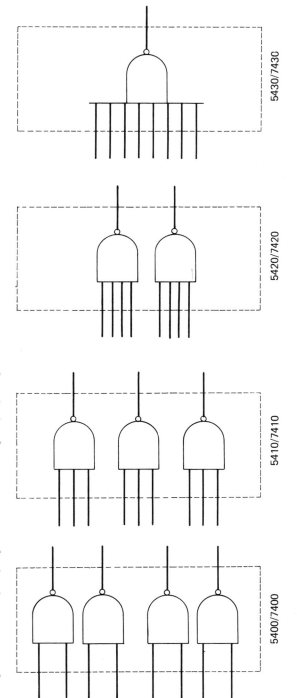

(b)

Fig. 12-3 (b) Various *NAND* gate packages. (Courtesy Sprague Electric Co.)

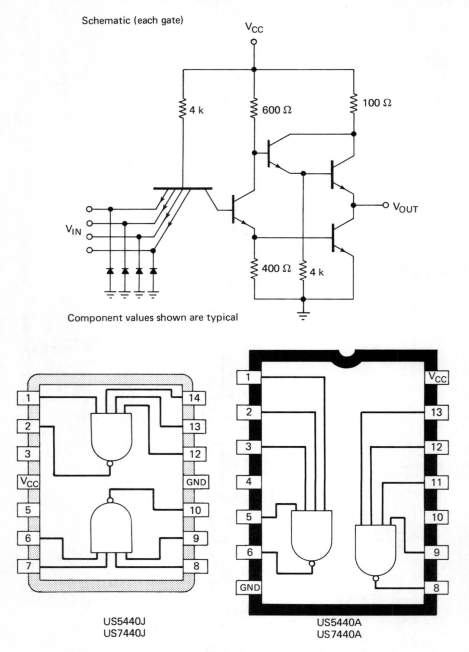

Fig. 12-4 A Dual 4-input *NAND* buffer gate. (Courtesy Sprague Electric Co.)

US5402 and US7402
Quad 2-input NOR gate

Schematic (each gate)

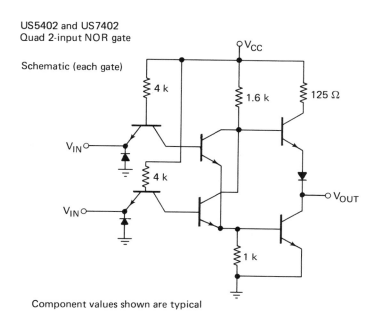

Component values shown are typical

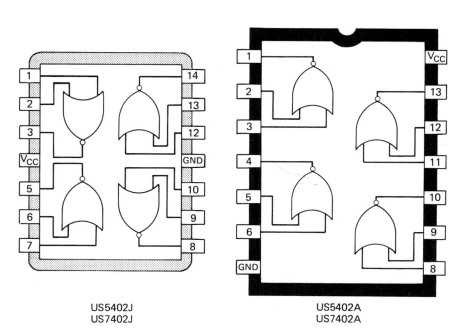

US5402J
US7402J

US5402A
US7402A

Fig. 12-5 A 2-input *NOR* gate. (Courtesy Sprague Electric Co.)

AND gates:
 5408/7408 Quad 2-input
 5409/7409 Quad 2-input—Open collector output
 5411/7411 Triple 3-input

These AND gates utilize the same basic construction as the NAND gates, except that an extra internal inverter stage is added to obtain the AND logic function. These devices are especially useful in high-speed arithmetic sections where it is necessary to avoid excessive propagation delays.

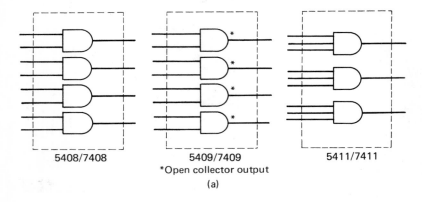

5408/7408   5409/7409   5411/7411
           *Open collector output
                (a)

OR gates:
 5418/7418 Triple 3-input
 5432/7432 Quad 2-input

These OR gates utilize the same basic construction as the NOR gates, except that an extra internal inverter stage is added to obtain the OR logic function.

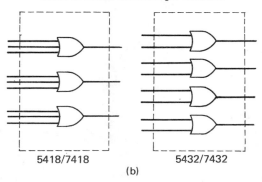

5418/7418   5432/7432
         (b)

Fig. 12-6 (a) TTL *AND* gate packages. (b) TTL *OR* gate packages. (Courtesy Sprague Electric Co.)

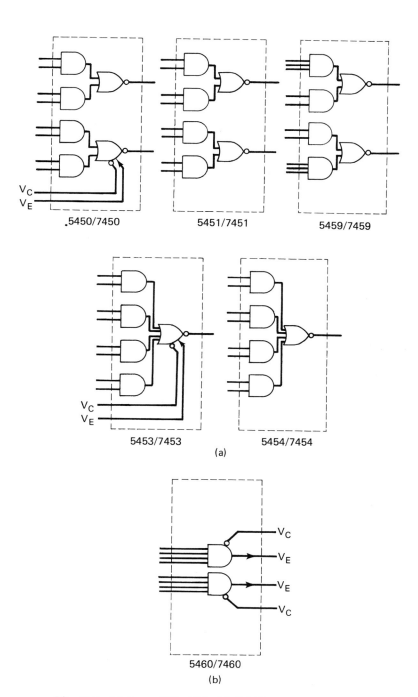

Fig. 12-7 (a) Some TTL *AND-OR-INVERT AOI* packages. (b) A dual 4-input expander for expanding the *AND* function of the 5450/7450 and 5453/7453 gates. (Courtesy Sprague Electric Co.)

US5405 and US7405
Hex inverter—Open collector output

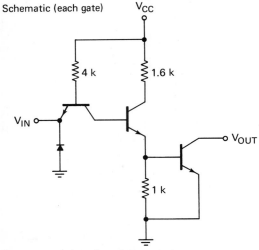

Schematic (each gate)

Component values shown are typical

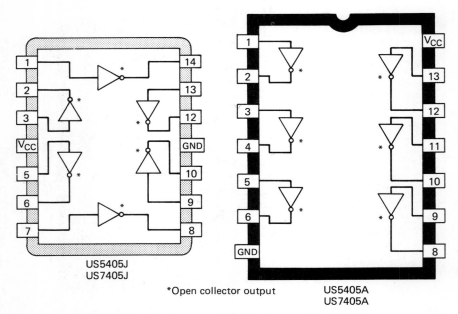

*Open collector output

US5405J
US7405J

US5405A
US7405A

Fig. 12-8 A hex-inverter. (Courtesy Sprague Electric Co.)

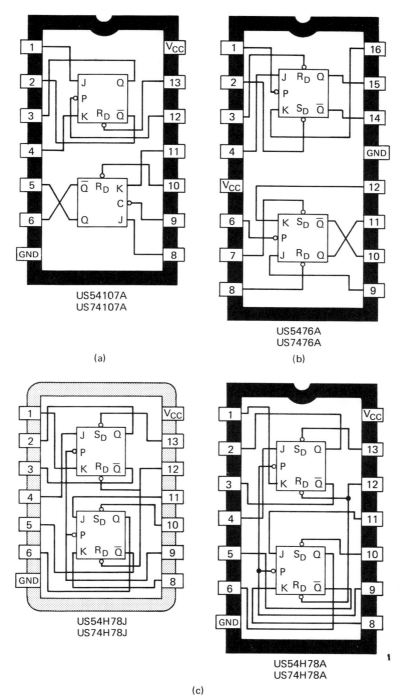

Fig. 12-9 Some master-slave JK flip-flop packages. (Courtesy Sprague Electric Co.)

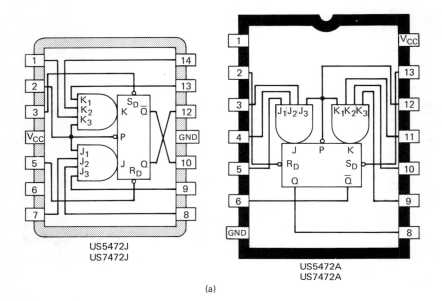

(a)

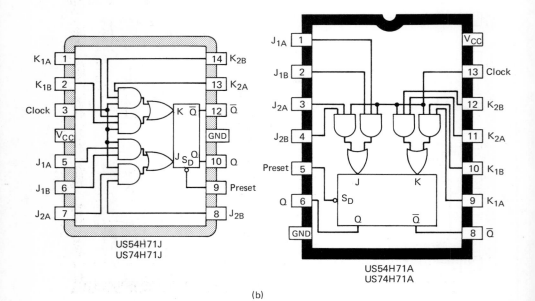

(b)

Fig. 12-10 (a) A master-slave JK flip-flop with multiple J and K inputs for synchronous operation. (b) A JK flip-flop with AND-OR synchronous inputs. (Courtesy Sprague Electric Co.)

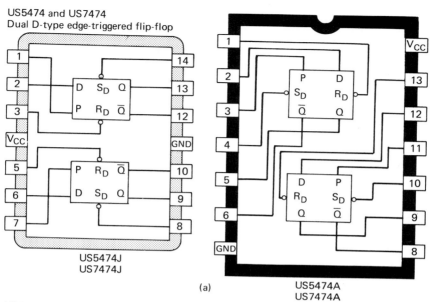
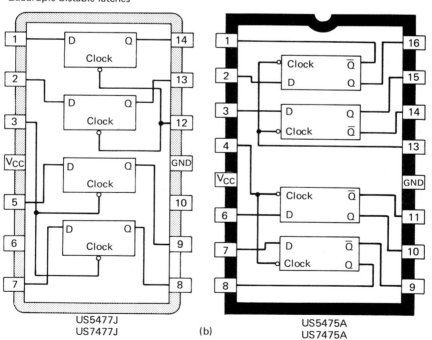

Fig. 12-11 Some D-type flip-flop packages. (Courtesy Sprague Electric Co.)

a direct set S_D input for asynchronous operation. It can be cleared by applying inputs to make $J = 0$ and $K = 1$ while a clock pulse is applied. The 54H71/74H71 is available in both the "J" and "A" package types.

The 5474/7474 packages shown in Fig. 12-11(a) contain two D-type, edge-triggered flip-flops that respond to the positive edge of the clock pulse. They are provided with direct set S_D and direct reset R_D inputs and are available in both the "J" and "A" package types.

The 5475/7475 packages of Fig. 12-11(b) contain four D-type flip-flops. Only synchronous operation is possible. It is available in the 16-pin "A" package. The 5477/7477 packages contain four D-type flip-flops that have only Q outputs. They come in the 14-pin "J" package.

12-2 A Monostable Package

The 8162 monostable multivibrator of Fig. 12-12 is available in both the "A" and "J" package types. The circuit has complementary outputs with 3-k pull-up resistors. Optional 500-Ω pull-up resistors are included. They may be used to maintain rise times when driving heavy capacitance loads. The 8162 has a 30-pF timing capacitor and an optional 1.5-k timing resistor. The output pulse may be varied over the range of 80 ns to 2 s by making the appropriate connections at the C_T, R_T, R_Y, and $R_{\bar{Y}}$ terminals. The following equations should be used to obtain a desired pulse width:

1. $PW \approx 0.85(C_x + C_{int})(10^{-3} \text{ s}/\mu\text{F})$

   with internal resistor R_x connected to V_{CC}.

2. $PW \approx \dfrac{0.85 R'_x(C_x - C_{int})(10^{-3} \text{ s}/\mu\text{F})}{1.5 \text{ k} + R'_x}$

   with external resistor $R'_x > 1$ k connected between R_T and V_{CC} so that it is in parallel with R_x.

3. $PW \approx \dfrac{0.85 R'_x(C_x + C_{int})(10^{-3} \text{ s}/\mu\text{F})}{1.5 \text{ k}}$

   with external resistor $R'_x (0.5 \text{ k} < R'_x < 4.7 \text{ k})$ connected between R_T and V_{CC} and R_x not connected.

where

PW is the pulse width. PW tolerance is ± 25 percent with R_x connected (unit-to-unit variations) and ± 10 percent with R'_x connected.
R_x is a 1.5-k internal resistor.

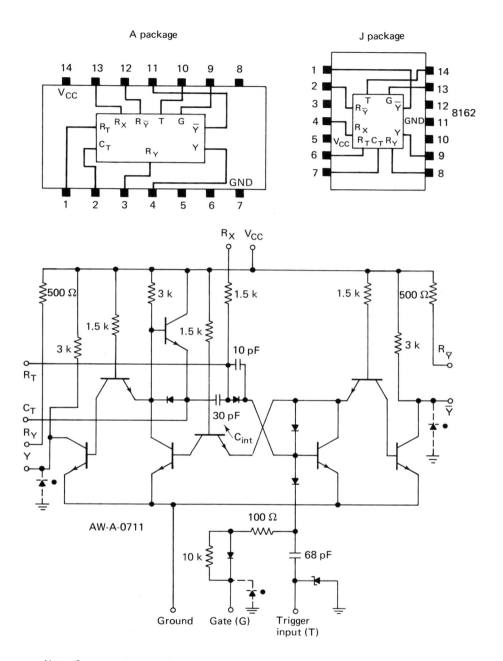

Note: Component values shown are typical • Isolation diodes

Fig. 12-12 A monastable multivibrator package and circuit. (Courtesy Signetics Corporation.)

C_x is an external capacitance in μF connected between C_T and R_T so that it is in parallel with C_{int}.
R'_x is an external resistor connected between R_T and V_{CC} so that it is in parallel with R_x.
With R_x connected, $PW \approx 0.85$ ms/μF of C_x.

The 8162 one-shot, can provide a duty cycle up to 75 percent. The output stage is isolated from the timing stage, resulting in good fall times even with wide pulse widths.

12-3 Summary

The reasons that the various digital circuits come in so many different packages can be seen from the preceding discussions of the TTL packages. One user may require more inputs to the gates than another. As shown in Fig. 12-3(b), if only two inputs are needed, four gates can be included in a 14-pin package. Eight pins are used for the inputs, 4 pins for the outputs, 1 pin for V_{CC}, and 1 pin for a ground connection. If three-input gates are needed, only three gates can be included in a 14-pin package. The 14-pin package can also hold two four-input gates or one eight-input gate. Eight pins are used for the inputs, 4 pins for the outputs, 1 pin for V_{CC}, and 1 pin for a ground connection. If three-input gates are needed, only three gates can be included in a 14-pin package. The 14-pin package can also hold two four-input gates or one eight-input gate.

With flip-flops, the number of inputs to the gates that drive the synchronous inputs can be increased if one of the asynchronous inputs R_D or S_D is eliminated or if common R_D and/or S_D inputs are used. Also, as shown in Fig. 12-10, if gated J and K inputs are provided, the package can hold only one flip-flop. But if no internal gates are included, the package, as shown in Fig. 12-9, can hold two flip-flops.

12-4 Complex Arrays

Packages containing complex arrays, such as Divide-by-N Ripple Counters, Shift Registers, Full Adders, NBCD-to-Decimal Decoder/Drivers, NBCD-to-Seven Segment Decoder/Drivers, Binary-to-Octal Decoders, and Digital Multiplexers, are also available. Examples of these packages are discussed in this section. Applications of nearly all the circuits previously discussed in this text are seen in these complex arrays.

Complex Arrays / 375

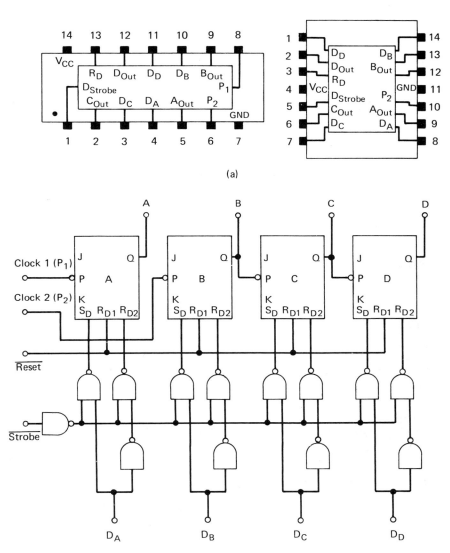

Fig. 12-13 A 4-bit (16-state) ripple counter package and logic diagram. (Courtesy Signetics Corporation.)

A 4-Bit Binary Ripple Counter. A 4-bit (16-state) ripple counter 8281 is shown in Fig. 12-13. It consists of four clocked JK flip-flops with direct set S_D and reset R_D inputs. The first flip-flop is separated from the others to increase flexibility. For example, the 8281 may be connected as a divide-by-two, eight, or sixteen counter. To divide by two, only flip-flop A is used. To divide by eight, only B, C, and D are used, with the input applied to the Clock 2 terminal. To divide by sixteen, the input is applied to the Clock 1 terminal and the output of the A flip-flop must be externally connected to the B input (Clock 2).

The counter has a $\overline{\text{Strobe}}$ input to permit asynchronous parallel-entry capability. This allows it to be preset to any desired count. A 1 or 0 at a data input (D_A, D_B, D_C, or D_D) is transferred to the associated output when a 0 level is applied to the $\overline{\text{Strobe}}$ input. The preset capability permits the counter to be used as an event counter, a fixed quantity totalizer, an arbitrary length frequency divider, and as a storage register. It also has a common reset input R_D, to further increase flexibility. A 0 level to the $\overline{\text{Reset}}$ input sends all outputs to the 0 level. The $\overline{\text{Reset}}$ and $\overline{\text{Strobe}}$ inputs should be at the 1 level (inactive) when normal counting is taking place. The flip-flops operate on the trailing edge of the clock.

There are intermediate counts produced during some clock pulse intervals. They may last from 10 to 80 ns in the 8281. The actual delay to the desired count depends on the next desired count. The intermediate counts for modified counters depend on the internal common clock connections and the internal gating. Each modified ripple counter must be considered separately. The intermediate counts for the 8281 are shown in Table 12-1. For example, when the count is 0000, it goes directly to 0001 after the first clock pulse to the A flip-flop. After the next clock pulse, A goes back to the 0 state, making the count temporarily 0000 again. However, when the Q output of FFA goes low it makes FFB go to the 1 state, producing the correct count 0010. The third clock pulse puts A back in the 1 state, sending the count directly to 0011.

In most cases intermediate counts are not important. Each output is considered separately, in terms of the number of pulses it produces in relation to the number of clock pulses applied to the input of the counter. Care should be taken, however, when decoding the outputs. The decoder must not be activated until after the intermediate counts are gone.

A Decade Counter. A divide-by-ten decade counter 8280 is shown in Fig. 12-14(a). The 8280 packages are the same as those shown for the 8281 in Fig. 12-13(a). One internal connection is made from the $\bar{Q}$ output of FFD to the J input of FFB. Another is made from the Q outputs of FFB and FFC through an AND gate to the J input of FFD. Note that the B and D flip-flops are clocked simultaneously. The sequence of counts for the decade counter is given in Fig. 12-14(b).

In addition to counting in the often used NBCD code, the 8280 can be connected to count in the divide-by-two or five modes or in the 5421 biquinary ($\div$ 2 and $\div$ 5) mode. In the biquinary mode, the input is applied to the

Table 12-1 The sequence of counts in the 4-bit ($\div$ 16) 8281 ripple counter. It is assumed that the count is 0 before the first clock pulse is applied.

Clock Pulse	Desired Decimal Count	Number of Intermediate States	Output DCBA	Approximate Delay to Desired Count
Start	0		0000	
1	1	None	0001	20 ns
2	2	One	0000	
			0010	40 ns
3	3	None	0011	20 ns
4	4	Two	0010	
			0000	
			0100	60 ns
5	5	None	0101	20 ns
6	6	One	0100	
			0110	40 ns
7	7	None	0111	20 ns
8	8	Three	0110	
			0100	
			0000	
			1000	80 ns
9	9	None	1001	20 ns
10	10	One	1000	
			1010	40 ns
11	11	None	1011	20 ns
12	12	Two	1010	
			1000	
			1100	60 ns
13	13	None	1101	20 ns
14	14	One	1100	
			1110	40 ns
15	15	None	1111	20 ns
16	0	Three	1110	
			1100	
			1000	
			0000	80 ns

Clock 2 terminal and the D output is fed back to the Clock 1 input. The weight of each flip-flop is $A = 5$, $D = 4$, $C = 2$, and $B = 1$. The counts produced in this mode are $ADCB$ (5421) = 0000, 0001, 0010, 0011, 0100, 1000, 1001, 1010, 1011, and 1100. A square wave is available from the A output when this code is used.

Like the 8281 16-state counter of Fig. 12-13, the 8280 has strobed parallel-entry capability. This permits it to be preset to a desired output. It also has the common direct reset R_{D_1} input so that it may be cleared.

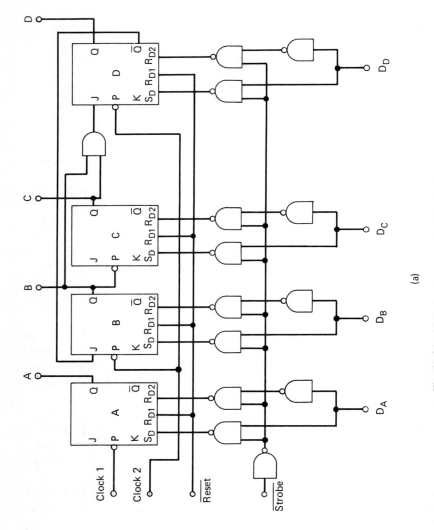

Fig. 12-14 (a) A decade counter logic diagram.

Clock pulse	Desired decimal count	Number of intermediate states	Output DCBA	Approximate delay to desired state
1	0		0000	
		None		
2	1		0001	
		One	0000	
3	2		0010	40 ns
		None		20 ns
4	3		0011	
		Two	0010	
			0000	
5	4		0100	60 ns
		None		20 ns
6	5		0101	
		One	0100	
7	6		0110	40 ns
		None		20 ns
8	7		0111	
		Two	0110	
			1100	
9	8		1000	60 ns
		None		20 ns
10	9		1001	
		One	1000	
11	0		0000	40 ns

(b)

Fig. 12-14 (b) Its sequence of counts. (*Cont.*)

An NBCD-to-Decimal Decoder/Driver. An NBCD-to-Decimal Decoder/Driver package is shown in Fig. 12-15(a). It is designed to directly drive gas-filled cold-cathode indicator lamps. Its logic diagram and truth table are shown in Fig. 12-15(b) and (c). The operation of the circuit will be explained for the 0001 input condition.

In the logic diagram of Fig. 12-15(b) the 1 input at A is inverted by gate 1, applying a 0 to the inverted inputs of gates 14, 16, 18, 20, and 22. The 0 out of gate 1 is inverted by gate 5, applying a 1 to the inverted inputs of gates 13, 15, 17, 19, and 21. This produces a 1-level output from these gates cutting off Q_{35}, Q_{37}, Q_{39}, Q_{41} and Q_{43}.

The 0 inputs at B, C, and D are inverted by gates 2, 3, and 4, making all inputs to gate 9 high. This applies a 1 level to the top input of gate 14 which, along with the 0 at the inverted input, produces a 0 level at the emitter of Q_{36}. By following the 1-level outputs of gates 2, 3, and 4 through their other paths it will be seen that Q_{38}, Q_{40}, Q_{42}, and Q_{44} are cut off. With an input of 0001, only Q_{36} has a 0 at its emitter and is able to conduct. All other outputs are off.

An NBCD-to-Seven Segment Decoder/Driver. The package shown in Fig. 12-16(a) contains an NBCD-to-Seven Segment Decoder/Driver. The

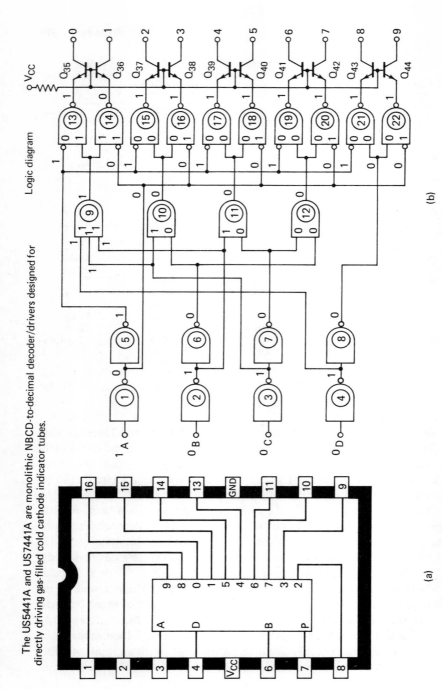

Fig. 12-15 An NBCD-to-Decimal Decoder/Driver package with (b) its logic diagram. (Courtesy Sprague Electric Co.)

Truth table

NBCD Input				Output on*
D	C	B	A	
0	0	0	0	0
0	0	0	1	1
0	0	1	0	2
0	0	1	1	3
0	1	0	0	4
0	1	0	1	5
0	1	1	0	6
0	1	1	1	7
1	0	0	0	8
1	0	0	1	9

*All other outputs are off.

(c)

Fig. 12-15 (c) Its truth table. (Courtesy Sprague Electric Co.) (*Cont.*)

US5446A and US7446A output transistors are designed for 30-V applications, and the US5447A and US7447A are designed for 15-V applications. The segment identification, truth table, and logic diagram are given in Fig. 12-16(b)–(d). The segments are on when the corresponding output is at the 0 level.

Each of the decoder/drivers has automatic leading and/or trailing-edge zero-blanking controls (ripple-blanking input RBI and ripple-blanking output RBO). An overriding blanking input BI is provided to permit control of the lamp intensity or to inhibit the outputs. A lamp test LT is also available. It may be used whenever the BI/RBO node is at the 1 level.

Wired logic (see Sec. 7-3) is used in the blanking input BI and/or the ripple-blanking output RBO circuity. The blanking input must be floating or at the 1 level when outputs 0 through 15 are desired. The ripple-blanking input must be floating or at the 1 level during the decimal 0 input.

When a 0 level is applied to the blanking input BI, all segment outputs go to the 1 level, regardless of the state of any other input. When a 0 level is applied to the ripple-blanking input and A, B, C, and D are also low, all segment outputs go to the 1 level and the ripple-blanking output goes to the 0 level (response condition).

When the blanking input/ripple blanking output BI/RBO is floating or at the 1 level and a 0 level is applied to the lamp-test input LT, all segment outputs go to the 0 level.

Binary Full Adders. Each of the packages shown in Fig. 12-17(a) contains a Binary Full Adder capable of adding 2-bit numbers. Outputs are

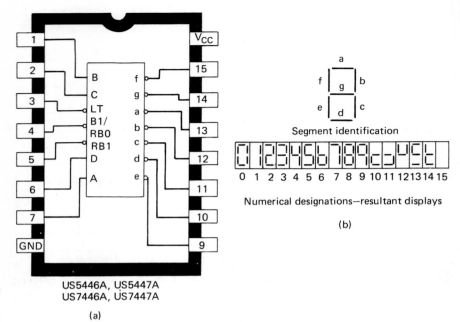

Truth table

Decimal or function	Inputs							Outputs						
	LT	RB1	D	C	B	A	B1/RB0	a	b	c	d	e	f	g
0	1	1	0	0	0	0	1	0	0	0	0	0	0	1
1	1	x	0	0	0	1	1	1	0	0	1	1	1	1
2	1	x	0	0	1	0	1	0	0	1	0	0	1	0
3	1	x	0	0	1	1	1	0	0	0	0	1	1	0
4	1	x	0	1	0	0	1	1	0	0	1	1	0	0
5	1	x	0	1	0	1	1	0	1	0	0	1	0	0
6	1	x	0	1	1	0	1	1	1	0	0	0	0	0
7	1	x	0	1	1	1	1	0	0	0	1	1	1	1
8	1	x	1	0	0	0	1	0	0	0	0	0	0	0
9	1	x	1	0	0	1	1	0	0	0	1	1	0	0
10	1	x	1	0	1	0	1	1	1	1	0	0	1	0
11	1	x	1	0	1	1	1	1	1	0	0	1	1	0
12	1	x	1	1	0	0	1	1	0	1	1	1	0	0
13	1	x	1	1	0	1	1	0	1	1	0	1	0	0
14	1	x	1	1	1	0	1	1	1	1	0	0	0	0
15	1	x	1	1	1	1	1	1	1	1	1	1	1	1
B1	x	x	x	x	x	x	0	1	1	1	1	1	1	1
RB1	1	0	0	0	0	0	0	1	1	1	1	1	1	1
LT	0	x	x	x	x	x	1	0	0	0	0	0	0	0

(c)

Fig. 12-16 (a) AN NBCD-to-Seven Segment Decoder/Driver. (b) The segment identification. (c) The truth table. (Courtesy Sprague Electric Co.)

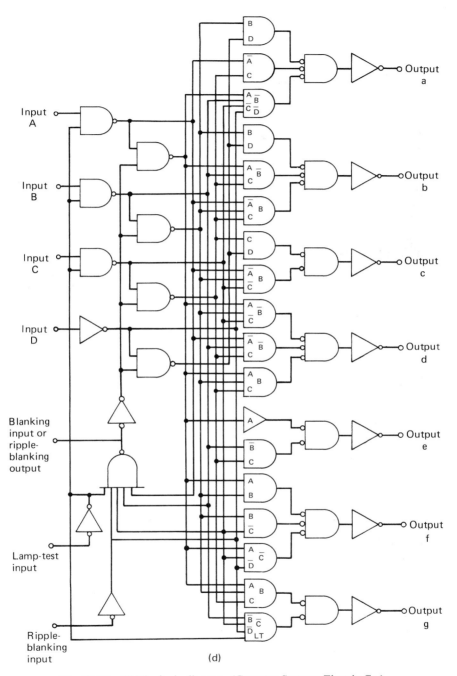

Fig. 12-16 (d) The logic diagram. (Courtesy Sprague Electric Co.) (*Cont.*)

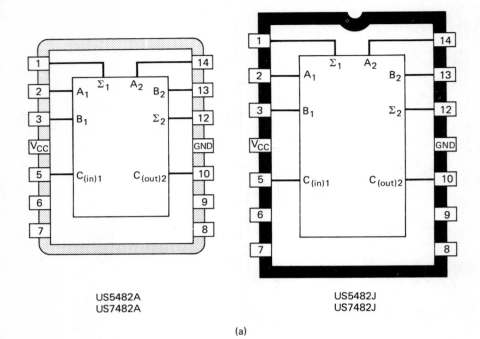

Input				Output					
				For $C_{(in)1} = 0$			For $C_{(in)1} = 1$		
A_1	B_1	A_2	B_2	Σ_1	Σ_2	$C_{(out)2}$	Σ_1	Σ_2	$C_{(out)2}$
0	0	0	0	0	0	0	1	0	0
1	0	0	0	1	0	0	0	1	0
0	1	0	0	1	0	0	0	1	0
1	1	0	0	0	1	0	1	1	0
0	0	1	0	0	1	0	1	1	0
1	0	1	0	1	1	0	0	0	1
0	1	1	0	1	1	0	0	0	1
1	1	1	0	0	0	1	1	0	1
0	0	0	1	0	1	0	1	1	0
1	0	0	1	1	1	0	0	0	1
0	1	0	1	1	1	0	0	0	1
1	1	0	1	0	0	1	1	0	1
0	0	1	1	0	0	1	1	0	1
1	0	1	1	1	0	1	0	1	1
0	1	1	1	1	0	1	0	1	1
1	1	1	1	0	1	1	1	1	1

(b)

Fig. 12-17 (a) 2-bit binary full adder packages with (b) their truth table.

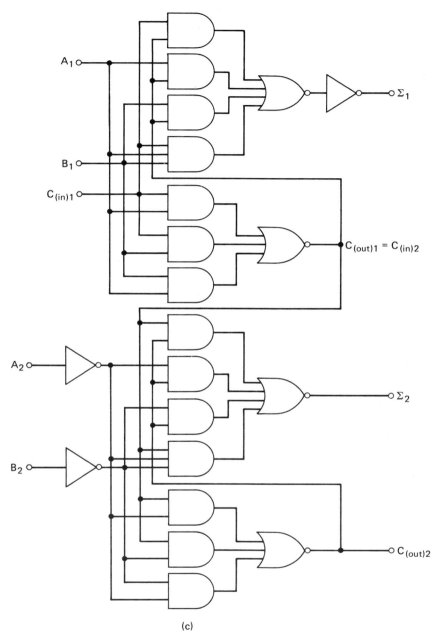

Fig. 12-17 (c) Logic diagram. (Courtesy Sprague Electric Co.) (*Cont.*)

provided for the 1s and 2s sums Σ_1 and Σ_2 and the 2s carry-out $C_{(out)2}$. The truth table and logic diagram are given in Fig. 12-17(b) and (c).

The Boolean equations for the sum and carry outputs of a 1-bit Full Adder are:

$$f_S = A\bar{B}\bar{C}_{(in)} + \bar{A}B\bar{C}_{(in)} + \bar{A}\bar{B}C_{(in)} + ABC_{(in)} \qquad (12\text{-}1)$$

and

$$f_{C(out)} = AB + AC_{(in)} + BC_{(in)} \qquad (12\text{-}2)$$

The design procedure for Full Adders can be found in most logic design text books. The sum will be a 1 if any one of the inputs is a 1 or if all three inputs are 1s. The carry-out will be a 1 if at least two of the inputs are 1s. If the logic diagram of Fig. 12-17(c) is analyzed and the Boolean equations for each output are written, it will be determined that

$$f_{\Sigma_1} = A_1\bar{B}_1\bar{C}_{(in)1} + \bar{A}_1B_1\bar{C}_{(in)1} + \bar{A}_1\bar{B}_1C_{(in)1} + A_1B_1C_{(in)1} \qquad (12\text{-}3)$$

and

$$f_{C(out)1} = \overline{A_1B_1 + A_1C_{(in)1} + B_1C_{(in)1}} \qquad (12\text{-}4a)$$
$$= (\bar{A}_1 + \bar{B}_1)(\bar{A}_1 + \bar{C}_{(in)1})(\bar{B}_1 + \bar{C}_{(in)1})$$
$$f_{C(out)1} = \bar{A}_1\bar{B}_1 + \bar{A}_1\bar{C}_{(in)1} + \bar{B}_1\bar{C}_{(in)1} \qquad (12\text{-}4b)$$

where $C_{(in)1}$ represents the carry-in of the 1s adder, and $C_{(out)1}$ is the carry-out of the 1s adder and the carry-in of the 2s adder.

Note that Eq. (12-3) is equal to Eq. (12-1), but Eq. (12-4a) is the complement of Eq. (12-2). However, the carry-out $C_{(out)1}$ is only an internal output. By applying $C_{(out)1}$ to the carry-in input of the 2s Full Adder while at the same time omitting the inverter in the sum output line and adding inverters in the A_2 and B_2 input lines, the correct Σ_2 and $C_{(out)2}$ outputs are obtained. The analysis is easier if the form given in Eq. (12-4b) is used. The Σ_2 and $C_{(out)2}$ equations become

$$f_{\Sigma_2} = A_2\bar{B}_2\bar{C}_{(in)2} + \bar{A}_2B_2\bar{C}_{(in)2} + \bar{A}_2\bar{B}_2C_{(in)2} + A_2B_2C_{(in)2} \qquad (12\text{-}5)$$

and

$$f_{C(out)2} = A_2B_2 + A_2(C)_{(in)2} + B_2C_{(in)2} \qquad (12\text{-}6)$$

where $C_{(in)2}$ represents the true carry-in of the 2s. Note that Eq. (12-5) corresponds to Eq. (12-1) and Eq. (12-6) corresponds to Eq. (12-2).

The US5482/7482, which uses TTL circuitry, is designed for medium-to-high speed, multiple-bit, parallel-add/serial-carry applications. Propagation delays of about 8 ns per bit are typical. The single-inversion, serial-carry circuit within each bit minimizes the need for extensive "*look-ahead*" and carry-cascading circuits.

A 4-bit Full Adder (US5483/7483 Sprague) is in a 16-pin package. It consists of two cascaded 2-bit Full Adders. The $C_{(out)2}$ output of the first 2-bit Full Adder is internally connected to the $C_{(in)1}$ input of the second Full Adder. It therefore becomes the carry-in of the 4s adder. Nine of the pins are used to apply the inputs A_1, B_1, $C_{(in)1}$, A_2, B_2, A_4, B_4, A_8, and B_8. Five pins are used to provide the four sum outputs ($\Sigma_1, \Sigma_2, \Sigma_4$, and Σ_8) and one pin for the carry-out of the 8s adder $C_{(out)8}$. The remaining two pins are for the V_{CC} and ground connections.

An 8-Input Digital Multiplexer. The package shown in Fig. 12-18(a) contains an 8-input Digital Multiplexer. It is the logical equivalent of a single-pole, 8-position switch, the position of which is determined by a 3-bit input address.

The 8230/31 of Fig. 12-18(b) has an *INHIBIT* input. When this input is at the 0 level, the f output is the same as the I input specified by the address inputs A_4, A_2, and A_1. For example, in the truth table of Fig. 12-18(c), if the address inputs A_4, A_2, and A_1 are, respectively, 1, 1, and 0 and the *INHIBIT* input is a 0, the f output is the same as the I_6 input. If the address inputs are $011 = 3$, the f output is the same as the I_3 input. If the *INHIBIT* input is at the 1 level, the $\bar{f}$ output is unconditionally high and the f output is unconditionally low.

The 8231 is the same as the 8230 except that it has an open collector $\bar{f}$ output. The 8232 is the same as the 8230 except for the effect of a high-level *INHIBIT* input. When the *INHIBIT* input is at the 1 level in the 8232, both outputs are unconditionally low.

A 2-Input, 4-Bit Digital Multiplexer. The package shown in Fig. 12-19(a) is a 2-Input, 4-Bit Digital Multiplexer. The logic diagram and truth tables are given in Fig. 12-19(b) and (c). The 8267 has an open-collector output. This multiplexer is used at the inputs of adders and registers and in parallel data handling applications. It is able to select from two 4-bit input sources: $A = A_0, A_1, A_2$, and A_3; and $B = B_0, B_1, B_2$, and B_3. The selection is controlled by the S_0 input. If $S_0 = 0, f_n = B_n$, regardless of the S_1 input. If $S_0 = 1$ and $S_1 = 0, f_n = \bar{A}_n$. The *INHIBIT* state $S_0 = S_1 = 1$, which makes $f_n = 1$, can be used to facilitate transfer operations in an arithmetic unit.

Conditional complementing may be obtained by tying the two inputs A_n and B_n together. This produces the *TRUE/COMPLEMENT* function, which is needed in conjunction with ADDER elements to perform *ADDITION/SUBTRACTION*. If $S_0 = 0$ while $S_1 = 0$, the output f_n is the same as the input. If $S_0 = 1$ while $S_1 = 0$, the output f_n is the complement of the input.

The schematic of the 8266/8267 is given in Fig. 12-20. Note that it uses TTL circuitry. The outputs are the totem-pole type and wired-logic is used in some of the internal gates.

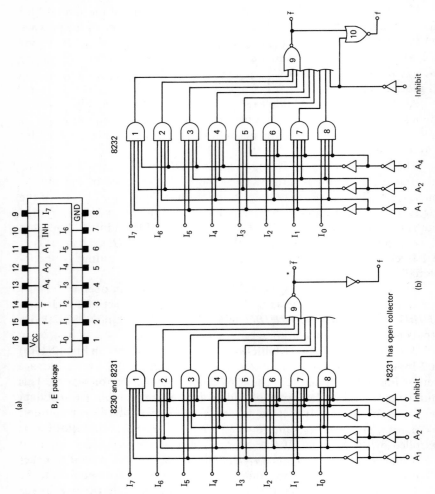

Fig. 12-18 (a) An 8-input digital multiplexer package with (b) its logic diagram

Truth table

Address			Data inputs								INH	f	Output	
A_4	A_2	A_1	I_7	I_6	I_5	I_4	I_3	I_2	I_1	I_0			8230 8231 $\bar{f}$	8232 $\bar{f}$
0	0	0	x	x	x	x	x	x	x	1	0	1	0	0
0	0	1	x	x	x	x	x	x	1	x	0	1	0	0
0	1	0	x	x	x	x	x	1	x	x	0	1	0	0
0	1	1	x	x	x	x	1	x	x	x	0	1	0	0
1	0	0	x	x	x	1	x	x	x	x	0	1	0	0
1	0	1	x	x	1	x	x	x	x	x	0	1	0	0
1	1	0	x	1	x	x	x	x	x	x	0	1	0	0
1	1	1	1	x	x	x	x	x	x	x	0	1	0	0
0	0	0	x	x	x	x	x	x	x	0	0	0	1	1
0	0	1	x	x	x	x	x	x	0	x	0	0	1	1
0	1	0	x	x	x	x	x	0	x	x	0	0	1	1
0	1	1	x	x	x	x	0	x	x	x	0	0	1	1
1	0	0	x	x	x	0	x	x	x	x	0	0	1	1
1	0	1	x	x	0	x	x	x	x	x	0	0	1	1
1	1	0	x	0	x	x	x	x	x	x	0	0	1	1
1	1	1	0	x	x	x	x	x	x	x	0	0	1	1
x	x	x	x	x	x	x	x	x	x	x	1	0	1	0

x = don't care

(c)

Fig. 12-18 (c) its truth table. (*Cont.*)

Shift Registers. A shift register consists of a series of flip-flops and is used for temporary storage of digital data. Shift registers come in four basic categories: (1) serial-in, serial-out, (2) serial-in, parallel-out, (3) parallel-in, serial-out, and (4) parallel-in, parallel-out.

An 8-Bit Serial-In, Serial-Out Register. The 5491/7491 packages of Fig. 12-21(a) contain high-speed monolithic 8-bit shift registers. The register, shown in Fig. 12-21(b), consists of eight RS master-slave flip-flops with input gating and a clock driver. Information is shifted from the data input to the Q_H output after eight clock pulses.

Information is applied to either input of the control gate. The other input of the control gate must be at the 1 level in order for the data to be transferred. In the truth table of Fig. 12-21(c) it is assumed that data information is applied to the X input and that the Y input is used as a data-enable control. Note that if the data-enable control is low, the Q_H output is low, regardless of the data input. If the data-enable control is high, the output is the same as the data input after eight clock pulses. The information is stored in *FFA* after the first clock pulse, in *FFB* after the second pulse, in *FFC* after the third pulse, and so on, until it is in *FFH* after the eighth clock pulse.

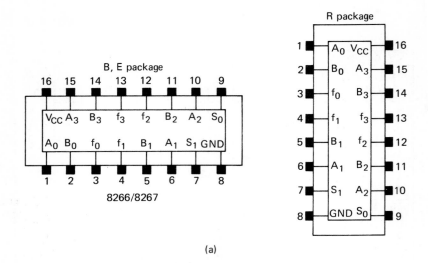

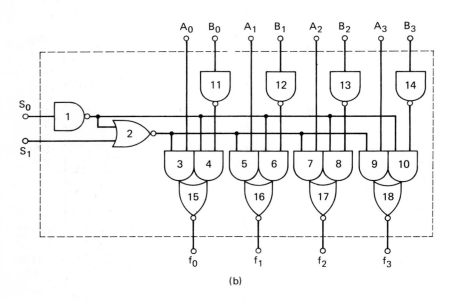

Fig. 12-19 (a) A 2-input, 4-bit digital multiplexer package with (b) its logic diagram and (c) its truth table.

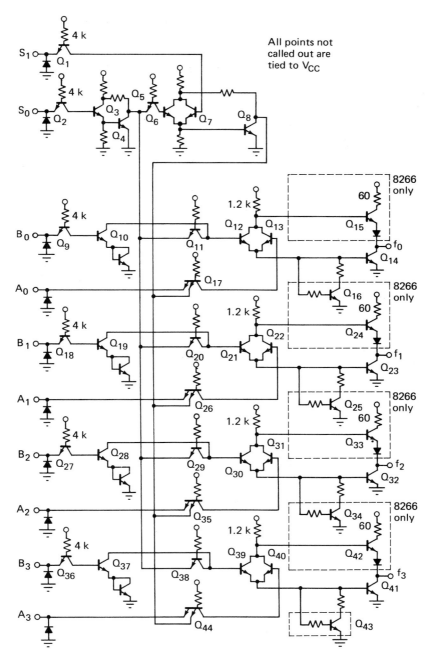

Fig. 12-20 The schematic of the 2-input, 4-bit digital multiplexer of Fig. 12-19.

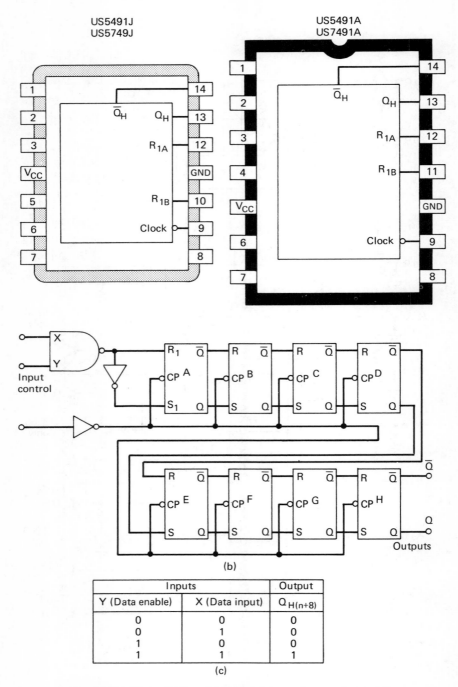

Fig. 12-21 (a) An 8-bit serial-in, serial-out register package with (b) its logic and (c) its truth table. (Courtesy Sprague Electric Co.)

A 4-bit Serial and Parallel Shift Register. The 8270/8271* packages shown in Fig. 12-22(a) contain a 4-bit shift register that has both serial and parallel synchronous entry capability with serial and parallel outputs. Serial data information is applied to the D_S input, and the inputs D_A, D_B, D_C, and D_D are used for parallel entry. The logic diagram is shown in Fig. 12-22(b).

The 8270/8271 has three modes of operation: (1) it stores ($HOLD$) information even with the clock running, (2) it permits parallel entry of data, and (3) serial information is shifted to the right. If the $HOLD$ mode is not needed, the load input may be tied to a 1 level and the shift input used as the mode control. The effect of the load and shift inputs is given in the truth table of Fig. 12-22(c).

The 8271 has a common asynchronous clear capability which is activated by applying a 0 level to the $\bar{R}_D$ input. The clock and control inputs are inhibited while the $\bar{R}_D$ input is low. The 8271 also has a $FALSE$ output $\bar{D}$ from the fourth flip-flop.

---

*Signetics Corporation, *Designing with MSI* (Medium Scale Integration), Vol. 1, Sec. 4, pp. 1–2.

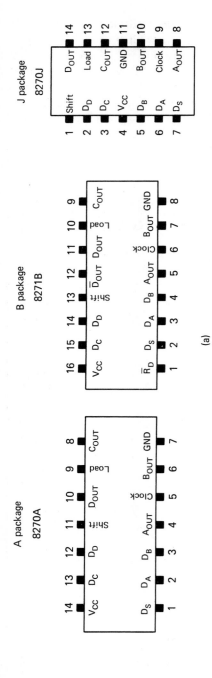

Fig. 12-22 (a) 4-bit serial and parallel shift register packages. (Courtesy Signetics Corporation)

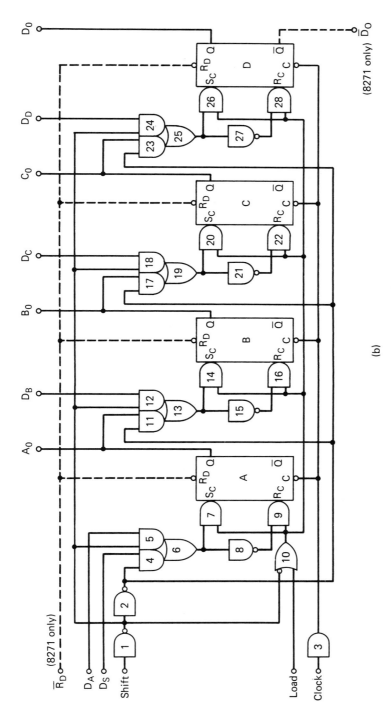

Fig. 12-22 (b) Their logic diagram. (Courtesy Signetics Corporation) (*Cont.*)

Shift	Load	Mode
0	0	Store (with clock running)
0	1	Synchronous parallel entry
1	0	Shift right
1	1	Shift right

(c)

Fig. 12-22 (c) Their truth table. (Courtesy Signetics Corporation) (*Cont.*)

Problems

12-1 How many of each of the following gates can be included in a 14-pin flat-pack?
 (a) 2-input *NOR* gates.
 (b) 3-input *OR* gates.
 (c) 4-input *NAND* gates.
 (d) 2-wide, 2-input *AOI* gates.
 (e) Inverters.

12-2 How many of each of the following flip-flops with Q and $\bar{Q}$ outputs can be included in a 14-pin dual-in-line package?
 (a) JK flip-flops with separate clock and separate R_D inputs.
 (b) JK flip-flops with separate clock, common R_D, and common S_D inputs.
 (c) JK flip-flops with common clock, common R_D, and separate S_D inputs.
 (d) JK flip-flops with separate clock and separate R_D and S_D inputs.
 (e) JK flip-flops with 2-wide, 2-input *AOI* J and K inputs with separate R_D and common clock inputs. How could this flip-flop be preset in the 1 state?
 (f) D-type flip-flops with separate clock and separate R_D and S_D inputs.
 (g) D-type flip-flops with common clock, common R_D, and common S_D inputs.

12-3 Determine the sequence of counts produced by the counters of Fig. 12-23. Assume that all flip-flops are initially in the 0 state and operate on the trailing edge of the clock. The output of the Reset R_0 gates must go low in order to activate the direct reset R_D inputs.

12-4 The synchronous ripple counter of Fig. 12-24, which uses flip-flops connected in the Toggle (T) configurations, can be made to act as an up counter or a down counter. It counts up when the UP (U) line is high and down when the DOWN (*D*) line is high. Both lines must not be high at the same time, but if both are low simultaneously, the circuit simply stops counting. It can be made to switch from an up counter to a down counter, or vice-versa, without producing an error. Verify the above statements by first assuming that all flip-flops are initially in the 0 state, *U* is high, *D* is low, and several clock pulses are applied. Then make *D* high, *U* low, and apply several more clock pulses. Finally, set *U* and *D* both at the 0 level and apply a clock pulse.

12-5 Referring to the circuit of Prob. 12-4, determine the Boolean equation for each *T* input $f_{T_A}, f_{T_B},$ and f_{T_C} in terms of *U, D, A, B, C,* and *P*. Let the *Q* outputs of the three flip-flops equal *A, B,* and *C,* and the $\bar{Q}$ outputs equal $\bar{A}, \bar{B},$ and $\bar{C}$.

12-6 Determine the output (1 or 0) of each gate and the state (1 or 0) of each output line in Fig. 12-15(b) for the NBCD input $0111 = 7_{10}$.

12-7 Verify that the seven-segment indicator will read 3 if NBCD 0011 is applied to the logic diagram of Fig. 12-16(d).

12-8 Verify that $\Sigma_1 = 1, \Sigma_2 = 1,$ and $C_{(out)2} = 0$ if the inputs to Fig. 12-17(c) are $A_1 = 1, B_1 = 0, A_2 = 0, B_2 = 1,$ and $C_{(in)1} = 0$.

12-9 Show how a 4-bit Full Adder can be made by using two of the 5482 packages of Fig. 12-17(a).

12-10 Verify that the *f* output of the 8230/31 of Fig. 12-18(b) is the same as (a) the I_5 input if $A_4 = 1, A_2 = 0,$ and $A_1 = 1$ and (b) the I_3 input if $A_4 = 0, A_2 = 1,$ and $A_1 = 1$ when the *INHIBIT* input is at the 0 level.

12-11 Verify that the *f* output of the 8230/31 of Fig. 12-18(b) is at the 0 level and both outputs of the 8232 are at the 0 level when their *INHIBIT* inputs are high.

12-12 Verify the truth table of the 2-input, 4-bit Digital Multiplexer of Fig. 12-19.

12-13 Verify the truth table of the 8-bit shift register of Fig. 12-21.

12-14 Verify the truth table of the 4-bit shift register of Fig. 12-22.

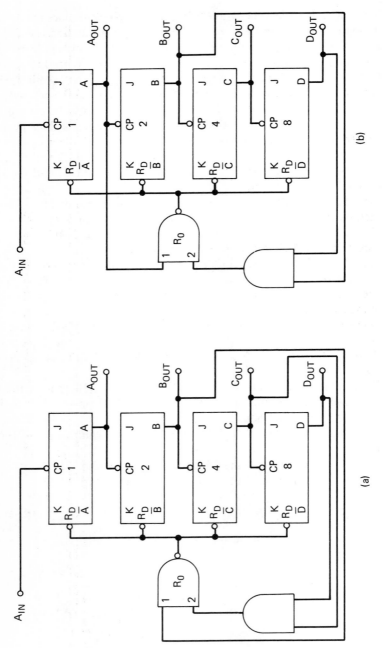

Fig. 12-23 (a) (b) (Courtesy Sprague Electric Co.)

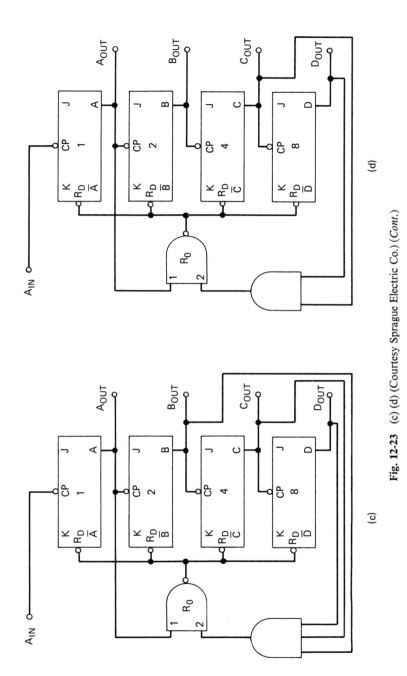

Fig. 12-23 (c) (d) (Courtesy Sprague Electric Co.) (*Cont.*)

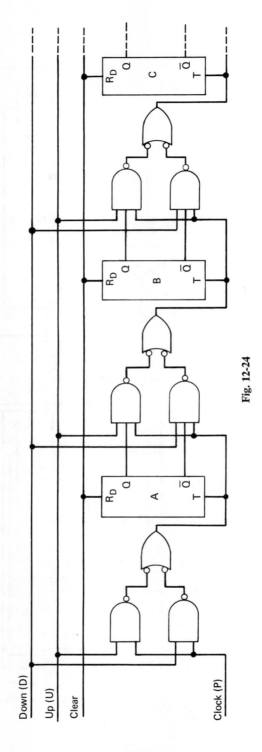

Fig. 12-24

INDEX

A

Absorption theorem, 112
Acceptance levels, 167
Addend, 94
Alpha, 41, 46
AND expander, 235-36
AND function, 107
AND gates:
   diode, 128-39, 153-54
   IC packages, 357, 366
   TTL, 230-32
AND-OR-INVERT AOI gate, 233-35, 357, 367
Associative laws, 112
Astable Multivibrator, 256-73
Asynchronous counters, 301, 304-6, 375-79
Attenuators, 29
Augend, 94
Avalanche breakdown, 38

B

Back clamp, 62-65
Baker clamp, 65
Bare-collector output, 184, 196
Barrier capacitance, 39
Base, 84
Base clamping, 68
Beta, 45-46

Binary arithmetic, 93-97
   addition, 93
   division, 97
   multiplication, 96
   subtraction, 94
   using complements, 99-101
Binary full adders, 381, 384-87
Binary system, 85
Bipolar transistor, 68
Biquinary codes, 106
   50-43210, 106
   5421, 106
   4'421, 106
   2'421, 106
Bistable IC packages, 359, 369-72
Bistable multivibrator, 257, 288-315, 299-301, 293-99, 309-15, 359, 369-72
Boole, George, 106
Boolean algebra, 106-16
   absorption theorem, 112
   AND, 107
   associative laws, 112
   complementary absorption theorem, 113-14
   De Morgan's theorem, 114-15
   distributive laws, 110-11
   negation theorem, 112
   NOT, 107
   OR, 107

Boolean algebra (*cont.*)
  propositions, 107
  redundancy theorem, 115-16
Borrow-in, 94
Borrow-out, 95
Breakdown voltages of transistors, 51
Breakdown voltage of diodes, 38
Buffer, 184-87

C

Carry-in, 94
Carry-out, 94
Cascaded diode gates, 148-52
Clamping diodes, 226-27
Clamping in internal gates, 320-23
Clock pulse generator, 328-30
CMOS, 346-50
  inverter, 347-50
  NAND gate, 350
  noise margins, 347-49
  NOR gate, 350
Codes, 102-6
Collector-base coupled astable multivibrator, 256-73
Collector-base coupled monostable multivibrator, 273-86
Collector clamp, 62
Collector-clamped inverter, 165-67
Collector dissipation, 45
Common base amplifier, 42-43
Common collector amplifier, 42-43
Common emitter amplifier, 43-48
  characteristic curves, 44
  large-signal current amplification factor, 45
  signal phase relationships, 46-48
  small-signal current amplification factor, 45-46
Commutating (speed-up) capacitor, 58-62
Compensated attenuator, 29, 30
Complementary absorption theorem, 113-14
Complements, 98
  arithmetic with, 99-101
  eights system, 98
  nines system, 98
  ones system, 98
  radix minus one (R-1) system, 98
  radix (Rs) system, 98
  sevens system, 98
  tens system, 98
  twos system, 98

Complex arrays, 374-96, 398-400
  binary full adders, 381, 384-87
  decade counter, 376-79
  digital multiplexes, 387-89
  four-bit ripple counter, 375-77
  NBCD-to-Decimal Decoder/Driver, 379-81
  NBCD-to-Seven Segment Decoder/Driver, 379, 381-83
  shift registers, 389, 392-96
Conversion techniques, 86-93
Counters, 301-7, 375-79
Current-controlled logic CCL, 187-95
Current-mode logic CML, 187-95
Cutin voltage:
  of diodes, 34
  of transistors, 51
Cutoff, 48-51
  in FETs, 72
  I_{CO}, 48-51
  $V_{BE_{co}}$, 50

D

Darlington amplifier, 248-50
dc noise margins, 167
Decimal codes, 102-6
  biquinary codes, 106
  excess three XS3, 104
  natural binary coded decimal NBCD (8421), 102-4
  74210, 105
Delay circuit, 284-85
Delay time t_d, 54
De Morgan's theorems, 114-15
Depletion mode, 70
Differentiator circuits, 27
Digital multiplexers, 387-89
Diode, 33-39
  avalanche breakdown, 38
  barrier capacitance, 39
  breakdown voltage, 38
  cutin voltage V_γ, 34
  equivalent circuit, 38
  forward bias, 33
  forward recovery time, 39
  leakage current I_{CO}, 35-38
  reverse bias, 35
  reverse recovery time, 39
  signal characteristics, 39
  Zener breakdown, 38
  Zener diode, 38

Diode logic:
  AND gate, 128-39
  cascaded gates, 148-52
  OR gates, 139-50
Diode-transistor logic DTL, 180-87
Distributive laws, 110-11
Dividend, 97
Divisor, 97
Double-diode back clamp, 65
D-type flip-flop, 297, 312-15
Dual-in-line package DIP, 357-59

E

Eccles-Jordan MV, 257
Eights complement, 98
8421 Code, 102-4
Emitter-coupled logic ECL, 187-95
Emitter-follower output transient response, 245-50
Emitter-to-emitter transistor, 213-14
Encoders, 119
Enhancement mode, 70, 74
EQUIVALENCE function, 118-19, 235, 345-46
Error-detecting codes, 105, 106
Excess three code, 104
EXCLUSIVE-OR function, 116-18, 123, 234-35, 345-46

F

Fall time t_f, 56
  of diode AND gates, 137-39, 153-54
  of diode OR gates, 146-48, 154-55
  of output circuits, 236-52
Fan-in, 167
Fan-out, 167
FET logic circuits, 342-50
  CMOS inverter, 347-50
  CMOS NAND gate, 350
  CMOS NOR gate, 350
  EQUIVALENCE circuit, 345-46
  EXCLUSIVE-OR circuit, 345-46
  inverting buffer, 345
  NAND gate, 344-45
  non-inverting buffer, 345
  NOR gate, 343-44
  RS flip-flop, 345, 347
FET switch, 76-77
Field-effect transistors, 68-77, 342-50
  construction, 69, 73
  cutoff, 72
  depletion mode, 70, 74

Field-effect transistors (cont.)
  enhancement mode, 70, 74
  FET switch, 76, 77
  input capacitance, 77
  input impedance, 70, 74
  insulated gate FET (IGFET), 73-76
  junction FET, 69-73
  metal oxide semiconductor FET (MOSFET), 73-76
  output impedance, 76
  pinch-off, 69
  volt-ampere characteristics, 71-72
Filtering, 24
Flat pack, 357, 360-61
Flip-flop, 257, 288-315, 293-301, 309-15, 359, 369-72
Flip-flop IC packages, 359, 369-72
Forbidden numbers, 102, 105
Forward recovery time, 39
Frequency division in:
  astable MV, 270-71
  bistable MV, 301-6
Frequency of multivibrator, 265-66
Full adders, 381, 384-87

H

Hexadecimal, 91-93
Hex-inverters, 359, 368
Hysteresis voltage, 333-34, 338

I

IMPLICATION gates, 121, 123, 124
INHIBIT gates, 121, 123, 124
Input clamping diodes, 226-27
Insulated-gate FET (IGFET), 73-76
  construction, 73
  depletion mode, 74
  enhancement mode, 74
  input impedance, 74
Integrated-circuit packages, 356-95
Integrated-circuit packaging, 356-61
Integrator circuit, 28
Interface elements, 352-54
Internal gates of complex arrays, 320-23
Intrinsic value, 84
Inverse transistor, 214
Inverter, 159-68, 229-30, 345, 347-50, 358, 368
Inverter interface elements, 352-54
Inverter output transient response, 236-45

J

JK flip-flop, 296-303
Junction FET, 69-73

L

Lamp drivers, 325-27, 379-81
Large-scale integration (LSI), 77
Large-signal current amplification factor $\beta_{dc} h_{FE}$, 45
Leakage current I_{CO}:
 of amplifiers, 41, 48-51
 of diodes, 35-38
Logical design, 116-21
Logic flip-flop:
 basic circuit, 293-94
 IC packages, 359, 369-72
 integrated circuit JK, 299-301
 practical circuits, 297-99
 relationships between various types, 314-15
 types, 294-97
Low-capacitance probe LCP, 29, 30
Lower acceptance level, 168
Lower threshold voltage of Schmitt triggers, 332-33, 336-38
Low-pass filter, 28

M

Mark-to-space ratio of Schmitt triggers, 338-40
Master-slave flip-flop, 309-12
Metal oxide semiconductor FET (MOSFET), 73-76
 construction, 73
 depletion mode, 74
 enhancement mode, 74
 input impedance, 74
Millman's theorem, 12-16
Minuend, 94
Modified collector-coupled astable MV, 268-70
Monostable multivibrator, 256-57, 273-86, 372-74
Multi-emitter transistor, 204-7
Multiplexes, 387-89
Multiplicand, 96
Multiplier, 96
Multivibrators, 256

N

NAND gate interface element, 352-54
NAND gates, 121-23, 172-73, 180-87, 207-27, 344-45, 350-57, 362-64
Natural binary coded decimal NBCD (8421), 102-4
NBCD-to-decimal decoder/driver, 379-81
NBCD-to-seven segment decoder/driver, 379, 381-83
Negation theorem, 112
Negative logic, 130
Nines complement, 98
Noise margins, 167-68
Non-saturating techniques, 62-68
NOR gates, 121-23, 170-77, 227-29, 343-44, 350, 357, 365
NOR-OR, 187-95
Norton's theorem, 7-12
NOT function, 107
NOT gates, 159-68, 229-30
Number systems, 84

O

One-shot multivibrator, 256-57, 273-86
Ones complement, 98
Open-collector output, 184, 196
OR function, 107
OR gates:
 diodes, 139-48
 packages, 357, 366
 TTL, 233
OR-NOR, 187-95
Output circuit variations, 236-52
Output short-circuit current, 219-20
Overdrive, 56-62
Overdrive factor, 58

P

Packaging, 356-95
Parallel counters, 307
Parasitic transistor, 213-14
Partial products, 96
Phase splitter, 185
Piecewise linear diode equivalent, 38
Pinch-off, 69
Place value, 84
pn junction, 33-39
pnp astable MV, 271-73

Index / 405

Position value, 84
Positive logic, 128
Power gate, 184-87
Primitive functions, 121
Products, 96
Propagation time t_p, 56
Pulse width of monostable MV, 282
Pyramiding factor, 167

Q

Quotient, 97

R

Radix, 84
Radix complement, 98
Radix-minus-one complement, 98
RC attenuators, 29, 30
RC differentiator, 27
RC integrator, 28
RC time constants, 16-28
Recovery time of:
    astable MV, 268
    monostable MV, 284
Redundancy theorem, 115-16
Relay drivers, 327-28
Reset Set (RS) flip-flop, 294-95, 309, 312, 345, 347
Reset-Set-Toggle (RST) flip-flop, 295-96
Resistor-capacitor-transistor logic RCTL, 177
Resistor-transistor logic RTL, 170-77
Reverse bias, 35
Reverse recovery time, 39
Ripple counter, 301, 304-6, 375-79
Rise time t_r, 54
    of diode *AND* gates, 137-39, 153-54
    of diode *OR* gates, 146
    of output circuits, 236-52

S

Saturated inverter, 159-68
Saturation, 51-54
    $h_{FE_{min}}$, 54
    $I_{B_{min}}$, 53
    $I_{C_{sat}}$, 52
    $V_{BE_{sat}}$, 54

Saturation, (*cont.*)
    $V_{CE_{sat}}$, 52-54
Schift registers, 389, 392-96
Schmitt Trigger, 328, 330-42
    analysis, 333-41
    fast analysis, 341-42
    hysteresis, 333
    hysteresis voltage, 333-34, 338
    loop gain, 330-33
    lower threshold voltage V_{LT}, 332-33, 336-38
    mark-to-space ration, 338-40
    transfer characteristics, 330-33
    upper threshold voltage V_{UT}, 332-36
Schottky-clamped TTL gate, 324-25
Serial counter, 301, 304-6, 375-79
Sevens complement, 98
Short-circuit output current, 219-20
Signal characteristics, 39
Single-diode back clamp, 62-65
Sink loads, 167
Small-signal current amplification factor β, h_{fe}, 45-46
Source loads, 167
Speed-up (commutating) capacitor, 58-62
Storage time t_s, 56
Subtrahend, 94
Synchronization of the astable MV, 270-71
Synchronous counters, 307

T

Temperature stabilization, 49-51
Tens complement, 98
Thevenin's Theorem, 1-7, 10-12, 15, 16
Three-state logic TSL, 350-52
Time constants:
    of compensated attenuator, 29, 30
    long time constant, 27
    medium time constant, 27
    of RC circuits, 16-28
    short time constant, 27
    universal time constant chart, 17
Totem-pole output, 185, 207
Totem-pole output transient response, 250-52
Trailing-edge logic, 307-9
Transient analysis of:
    AND gates, 136-39, 153-54
    cascaded diode gates, 150-52

Transient analysis of: *(cont.)*
  OR gates, 145-48, 154-58
  output circuits, 236-52
Transient response:
  of output circuits, 236-52
  of RC attenuators, 29, 30
  of RC circuits, 16-28
  of TTL *NAND* gates, 222-26
Transient response of the transistor switch, 54-68
Transistor amplifier, 40-48
  alpha, 41-46
  beta, 45-46
  biasing, 41
  characteristic curves, 44
  common base, 42-43
  common collector, 42-43
  common emitter, 43-48
  currents, 41
  dc alpha a_{dc}, 41
  dc beta β dc h_{FE}, 45
    maximum collector dissipation rating, 45
  signal phase relationships of *CE*, 46-48
Transistor switch, 48-68, 76
  breakdown voltage, 51
  cutin voltage V_γ, 51
  cutoff, 48-51
  delay time t_d, 54-68
  fall time t_d, 56
  field-effect transistor FET, 76
  $h_{FE_{min}}$, 54
  $I_{B_{min}}$, 53
  $I_{C_{sat}}$, 52
    leakage current, 48-51
    non-saturating techniques, 62-68
    overdrive, 56-62
    overdrive factor, 58
    propagation time t_p, 56
    temperature stabilization, 49-51
    transient response, 54-68
    turn-off time t_{off}, 56
    turn-on time t_{on}, 56

Transistor switch *(cont.)*
  typical junction voltages, 54
  $V_{BE_{CO}}$, 50
  $V_{BE_{sat}}$, 54
  $V_{CE_{sat}}$, 52-54
Transistor-transistor logic, TTL, 204-36
  AND gate, 230-32
  AND-OR-INVERT AOI gate, 233-35
  INVERTER, 229-30
  JK flip-flop, 299-303
  NAND gate, 207-27
  NOR gate, 227-29
  OR gate, 233
Transmission-line driver, 328
Truth table, 114
TTL IC packaging, 356-95
Turn-off time t_{off}, 56
Turn-on time t_{on}, 56
Two-out-of-five code, 105
Twos complement, 98

U

Unipolar transistor, 68
Unit load, 167
Universal time constant chart, 17
Upper acceptance level, 168
Upper threshold voltage of Schmitt triggers, 332-36

W

Waveshaping, 24
Weights, 84
Wired *AND*, 195
Wired logic, 195-99
Wired *OR*, 195

X

XS3 Code, 104

Z

Zener breakdown, 38
Zener diode, 38